NANOLUBRICANTS

Tribology Series

Editors: M J Neale, M Priest and G Stachowiak

Stachowiak (ed)	Wear – Materials, Mechanisms and Practice	November 2005
Lansdown	Lubrication and Lubricant Selection, 3rd Edition	November 2003
Neale, Polak and Priest (eds)	Handbook of Surface Treatment and Coatings	May 2003
Sherrington, Rowe and Wood (eds)	Total Tribology – towards an integrated approach	December 2002
Kragelsky, Alisin, Myshkin and Petrokovets (eds)	Tribology – Lubrication, Friction and Wear	April 2001
Stolarski & Tobe	Rolling Contacts	December 2000
Neale and Gee	Guide to Wear Problems and Testing for Industry	October 2000
Martin and Ohmae	Nanolubricants	April 2008

Coming soon

Khonsari and Booser	Applied Tribology, 2nd Edition	April 2008

NANOLUBRICANTS

Edited by

Jean Michel Martin
École Centrale de Lyon, France

Nobuo Ohmae
Kobe University, Japan

John Wiley & Sons, Ltd

Telephone (+44) 1243 779777

Email (for orders and customer service enquiries): cs-books@wiley.co.uk
Visit our Home Page on www.wiley.com

Other Wiley Editorial Offices

John Wiley & Sons Inc., 111 River Street, Hoboken, NJ 07030, USA

Jossey-Bass, 989 Market Street, San Francisco, CA 94103-1741, USA

Wiley-VCH Verlag GmbH, Boschstr. 12, D-69469 Weinheim, Germany

John Wiley & Sons Australia Ltd, 42 McDougall Street, Milton, Queensland 4064, Australia

John Wiley & Sons (Asia) Pte Ltd, 2 Clementi Loop #02-01, Jin Xing Distripark, Singapore 129809

John Wiley & Sons Canada Ltd, 6045 Freemont Blvd, Mississauga, ONT, L5R 4J3

Wiley also publishes its books in a variety of electronic formats. Some content that appears in print may not be available in electronic books.

Library of Congress Cataloging-in-Publication Data

Martin, Jean Michel, 1948-
Nanolubricants / Jean Michel Martin, Nobuo Ohmae.
p. cm.
Includes index.
ISBN 978-0-470-06552-5 (cloth)
1. Lubrication and lubricants. 2. Nanoparticles. 3. Nanotechnology. 4. Metal clusters.
I. Ohmae, Nobuo. II. Title.
TJ1075.M278 2008
621.8′9–dc22 2008002530

British Library Cataloguing in Publication Data

A catalogue record for this book is available from the British Library

ISBN 978-0-470-06552-5 (H/B)

Typeset in 10/12pt Times by Aptara Inc., New Delhi, India
Printed and bound in Great Britain by Antony Rowe Ltd, Chippenham, Wiltshire

Contents

Preface

The use of nanoparticles as lubricants is a recent idea. The replacement of organic molecules by tiny particles of solid material is not straightforward and has only recently been recognized as feasible. There are two reasons for this: first, nanoparticles are not easy to synthesize; and second, colloidal solutions are intrinsically unstable. Nevertheless, these difficulties have been overcome, and descriptions of many kinds of nanolubricants can now be found in the literature. The main advantages of using nanolubricants are that they are relatively insensitive to temperature and that tribochemical reactions are limited, compared to traditional additives.

In Chapter 1 of this book, the general principles of colloidal lubrication are explained and the stability of sols and micellar systems is discussed.

Chapter 2 is devoted to nanoparticles made of metal dichalcogenides. Solid lubricants of this kind have long been in use, albeit not in the form of such tiny particles. Fullerene-like structures are very attractive and are now used in practical situations. The production of such materials is only briefly described, while their mechanisms of action are addressed in much more detail, using advanced analytical tools. Other particles such as nanowires are also studied.

In Chapter 3, carbon nanotubes and onions are evaluated as potential antiwear and friction-reducing additives. Similarly to Chapter 2, analytical tools including surface analysis are found to be very useful for identifying the mechanism of action of nanotubes and onions as lubricants. Interestingly, the formation of lubricious iron oxides contributes to the tribological performance of these additives.

Chapter 4 focuses on micellar systems. The nature of these particles is described, and the complex mechanisms that are responsible for their efficiency are discussed.

Chapter 5 is devoted to metallic nanolubricants. The idea of using metals (and particularly metals with low melting points) is not new. However, the synthesis of particles of the order of tens of nanometres in size is quite attractive because of the lack of corrosion effects.

The book concludes with an extensive discussion in Chapter 6 of boron-based solid nanolubricants, which are extensively used for a variety of purposes. As future tribological systems need to operate in ever more stringent conditions, boric acid and other solid lubricants offer unique opportunities and are expected to become more and more popular.

Jean Michel Martin and Nobuo Ohmae

List of Acronyms

AES	Auger electron spectroscopy
AFM	atomic force microscopy
CVD	chemical vapour deposition
DAC	diamond anvil cell
DLC	diamond-like carbon coatings
DOS	density of states
EDS	energy dispersive X-ray spectroscopy
EELS	electron energy-loss spectroscopy
EFTEM	energy filtered transmission electron microscopy
ELNES	energy-loss near edge structure
ESI	electron spectroscopic imaging
EXAFS	extended X-ray absorption fine structure
HAADF	high-angle annular dark field
HRTEM	high-resolution transmission electron microscopy
MEMS	microelectromechanical system
MoDTC	molybdenum dithiocarbamate
PEEM	photoemission electron microscopy
SAED	selected area electron diffraction
SEM	scanning electron microscopy
SFM	surface force microscopy
TEM	transmission electron microscope
TEY	total electron yield
TGA	thermogravimetric analysis
TOF-SIMS	time-of-flight secondary ion mass spectroscopy
UHV	ultrahigh vacuum
XANES	X-ray absorption near edge structure
XPS	X-ray photoelectron spectroscopy
XRD	X-ray diffraction
ZnDTP	zinc dithiophosphate

1

Colloidal Lubrication: General Principles

Jean Michel Martin and Nobuo Ohmae

Nanoparticles can be considered as potential lubricant additives. They present several major advantages compared to organic molecules currently used as lubricant additives, for example:

- Their nanometre size allows them to enter the contact area easily.
- They are often efficient at ambient temperature. Thus, no induction period is necessary to obtain interesting tribological properties.
- Several types of nanoparticles can be envisaged as lubricant additives. The chapters of this book describe different possible nanolubricants: nested nanoparticles (fullerenes, onions and nanotubes) made on metal dichalcogenides or carbon, reverse overbased micelles, metallic nanoparticles and boron-based nanolubricants. In some cases, the authors chose to study these particles because their structure is close to well-known lamellar structures in tribology (for example inorganic fullerenes of MoS_2). However, dispersion of nanoparticles in oil is not always simple. Thus, in a first part, some rules will be shown about the dispersion of solid nanoparticles in a liquid.

Another way to use nanoparticles in tribology is by embedding them inside a coating. Nanoparticles can improve the tribological of the coating itself, but nanoparticles can also be released by the coating during wear and can thus lubricate in the same way as if they were in the oil.

1.1 Stability of Colloids Dispersed in a Base Oil

A colloidal sol is generally a substance (liquid or gel) containing dispersed solid particles. To obtain a homogeneous and stable system, the size of the nanoparticles has to be small, say below 100 nm.

Nanolubricants Edited by J. M. Martin and N. Ohmae

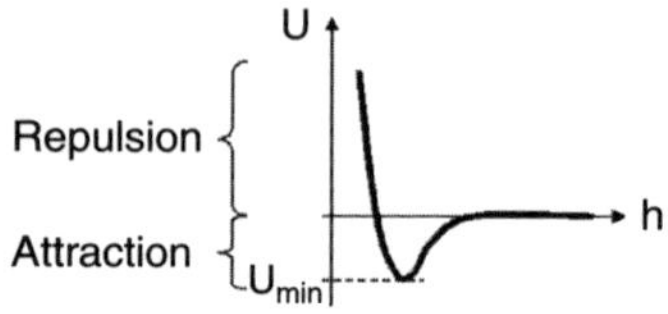

Figure 1.1 Lennard–Jones force describing attraction and repulsion between two molecules as a function of the distance between them

A sol is composed of two phases: a continuous phase and a dispersed phase. In the present case, the continuous phase is liquid (oil) and the dispersed phase is a solid (nanoparticles). A good stability of dispersion is not easy to obtain, and several phenomena lead to instability:

- aggregation (or flocculation) of the particles;
- sedimentation of particles and aggregates of particles.

Lennard–Jones forces are often used to describe intermolecular interactions. They result from steric repulsion at very short distance and attractive van der Waals forces at short distance (Figure 1.1).

At a short distance (below 50 nm), two particles attract themselves due to van der Waals forces. This force is expressed by the following relation:

$$E_{\text{vdw}} = -\frac{A_r}{12h} \tag{1}$$

where A_r is the Hamaker constant and h is the distance between particles. The Hamaker constant is equal to $\pi^2 C\rho^2$ (C is a constant that is characteristic of the material constituting the particles and of the dispersion medium). These forces become more significant when the distance h is below 10 nm.

Particles in suspension in a liquid medium are subjected to three kinds of forces: (a) gravitation forces the particles to fall down, (b) viscosity of the liquid decreases the speed of their displacement and (c) Archimedes force is opposed to gravitation forces in this case. By applying fundamental relation of dynamics, the expression of the steady state sedimentation speed of a particle is obtained:

$$V = \frac{2}{9}\frac{R^2}{\eta}(\rho - \rho')g \tag{2}$$

with

R radius of the particle
η dynamic viscosity of the liquid
ρ density of the particles
ρ' density of the liquid

It is important to notice that the speed varies in proportion to the square of the radius of the particle. Therefore, particles with large radius will sediment much faster than small ones. Due to the flocculation process described before, aggregated particles will present a larger size than individual ones and sedimentation is then faster.

Fortunately, other effects can allow for a better stability of a suspension. The Brownian motion is certainly the most interesting. It corresponds to chaotic movement of particles due to shocks of liquid molecules on its surface. For small particles, the relation of Stokes–Einstein gives the average distance x covered by a particle during a certain time t as

$$x = \sqrt{\frac{2kTt}{6\pi\eta R}} \tag{3}$$

with

R radius of the particle
η dynamic viscosity of the liquid
T temperature
k constant of Boltzmann ($k = 1.38 \times 10^{-23}$ J/K)
t time

Some nanoparticles are difficult to disperse in oil. For example, inorganic fullerenes (studied in Chapter 2) are mineral particles with a spherical or cylindrical shape and a size lying between 50 and 150 nm. They do not present any affinity with oil and their dispersion is thus difficult without surfactants.

To study the dispersion of such nanoparticles in base oil, they were mixed at different concentrations in poly-alpha-olefin (PAO) base oil using an ultrasonic bath (concentrations are expressed here in weight percent). By using formulae (2) and (3), the distance covered by one particle in 1 s can easily be compared by sedimentation and by Brownian motion, respectively (Table 1.1). The sedimentation force is oriented vertically whereas the distance covered by Brownian movement has no preferential direction. For this calculation, the density of the lamellar compounds was fixed at 5.06 g/cm^3 for MoS_2, 7.6 g/cm^3 for WS_2 and 4.4 g/cm^3 for NbS_2 (according to the data from the supplier). The value used for inorganic fullerene (IF)

Table 1.1 Dispersion properties of IF-MS_2 and corresponding lamellar structures at 20 °C

	Average diameter (nm)	Sedimentation: distance covered in 1 s (nm)	Brownian movement in 1 s (nm)
IF-WS_2	140	0.2	150
	90	0.1	188
2H-WS_2	**1000**	**55**	**39**
IF-MoS_2	40	0.01	282
2H-MoS_2	**2000**	**140**	**28**
IF-NbS_2	60	0.02	230
3R-NbS_2	**1000**	**29**	**13**

Table 1.2 Dispersion properties of IF-MS_2 and corresponding lamellar structures at 70 °C

	Average diameter (nm)	Sedimentation: distance covered in 1 s (nm)	Brownian movement in 1 s (nm)
IF-WS_2	140	2	503
	90	0.9	628
2H-WS_2	1000	531	133
IF-MoS_2	40	0.1	942
2H-MoS_2	2000	1329	94
IF-NbS_2	60	0.2	770
3R-NbS_2	1000	280	133

is 15 % lower than in the 2H structure to take into account the cavity generally present in the centre of IF particles.

The results show the advantage of using smaller IF-MS_2 particles instead of the corresponding 2H structure. The sedimentation speed is much higher in the case of the lamellar structure (which is a few tenths of a micrometre in size). Thus, calculations predict much better stability of IF suspensions, the distance due to sedimentation being lower than that of the Brownian motion. However, flocculation of particles could explain the poor stability of the suspensions because of the absence of any polar molecules. Calculations show that an aggregate of particles with a diameter of 650 nm will cover about the same distance by sedimentation and by Brownian motion.

An increase in the temperature provides a beneficial effect on the increased Brownian motion. However, it drastically decreases the viscosity of the oil and thus sedimentation of particles becomes faster. A comparison of results in Tables 1.1 and 1.2 illustrates this phenomenon, which is of course more important in the case of larger particles.

The stability of dispersions of PAO + IF − MS_2 was determined optically by measuring the time necessary to observe the total sedimentation of fullerenes at 20 °C (Table 1.3). These results confirm the importance of the size of the particle on the stability of the dispersion. The best stability is obtained for the oil/IF-MoS_2 dispersion, having a smaller particle size (about 40 nm).

In the case of nanotubes, the problem of dispersion is more complex and there are no simple relations to express the various trends. Generally nanotubes are intermingled and tend to form bundles, leading to a very difficult dispersion. The length of the nanotubes is certainly a crucial parameter.

Table 1.3 Sedimentation times for different nanoparticle dispersions at 20 °C. The times were determined optically

	IF-MoS_2 (40 nm)	IF-NbS_2 (60 nm)	IF-WS_2 (90 nm)	IF-WS_2 (140 nm)
Sedimentation time (hours)	10	8	6.5	5.5

Stability of a colloidal solution can be improved:

- by using steric repulsion between particles: functionalization by polymers;
- by using electrostatic repulsion: adsorption of ionic surfactants.

In the case of nanotubes, several techniques were proposed: functionalization by organic molecules [1] and the use of surfactants like sodium dodecyl sulfate (SDS) to disperse them in water. This kind of dispersion led to the manufacture of fibres made of nanotubes [2]. Nanotubes can also be cut to avoid their intermingling and thus improving their dispersion.

For the dispersion of IF-WS_2, a technique using small metallic balls has been developed by Nanomaterials Company (http://www.apnano.com). Vigorous stirring by using small metal balls makes it possible to obtain better dispersions compared with an ultrasonic bath. Moshkovith *et al.* studied the dispersion parameters on the tribological properties of IF-WS_2 and, in particular, the size of aggregates (large aggregates >2 μm and small aggregates <1 μm) as a function of the mixing time of the dispersion [3]. The increase of the mixing time leads to a decrease in the aggregate size. Friction results are then more reproducible.

1.2 Lubrication by Micellar Systems

Detergents in oils form micelles, which are aggregates of surfactant molecules. Polar molecules are made of a polar part (hydrophilic head) and a nonpolar hydrophobic part. According to the nature of the solvent (polarity), they can form micelles or reverse micelles. Hydrocarbon chains create a steric barrier, preventing the particles from aggregation. Overbased micelles have a mineral core made, for example, of calcium carbonate $CaCO_3$ (Figure 1.2).

Owing to their lipophilic chains and their very small size, micelles are very easily dispersed in base oils and dispersion is stable for a long time. Overbased micelles, used initially as detergents, also present antiwear properties (see Chapter 4). Overbased micelles can be used in several ways: (a) as thin coatings deposited on a solid surface, (b) as a colloidal dispersion in apolar medium (oil, grease, etc.) and (c) in gas phase lubrication.

The ability of overbased micelles to lubricate to some extent compared with molecules can be explained on the basis of the following features:

1. Micelles can adsorb physically on any surface (metal, ceramics, carbon-rich materials, etc.) because physical laws (van der Waals forces) mainly govern the attractive forces. In particular, they can adsorb at low temperature.

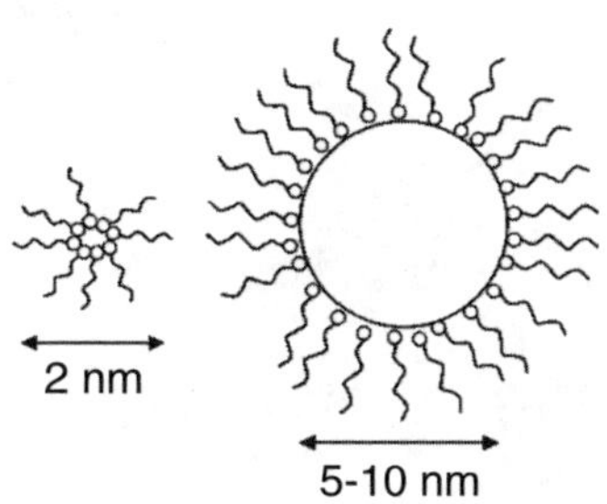

Figure 1.2 Schematic representation of an inverse micelle and overbased micelle structure

2. The adsorbed micelles are so small that they can hardly be removed from the surface by the moving liquid.
3. The size of the micelles is so small that they can easily enter a macroscopic sliding contact in boundary conditions but do not perturb the hydrodynamic regime significantly.
4. The lubrication effect can be generated by the chemical nature of both constituents of the micelles (the core and the shell) and even by the synergistic effect between the two. Moreover, tribochemical reactions in the contact area can create a new compound as the third body.

Overbased micelles are generally obtained by a polyphasic complex reaction. They are prepared by carbonating a metal or by hydroxide, which is dispersed in a polar mixture of surfactant, aromatic solvant and paraffinic oil [4,5]. Prior to bubbling carbon dioxide through the reaction vessel, polar promoters (alcohols, amines and water) are added. Thus, oxide or hydroxide particles can be converted into spherical colloidal particles in the size range of 2–20 nm. The colloidal dispersion obtained is centrifuged (typically 30 minutes at 3000g) in order to eliminate nonreacted micrometre-sized lime particles and large aggregates of crystallized micelles. Afterwards, the colloidal suspension is dialysized using pure heptane, to remove organic impurities and nonreacted micelles. At the end of dialysis, the heptane is evaporated and a yellow-brown solid varnish-like material is obtained, which can be stored and further used for any lubrication mode. The structure of the solid phase has been determined by small-angle electron diffraction in transmission electron microscopy (TEM) (see Figure 1.3). Preparation of a coating can be made using the standard procedure for Langmuir–Blodgett films. A monomicellar film can be produced on water and then transferred to a surface by dipping. This shows the amorphous character of the structure at the micellar level and the fact that colloids are packed together with the original structure they originally had in the solution (Figure 1.3). Another technique simply uses solvent evaporation after deposition of a drop of the solution on the surface. However, in this case, the thickness of the coating is not controlled. The fixation of micelles on surfaces has been attributed to van der Waals forces between the lipophilic end of surfactant chains and the solid surface, and also to the attraction of the mineral core by the substrate. The attractive forces as shown by TEM observation can

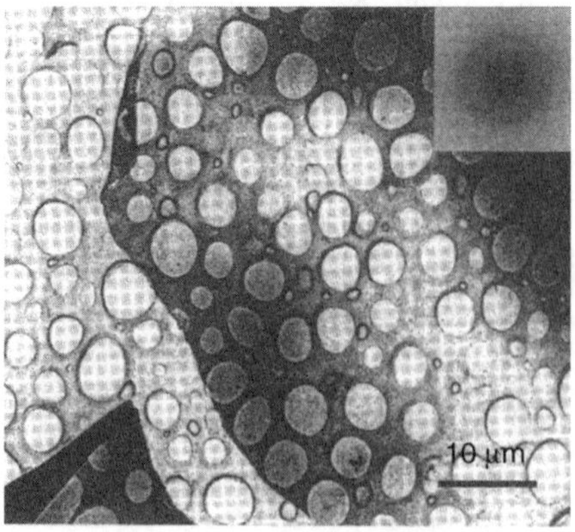

Figure 1.3 TEM image of Langmuir film made of overbased micelles

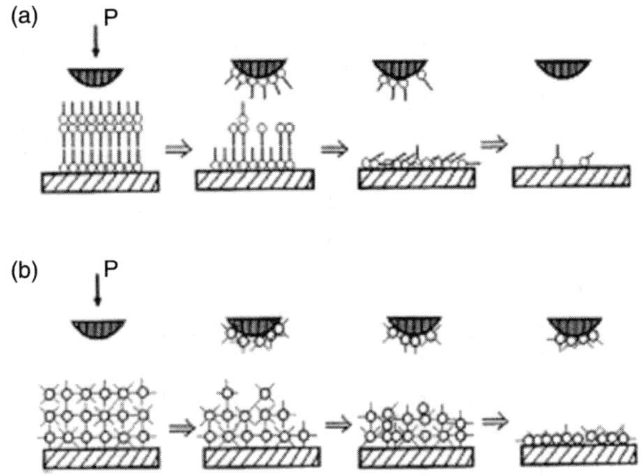

Figure 1.4 Lubrication mechanism of (a) fatty acid LB films and (b) MoS_2-DDP LB films. Reproduced by permission of Elsevier from Zhang *et al.* [8], © (2000) Elsevier

mechanically deform the mineral core when a film of micelles is deposited on a carbon film. Overbased micelles can be directly used as additive in a base oil. Due to the small size of the nanoparticle and the presence of the surfactant, the dispersion is very stable. Lubrication by such micellar systems are described in Chapter 4.

1.3 Lubrication by Metallic Nanoparticles

Metal nanoparticles are not easily dispersed in oil unless that they are of nanometre size. Zhou *et al.* studied the tribological properties of Cu nanoparticles dispersed in pure paraffin base oil [6]. To improve their compatibility with the base oil, the surface of Cu nanoparticles was modified with dialkyldithiphosphate (DDP). Zhang *et al.* used the same surface modification process with Mo-S nanocluster [7]. These nanoparticles can be used as additives in lubricant or as Langmuir–Blodgett (LB) films [8]. Zhang *et al.* compared the tribological behaviour of LB films of fatty acid and MoS_2-DDP (Figure 1.4). MoS_2-DDP films present better antiwear properties. These properties can be explained by a higher load-carrying capacity of MoS_2 nanocores. Chapter 5 is dedicated to metallic nanolubricants.

1.4 Colloids Embedded in a Coating

Nanoparticles can also be used as coatings or embedded in coatings. IF-MS_2, C_{60} and carbon nanotubes can be added into coatings that present interesting properties. Chhowalla and Amaratunga compared thin IF-MoS_2 films (1 μm thick) and 2H-MoS_2 films, and found a very low friction coefficient with IF-MoS_2 coatings, even in N_2 atmosphere [9]. These performances were explained by the absence of edges due to the nested structure of fullerenes. Joly-Pottuz compared the tribological properties of IF-MoS_2 and IF-WS_2 films, obtained by solvent evaporation or rubbing, under ultrahigh vacuum. Very low friction coefficients were obtained with all coatings (see Chapter 2). Recently, IF-WS_2 particles were deposited on diamond-like carbon (DLC) coatings and very a low friction coefficient was observed [10].

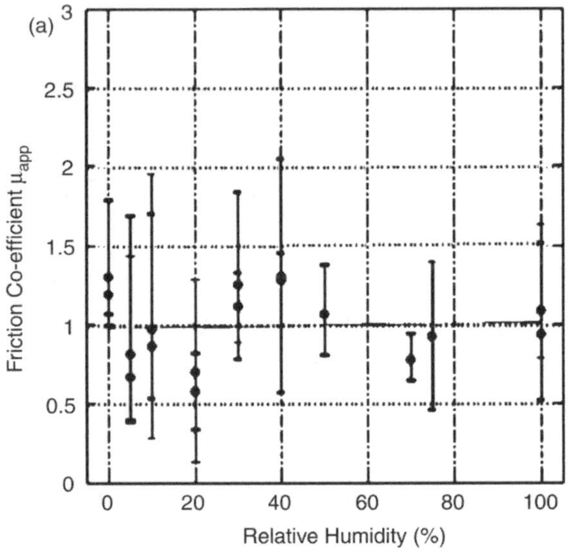

Figure 1.5 Influence of humidity on the friction coefficient of carbon nanotube film. With kind permission from Springer Science and Business Media from Turq *et al.* [14], © (2005) Springer Science and Business Media

Nakagawa *et al.* used molecular beam epitaxy (MBE) to deposit very thin coatings of C_{60}. Experiments by friction force microscopy (FFM) reveal a very low friction force [11]. Carbon nanotubes can be grown on surfaces to form film usually called 'forest' [12]. These films present particular properties. They are highly hydrophobic with a contact angle of 162°. The frictional behaviour of these films was evaluated against a gold tip [13]. High friction coefficients (1.2 to 1.5) were measured but no adhesion of the tip was observed. These properties can be attributed to the deformation properties of these films. Their deformation was followed by using a tribometer inside a scanning electron microscope (SEM). The influence of humidity on their tribological properties showed that friction is independent of the relative humidity (Figure 1.5) [14]. Miyoshi *et al.* also studied friction properties of multiwall carbon nanotube (MWNT) coatings. Friction coefficients of aligned MWNTs against steel were measured between 0.025 and 0.06 in air, and between 0.19 and 0.48 in vacuum [15].

Coating structures have changed during the last 25 years. A recent paper by Donnet and Erdemir gives an overview of the evolution of coatings [16]. Coatings are now multicomponent (three, four or more phases) and/or multilayered. These evolutions lead to an optimization of their properties: antiwear resistance and corrosion resistance. Addition of nanoparticles in coatings was also found to improve their properties. For example, addition of Al_2O_3 nanoparticles in polymer coatings leads to an improvement in their scratch and abrasive resistances [17].

Inorganic fullerenes of MoS_2 have been also added into coatings. Chen *et al.* compared tribological results obtained with Ni-P electroless composite coatings containing 2H-WS_2, graphite and IF-WS_2 [18]. Addition of graphite or 2H-WS_2 leads to a reduction of the friction coefficient, 0.067 and 0.062, respectively, compared to the one obtained for pure Ni-P coating (0.09). However, addition of IF-WS_2 leads to a strong friction reduction (0.03). No wear was observed on the Ni-P-(IF-WS2) coatings after friction tests (only slight scars). On the

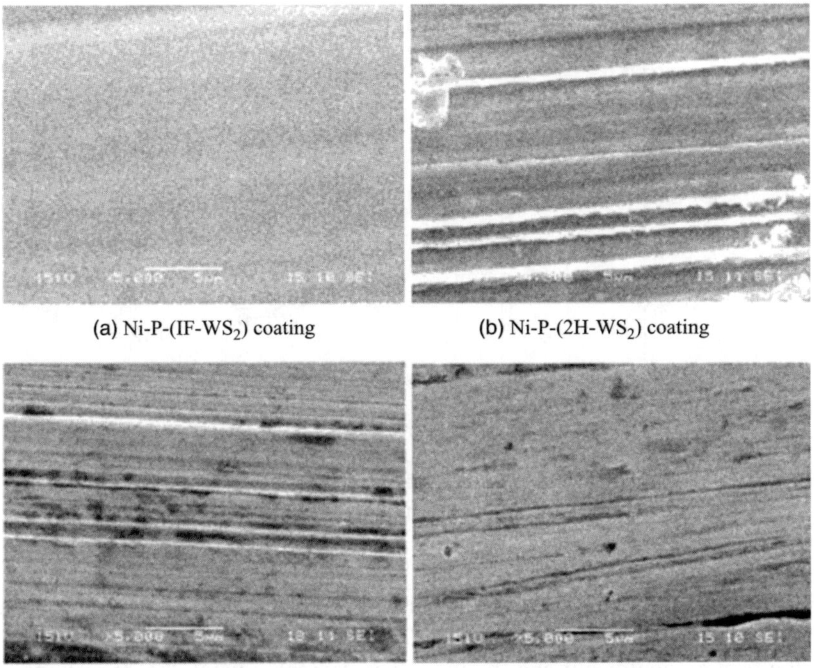

(a) Ni-P-(IF-WS_2) coating (b) Ni-P-(2H-WS_2) coating

(c) Ni-P-graphite coating (d) Ni-P coating

Figure 1.6 SEM images of worn surface of Ni-P coatings (d) and Ni-P coatings containing (b) 2H-WS_2, (c) graphite or (a) IF-WS_2 [18]

contrary, wear was observed with the three other coatings (see Figure 1.6). Interesting results were also observed with Ni-P-(IF-MoS_2) coatings tested in air and vacuum, while Ni-P-(2H-WS_2) presented a poor stability [19]. Katz *et al.* added IF-WS_2 nanoparticles in Ni-P coatings to lubricate orthodontic wires [20]. These coatings are envisaged to lubricate other medical devices (catheter and endoscope).

Friction results can not be explained by a rolling process of the fullerenes. Indeed, nanoparticles embedded in the coatings are unable to roll [21]. Rapoport *et al.* studied tribological properties of polymer nanocomposites with embedded IF nanoparticles [22]. The friction mechanism was found to be due to a transfer of IF from the polymer coatings to the counterface. The same mechanism was proposed for bronze powder composites impregnated with nanoparticles [23]. Coatings impregnated with IF-WS_2 present better properties (the transition to seizure is higher) than 2H-WS_2 impregnated ones. Platelets of 2H-WS_2 can not easily enter into the pores of the coatings and are accumulated at the metal surface. On the contrary, fullerenes enter the pores of the coatings easily and are gradually supplied inside the contact area. Thus, their effect on the tribological properties of the coatings is made longer. These two examples demonstrate that fullerenes containing coatings act by progressively releasing the particles in the contact area.

Nanotubes present exceptional mechanical properties and are now widely used to reinforce composites [24], particularly with polymers [25]. Tribological properties of composites

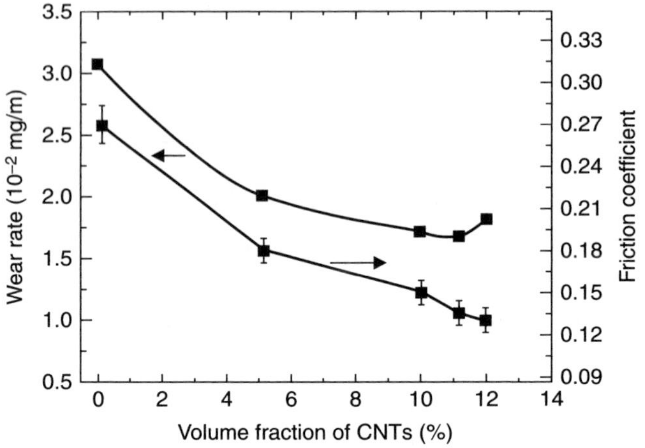

Figure 1.7 Variation of friction coefficient and wear loss of Ni-based composite with volume of carbon nanotubes embedded in the coatings. Reproduced by permission of Elsevier from Wang *et al.* [27], © (2003) Elsevier

containing nanotubes have also been widely studied. Several matrices can be considered: polyimide [26], Ni [27], Ni-P [28] and alumina [29,30]. A general trend is observed for all these matrices: the friction coefficient and the wear loss decreases when the percentage of carbon nanotubes in the coatings increases, as presented in Figure 1.7 for Ni-based carbon nanotube (CNT) composite. In the case of carbon/carbon composite, the addition of carbon nanotubes leads to a reduction in wear loss, but the friction coefficient slightly increases with the percentage of carbon nanotubes in the coatings [31].

The dispersion of nanotubes inside the matrix is not always easy. To solve this problem, Schittelhelm *et al.* proposed to deposit ta-C coatings by physical vapour deposition (PVD) on nanotubes previously deposited on Si substrates [32]. This paper shows that the amorphous diamond film provides scratch resistance for dispersed single-wall nanotubes (SWNTs). This technique can be envisaged to grow thin films containing carbon nanotubes. Recently, DLC coatings were also deposited on carbon nanotube forests (see Figure 1.8). Mechanical properties of these coatings were studied against the density of carbon nanotube forests (1 to $7 \times 10^9/cm^2$). The dynamic hardness and elastic modulus of the CNT/DLC composite films

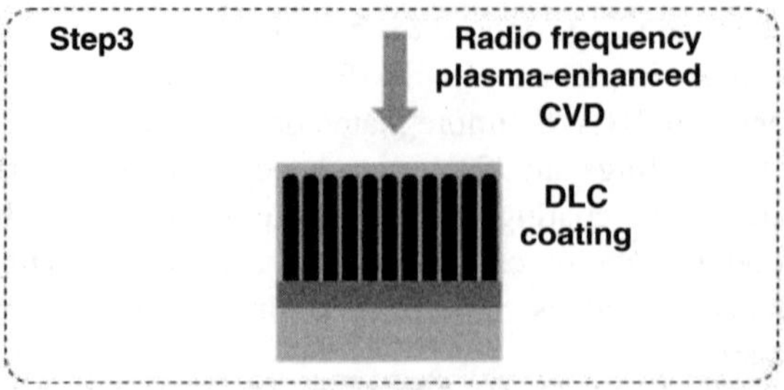

Figure 1.8 Schematic composition of CNT/DLC coatings [33]

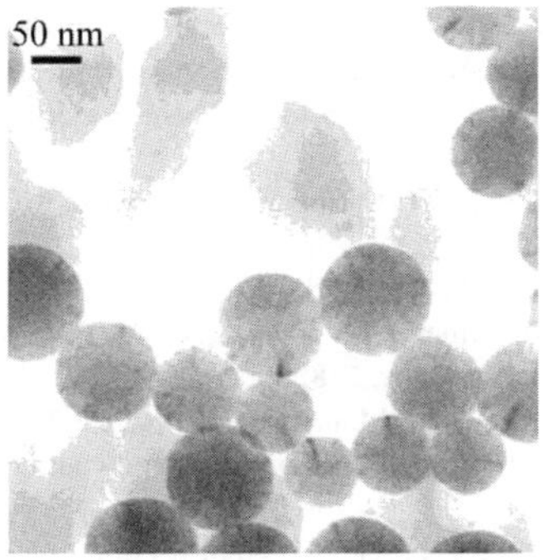

Figure 1.9 TEM image of carbon onions randomly distributed on a graphite film. Reproduced by permission of Elsevier from Cabioc'h *et al.* [36], © (2000) Elsevier

decrease linearly with increasing CNT density, because the DLC coating has a much higher hardness and elasticity than CNT films. The same friction coefficient of 0.23 was observed and was independent of the density of nanotubes. This friction coefficient corresponds to those obtained with pure DLC coatings (not deposited on CNT forests).

A way to synthesize carbon onions is by high-fluence ion implantation of carbon onions into metals (Cu, Ag). This is possible because of the very low solubility of carbon into copper or silver. The implanted carbon atoms precipitate into the bulk of the metal. A uniform graphite layer is formed with the carbon onions above [34]. Figure 1.9 presents a film observed after ion implantation of copper. Implantation parameters have an influence on the size of carbon onions: the mean diameter is 5 nm for a dose of $5 \times 10^{16}/\text{cm}^2$ and 20 nm for $2.5 \times 10^{17}/\text{cm}^2$ [35]. The growth mechanism of the graphite film and carbon onions was proposed by Cabioc'h *et al.* [34]. Tribological properties of these carbon films containing carbon onions were tested (after removal of silver by thermal annealing at 850 °C). Unfortunately, their thickness was too thin [37]. However, silver films containing carbon onions were tested, and friction results obtained were compared with those obtained on pure silver films. The presence of onions in silver films leads to an increase in lifetime of the coating by a factor of 15.

During the preparation of this chapter, such a new finding that carbon onion synthesized from diamond nanoparticle at 1600 °C shows a low friction coefficient of 0.005 in ultrahigh vacuum. Therefore, further application of nanolubricants to tribology is anticipated.

References

1. Chen, J., Hamon, M. A., Hu, H., Chen, Y. A., Rao, M., Eklund, P. C. and Haddon, R. C., Solution properties of single-wall carbon nanotubes, *Science*, **282**, 1998, 95–98.
2. Vigolo, B., Pénicaud, A., Coulon, C., Sauder, C., Pailler, R., Journet, C., Bernier, P. and Poulin, P., Macroscopic fibers and ribbons of oriented carbon nanotubes, *Science*, **290**, 2000, 1331.
3. Moshkovith, A., Perfiliev, V., Verdyan, A., Lapsker, I., Popovitz-Biro, R., Tenne, R. and Rapoport, L., Sedimentation of IF-WS_2 aggregates and a reproducibility of the tribological data, *Tribology International*, **40**, 2007, 117–124.
4. Le Suer, W. M., Patent 2247534, Lubrizol Corp.
5. Cleverley, J. H., Marsh, J. F. and Swietlik, J. M., Patent EP 0323087, Exxon Chemical Corp.
6. Zhou, J., Wu, Z., Zhang, Z., Liu, W. and Xue, Q., Tribological behavior and lubricating mechanism of Cu nanoparticles in oil, *Tribology Letters*, **8**, 2000, 213–218.

7. Zhang, Z., Xue, Q., Zhang, J., Synthesis, structure and lubricating properties of dialkyldithiophosphate-modifed Mo-S compound nanoclusters, *Wear*, **209**, 1997, 8–12.
8. Zhang, P., Xue, Q., Du, Z. and Zhang, Z., The tribological behavior of LB films of fatty acids and nanoparticles, *Wear*, **242**, 2000, 147–151.
9. Chhowalla, M. and Amaratunga, G. A. J., Thin films of fullerene-like MoS_2 nanoparticles with ultra-low friction and wear, *Nature*, **407**, 2000,164–167.
10. Joly-Pottuz, L. and Martin, J. M., unpublished work.
11. Nakagawa, H., Kibi, S., Tagawa, M., Umeno, M. and Ohmae, N., Microtribological properties of ultrathin C_{60} films grown by molecular beam epitaxy, *Wear*, **238**, 2000, 45–47.
12. Kinoshita, H., Kume, I., Sakai, H., Tagawa, M.and Ohmae, N., High growth rate of vertically aligned carbon nanotubes using a plasma shield in microwave plasma-enhanced chemical vapor deposition, *Carbon*, **42**, 2004, 2735–2777.
13. Kinoshita, H., Kume, I., Tagawa, M. and Ohmae, N., High friction of a vertically aligned carbon-nanotube film in microtribology, *Applied Physics Letters*, **85**, 2004, 2780–2781.
14. Turq, V., Ohmae, N., Martin, J. M., Fontaine, J., Kinoshita, H. and Loubet, J. L., Influence of humidity on microtribology of vertically aligned carbon nanotube film, *Tribology Letters*, **19** (1), 2005, 23–28.
15. Miyoshi, K., Street Jr, K. W., Vander Wal, R. L., Andrews, R. and Sayir, A., Solid lubrication by multiwalled carbon nanotubes in air and in vacuum, *Tribology Letters*, **19** (3), 2005, 191–201.
16. Donnet, C. and Erdemir, A., Historical developments and new trends in tribological and solid lubricant coatings, *Surface and Coatings Technology*, **180–181**, 2004, 76–84.
17. Wang, Y., Lim, S., Luo, J. L. and Xu, Z. H., Tribological and corrosion behaviors of Al_2O_3/polymer nanocomposite coatings, *Wear*, **260**, 2006, 976–983.
18. Chen, W. X., Xu, Z. D., Tenne, R., Rosenstveig, R., Chen, W. L., Gan, H. Y. and Tu, J. P., Wear and friction of Ni-P electroless composite coating including inorganic fullerene-WS_2 nanoparticles, *Advanced Engineering Materials*, **4**, 2004, 686–690.
19. Zou, T. Z., Tu, J. P., Zhang, S. C., Chen, L. M., Wang, Q., Zhang, L. L. and Ne, D. H., Friction and wear properties of electroless Ni-P-(IF-MoS_2) composite coatings in humid air and vacuum, *Materials Science and Engineering A*, **426**, 2006, 162–168.
20. Katz, A., Redlich, M., Rapoport, L., Wagner, H. D. and Tenne, R., Self-lubricating coatings containing fullerene-like WS_2 nanoparticles for orthodontic wires and other possible medical applications, *Tribology Letters*, **21** (2), 2006, 135–139.
21. Rapoport, L., Fleischer, N. and Tenne, R., Applications of WS_2 (MoS_2) inorganic nanotubes and fullerene-like nanoparticles for solid lubrication and for structural nanocomposites, *Journal of Materials Chemistry*, **15**, 2005, 1782–1788.
22. Rapoport, L., Nepomnyashchy, O., Verdyan, A., Popovitz-Biro, R., Volovik, Y., Ittah, B. and Tenne, R., Polymer nanocomposites with fullerene-like solid lubricant, *Avanced Engineering Materials*, 6, 2004, 44–48.
23. Rapoport, L., Lvovsky, M., Lapsker, I., Leschchinsky, W., Volovik, Y., Feldman, Y. and Tenne, R., Friction and wear of bronze powder composites including fullerene-like WS_2 nanoparticles, *Wear*, **249**, 2001, 150–157.
24. Barrera, E. V., Key methods for developing single-wall nanotube composites, *Chemistry and Materials Science*, **52** (11), 2007, 38–42.
25. Salvetat, J. P., Mechanical properties of individual nanotubes and composites, *Lecture Notes in Physics*, **667**, 2006, 439–493.
26. Cai, H., Yan, F. and Xue, Q., Investigation of tribological properties of polyimide/carbon nanotube nanocomposites, *Materials Science and Engineering A*, **364**, 2003, 94–100.
27. Wang, L. Y., Tu, J. P., Chen, W. X., Wang, Y. C., Liu, X. K., Olk, C., Cheng, D. H. and Zhang, X. B., Friction and wear behavior of electroless Ni-based CNT composite coatings, *Wear*, **254** (12), 2003, 1289–1293.
28. Chen, W. X., Tu, J. P., Wang, L. Y., Gan, H. Y., Xu, Z. D. and Zhang, X. B., Tribological application of carbon nanotubes in a metal-based composite coating and composites, *Carbon*, **41** (2), 2003, 215–222.
29. Lim, D. S., You, D. H., Choi, H. J., Lim, S. H. and Jang, H., Effect of CNT distribution on tribological behavior of alumina–CNT composites, *Wear*, **259** (1–6), 2005, 539–544.
30. Sun, J., Gao, L. and Jin, X., Reinforcement of alumina matrix with multi walled carbon nanotubes, *Ceramics International*, **31** (6), 2005, 893–896.
31. Lim, D.-S., An, J.-W. and Lee, H. J., Effect of carbon nanotube addition on the tribological behavior of carbon/carbon composites, *Wear*, **252** (5–6), 2002, 512–517.

32. Schittenhelm, H., Geohegan, D. B. and Jellison, G. E., Synthesis and characterization of single-wall carbon nanotube–amorphous diamond thin-film composites, *Applied Physics Letters*, **81**, 2001, 2097–2099.
33. Kinoshita, H., Kume, I., Sakai, H. and Ohmae, N., Synthesis and mechanical properties of carbon nanotube/diamond-like carbon composite films, *Diamond and Related Materials*, **16**, 2007, 1940–1944.
34. Cabioc'h, T., Jaouen, M., Rivière, J. P., Delafond, J. and Hug, G., Characterization and growth of carbon phases synthesized by high temperature carbon ion implantation into copper, *Diamond and Related Materials*, **6**, 1997, 261–265.
35. Cabioc'h, T., Jaouen, M., Denanot, M. F. and Bechet, P., Influence of the implantation parameters on the microstructure of carbon onions produced by carbon ion implantation, *Applied Physics Letters*, **73** (21), 1998, 3096–3098.
36. Cabioc'h, T., Jaouen, M., Thune, E., Guérin, P., Fayoux, C. and Denanot, M. F., Carbon onions formation by high-dose carbon ion implantation into copper and silver, *Surface and Coatings Technology*, **128–129**, 2000, 43–50.
37. Cabioc'h, T., Thune, E., Riviere, J. P., Camelio, S., Girard, J. C., Guérin, P., Jaouen, M., Henrard, L. and Lambin, P., Structure and properties of carbon onion layers onto various substrates, *Journal of Applied Physics*, **91** (3), 2002, 1560–1567.

2

Nanoparticles Made of Metal Dichalcogenides

Lucile Joly-Pottuz and Fabrice Dassenoy

2.1 Tribological Properties of 2H-MoS_2

Molybdenum disulfide (MoS_2) has been used for many years as a solid lubricant because of its interesting friction-reducing properties related to its crystalline structure. MoS_2 is a lamellar compound made of a stacking of S-Mo-S layers (Figure 2.1(a)). In each of them, the molybdenum atom is surrounded by six sulfur atoms located at the top of a trigonal prism (see Figure 2.1(b)). The distance between a molybdenum atom and a sulfur atom is equal to 0.241 nm, whereas the distance between two sulfur atoms from two adjacent layers is equal to 0.349 nm. This characteristic was often used to explain easy cleavage between the layers and therefore the lubricating properties of MoS_2. Nevertheless, in this chapter it will be seen that this point is debatable.

Atoms inside a layer are strongly bonded (covalent bonds) whereas the sulfur atoms between two consecutive layers are slightly bonded (van der Waals bonds). According to some researchers, these weak interlayer bonds are responsible for an easy slip between the layers and could thus explain the friction-reducing properties of MoS_2. On the contrary, Sokoloff showed that the slip of two layers one on another is not possible without the presence of dislocations [2]. In the case of a crystal without defects, low friction can only be obtained if the superimposed layers are in incommensurate conditions. For example, in the case of a monoclinical structure, Hirano *et al.* showed that this is possible with a 'magic' angle of rotation between two superimposed layers [3]. In the case of a hexagonal structure, the magic angle is equal to 30°.

The majority of MoS_2 films deposited in vacuum contains a nonnegligible quantity of oxygen (10 to 20 % in atomic concentrations). These oxygen atoms easily replace sulfur atoms in the structure of MoS_2, creating an isostructural phase called $MoS_{2-x}O_x$. This substitution is easily observed by X-ray diffraction with the displacements of interreticular

Nanolubricants Edited by J. M. Martin and N. Ohmae

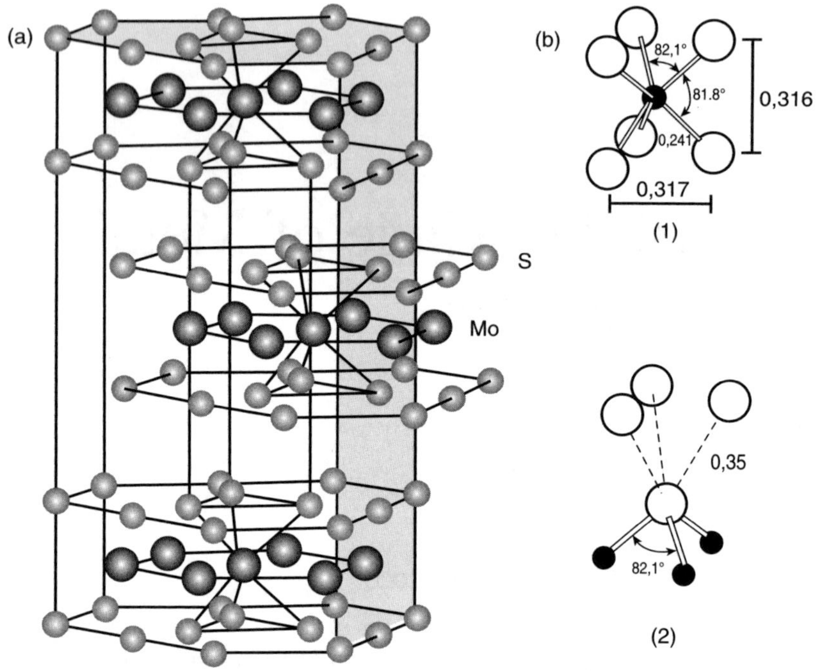

Figure 2.1 (a) Crystalline structure of 2H-MoS_2, (b) atom coordination: (1) configuration of MoS_6 trigonal prism, (2) surroundings of sulfur atoms. Reproduced by permission of Elsevier from Winer [1], © (1967) Elsevier

distances: dilation of the distances between (002) planes and contraction between (001) planes [4].

Fleischauer and Lince studied the effect of the oxygen contained in the layers on the friction properties of the material under ultrahigh vacuum [5]. Between 2 and 10 % of oxygen a decrease of the friction coefficient was observed (Figure 2.2). This evolution was explained by considering friction at an atomic level, and more precisely friction between two planes of sulfur atoms of two consecutive layers. The length of the molybdenum–oxygen bond is smaller than the molybdenum–sulfur bond. Thus the substitution of a sulfur atom by oxygen involves the formation of a 'depression' on the surface of the layer, which causes according to Fleischauer a kind of 'notch' (notch model) (see Figure 2.2). Consequently, friction between two layers whose surfaces are not perfectly smooth (at the atomic level) is higher. If the quantity of oxygen increases beyond 10 %, surfaces become more homogeneous in defects and friction strongly decreases. Beyond 15 % of oxygen in the structure, substitution is not possible any more: a molybdenum oxide structure is formed.

Martin *et al.* obtained a very low friction coefficient under ultrahigh vacuum with MoS_2 coatings containing less than 1 % of oxygen [6]. In this case, weak friction can only be explained by the presence of layers in incommensurate conditions. These superimposed layers with a rotational disorder were observed by high-resolution transmission electron microscopy (TEM) on very thin wear particles where only two layers can be observed (Figure 2.3).

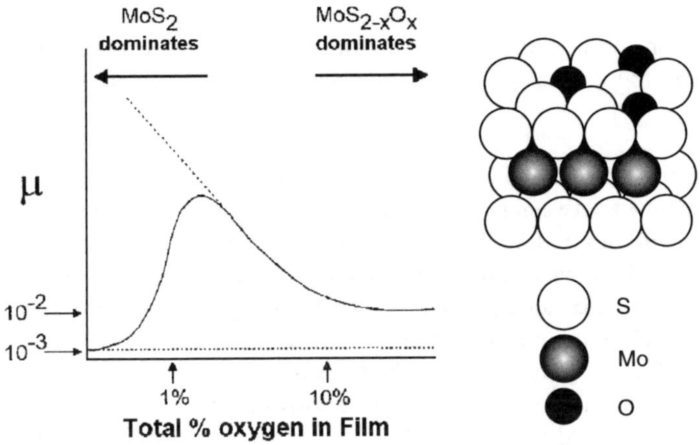

Figure 2.2 Influence of the oxygen content in MoS_2 films on the friction coefficient. Model of sulfur substitution by oxygen in the MoS_2 structure. Reproduced by permission of Elsevier from Fleishchauer and Lince [5], Figures 9 and 10, © (1999) Elsevier

Obtaining suprafriction with MoS_2 requires three conditions [6]:

- Formation of a transfer film on the antagonist surface. Its formation depends on the capacity of adhesion of MoS_2 on the metallic surface. This film is formed very quickly.
- Orientation of the (0001) planes parallel to the surface (direction of friction). It takes place in the transfer film, the coatings and the third body. This orientation takes place very early during the test and more quickly if the size of the crystals is small.

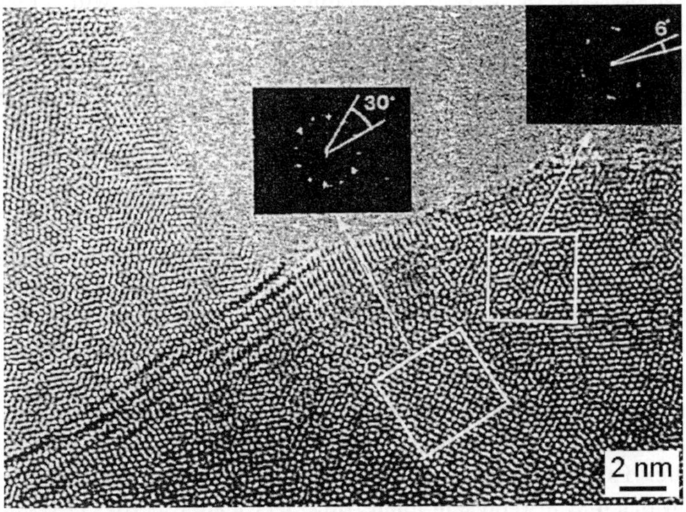

Figure 2.3 Wear particles after friction between steel and MoS_2 coatings, containing two superimposed stackings in incommensurate conditions (angle of 30° measured on optical diffractograms). Reprinted with permission from Martin *et al.* [6], Figure 4, © (1993) by the American Physical Society

- Absence of contaminants. Contaminants can be oxygen, carbon or water vapour. Oxygen is in oxide forms or in substitution of sulfur atoms in the crystal. Carbon comes from residual gas used during the deposit phase. Water vapour can contaminate the sample during the phase of storage.

The environment in which the layers of MoS_2 are used has an influence on their efficiency. In the case of a contact between two metal surfaces lubricated by an MoS_2 coating, the coefficient of friction increases with moisture up to 65 % RH (relative humidity). Then it decreases [7]. An increase in the contact pressure or sliding velocity leads to a reduction of the friction coefficient. At high contact pressures or high sliding speeds, there would be heating of the contact area leading to an evaporation of the water present in MoS_2 films [1]. Thus, the durability of the film is largely controlled by the water content.

In the case of a test performed under a dry atmosphere, MoS_2 is transferred layer by layer. Wear is therefore explained by an ejection of the matter out of the contact (debris observed around the contact area). In the case of considerable amounts of moisture, water leads to the cohesion of crystallites [8], thus preventing the transfer of MoS_2 or carrying out the removal of all the coating.

The increase of friction coefficient with moisture is generally assigned to the formation of molybdenum hydroxides or oxides. Nevertheless, Wahl and Singer monitored by *in vivo* Raman spectroscopy the wear of a layer of Pb-Mo-S rubbed against a glass ball and did not detect the formation of MoO_3, which is very easily detected by this technique [9]. They attributed the high friction to an increase in interfacial shearing. Water molecules can also react with dangling bonds at the edge of crystallites. This will avoid the sliding of one layer on the other, or their orientation inside the contact area.

To resume, the MoS_2 coating is very sensitive to the environment (water, gas, etc.) and its tribological properties are strongly dependent on its reaction with different compounds. This is due to the presence of edges on crystallites of MoS_2 films, which make the material chemically very reactive. To solve this problem, Cohen *et al.* synthesized a film named type II, made up of large crystals oriented parallel to the surface [10]. A reduction of dangling bonds at the interface makes the film much less reactive and leads to a reduction of both friction and wear.

2.2 IF-MoS_2 and IF-WS_2 Fullerene-like Nanoparticles

By analogy to carbon which can form closed structures (fullerenes and nanotubes), Tenne and co-authors synthesized the same type of closed structures with metal dichalcogenides, lamellar compounds of MX_2 formula [11, 12]. One of the principal advantages of these structures is the absence of edges of the crystals. These particles are called 'inorganic fullerene-like nanoparticles' or IF-MS_2, and can be described as several hollow spheres with different diameters (fullerenes) fitted together to form a structure that looks like an onion. Thereafter, we will use the term IF to describe the structure 'fullerene-like' and 2H for the corresponding lamellar structure. IFs were first studied for their tribological and mechanical properties. Recent results show that they also present interesting electrical properties [13], indicating a large range of applications for this new material.

In 1996, Feldman *et al.* described a model of growth of the IF-MoS_2 and IF-WS_2 [14]. Particles made up of metallic oxides are reduced in a gas mixture containing 5 % of H_2 and

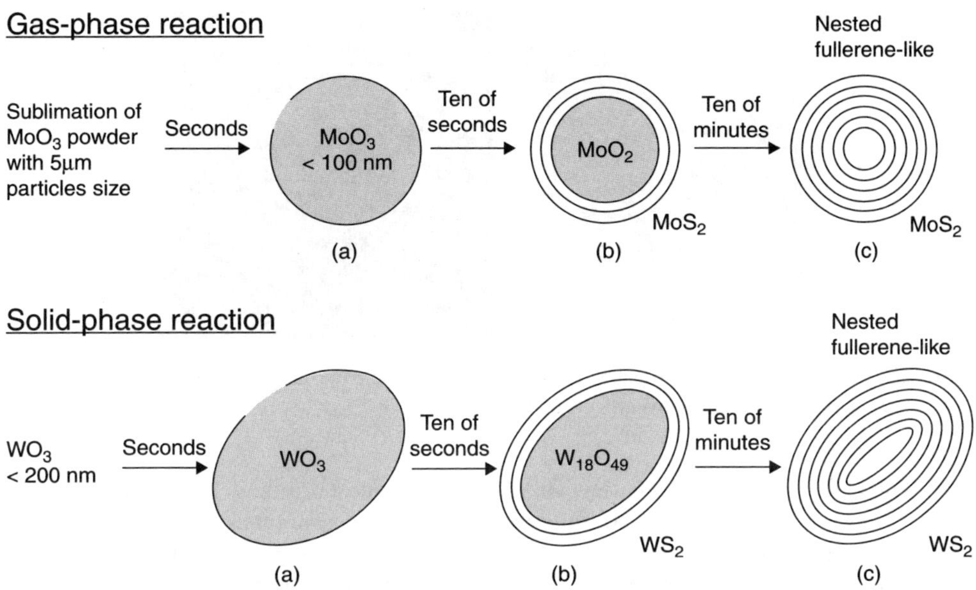

Figure 2.4 Schematic representation of the growth model of IF-MoS_2 and IF-WS_2. Reproduced by permission of American Chemical Society from Feldman *et al.* [14], Figure 1, © (1996) American Chemical Society

95 % of N_2 and then react with sufhydric acid (H_2S). The surface of the oxide particles MO_3 (M = Mo, W) reacts initially with H_2S, which leads to the formation of a layer of MS_2 around the oxide particle (Figure 2.4(a)). The formation of this surface layer avoids aggregation and coalescence of the nanoparticles, which could involve the formation of macroscopic entities and 2H-MS_2. The fast diffusion of dihydrogene in the core of the particles allows a complete reduction of oxide to form MoO_2 or $W_{18}O_{49}$ (Figure 2.4(b)). This core is then transformed slowly and gradually into sulfide (Figure 2.4(c)). When the reaction is finished, metallic oxide present at the centre of the IF structure has completely reacted. There is a hollow cavity in the centre of the particle since densities of oxide and sulfide are different (Figure 2.4(c)). This cavity is approximately equal to 15 % of the total volume of the particle. The size of the IF is directly related to the initial size of the oxide nanoparticles.

Nevertheless, synthesis methods of these two IFs are different. The synthesis of the IF-MoS_2 is a gas phase reaction, MoO_3 particles being volatile at 700 °C. For IF-WS_2, a solid-gas reaction occurs since WO_3 particles are not sublimable below 1000 °C. In 1998, Feldman *et al.* studied the kinetics of formation of these fullerenes [15], mainly the influence of the temperature and gas composition. The presence of H_2 in the gas mixture appears to be necessary even if the sulfidation of a metallic oxide does not require a reducing atmosphere. Regarding the synthesis of the IF-WS_2, Feldman *et al.* developed a new type of reactor to get a better contact between the solid particles and gases [16].

Hu *et al.* recently proposed a new method of synthesis of IF-MoS_2 by electric arc in water [17]. The IFs formed have a lower diameter (5 to 30 nm), but contain a core made of molybdenum. The presence of this core could affect the tribological properties of these nanoparticles,

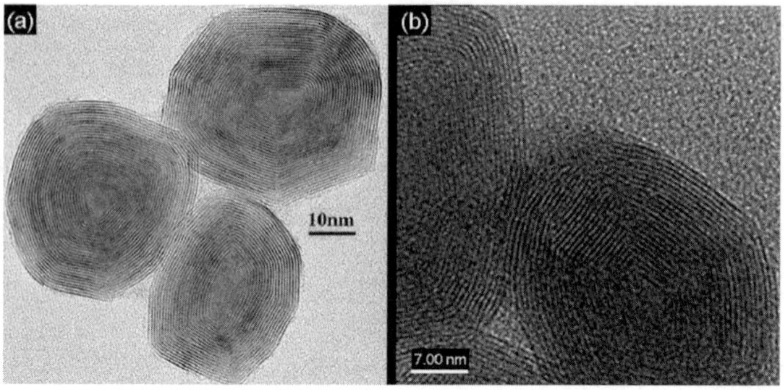

Figure 2.5 (a) HRTEM micrograph of IF-MoS_2 showing their partially faceted shape. (b) HRTEM image showing the interaction between two IF-MoS_2

in particular their elasticity. Recently, Brooks *et al.* proposed a new synthesis method using microwave-induced plasma [18]. Fullerenes of WS_2 and nanotubes of HfS_2 and ZrS_2 are produced by the reduction and the sulfidation of WO_3 or MS_3 (M = Zr, Hf) by microwave-induced plasmas of H_2S and H_2/N_2. The fullerenes prepared by the method nevertheless contain many defects. Thus, interest here is only in the nanoparticles synthesized by the method described by Feldman *et al.*

The authors studied IF-MoS_2 samples synthesized at Weizmann Institute of Science (Rehovot, Israël), by the laboratory of Professor Tenne. A size distribution made by TEM on 200 particles indicates that their mean diameter is close to 40 nm. Few IFs are perfectly spherical as in the model; the majority are partially faceted as shown in Figure 2.5(a). An aggregation of the nanoparticles by attractive van der Waals forces can also be observed (Figure 2.5(b)).

Cizaire studied in detail the structure of these IF-MoS_2 [19] and described them as an assembly of MoS_2 nanocrystals, of 10 nm length, joined on all edges to form irregular polyhedrons. An analysis by X-ray diffraction (XRD) showed a dilation of interlayers spacing from approximately 1 % (distance between the 0002 reticular planes). This expansion, initially observed in the case of closed structures made of carbon [20], is generally due to the presence of residual stresses in the curved layers. Moreover, the number of atoms increases from one layer to another so that the layers cannot be commensurable. Thus, an expansion along the *c* axis is necessary [14]. Raman spectroscopy performed with an argon laser (wavelength = 514.5 nm) allowed the hexagonal structure of nanocrystals to be determined, with Mo atoms in the trigonal position (the presence of the peak at 384 cm^{-1} was not observed in the octahedral structure) [19]. However, the resolution of these spectra is not good enough to distinguish the IF structure of the 2H structure.

Raman spectroscopy is nevertheless a very interesting technique to characterize metal dichalcogenide compounds. A Raman study of hexagonal MoS_2 highlighted the presence of three characteristic lines at 287, 383 and 409 cm^{-1}, corresponding respectively to the following modes: E_{1g}, E^1_{2g} and A_{1g}. A fourth mode, called the 'rigid layer mode' is also possible (E^2_{2g}) but difficult to observe since it appears at a wave number lower than 20 cm^{-1}. Recently, it was observed that this mode is blocked in the IF structure due to strains inside the structure which avoids the intralayer motion [21]. E_{1g}, E^1_{2g} and E^2_{2g} modes correspond to

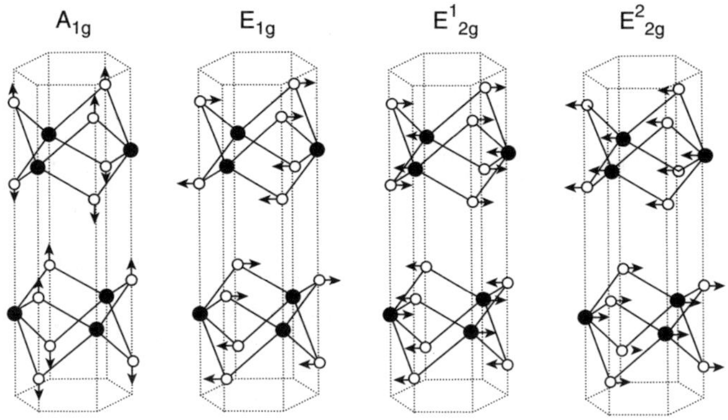

Figure 2.6 Raman active lattice vibration modes in 2H-MoS_2. Atoms vibrations are represented by arrows (molybdenum atoms are in grey and sulfur atoms in white)

vibrations of sulfur atoms (E_{1g}), and sulfur and molybdenum atoms (E^1_{2g} and E^2_{2g}) in basal plans (Figure 2.6). The A_{1g} mode corresponds to vibrations of sulfur atoms along the *c* axis.

The same transitions are observed with 2H-WS_2 (Table 2.1). Niobium and tantalum disulfide were not studied because their Raman signals were too weak. The weakness of the Raman signal of NbS_2 was observed by McMullan and Irwin [22]. Hirata and Ohuchi studied the influence of the temperature on the spectra intensity of IF-TaS_2 between 297 and 48 K. The signal intensity decreases with an increase in the temperature, and at ambient temperature the signal becomes too weak to study IF-TaS_2 by Raman spectroscopy [23]. Thus, Raman spectroscopy study was focused on IF-WS_2 and IF-MoS_2.

Resonant Raman conditions can be obtained by using a wavelength near the absorption range of the structure. 2H-MoS_2 and 2H-WS_2 are indirect gap semiconductors with two exciton absorption bands [26]. Their optical absorption spectra and those of their corresponding fullerenes and nanotube structures have already been studied and compared [27]. By using a laser wavelength of 632.817 nm, the energy of incident photons is equal to 1.96 eV, which corresponds to an absorption band of the 2H-MoS_2 structure. Resonant Raman conditions are of particular interest in the case of MoS_2: the intensity of the peak A_{1g} (409 cm^{-1}) increases (relative intensity of A_{1g} compared to E^1_{2g} (383 cm^{-1})) and new peaks appear, an asymmetrical peak at 460 cm^{-1} and peaks at 572, 599 and 641 cm^{-1}. The most intense peak appears at 460 cm^{-1} [28]. Frey *et al.* studied the deconvolution of the asymmetrical peak into two peaks at 455 and 466 cm^{-1}. The peak at 466 cm^{-1} corresponds to the A_{2u} mode not described by the theory of groups but appearing during Raman and infrared analyses. The ratio of intensity

Table 2.1 Raman active modes for the MS_2 lamellar structures

Mode	Atoms	Direction of Vibration	MoS_2 [24]	WS_2 [25]	NbS_2 [22]	TaS_2 [23]
E_{1g}	S	Basal plane	287	327	260	243
E^1_{2g}	Mo+S	Basal plane	383	351	304	306
A_{1g}	S	*c* axis	409	420	379	381

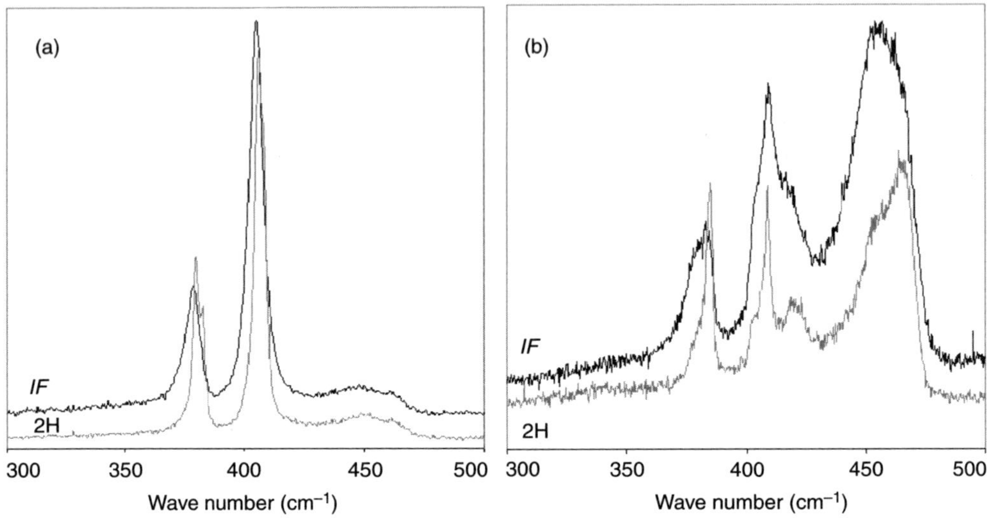

Figure 2.7 Raman spectra of IF-MoS_2 and 2H-MoS_2: (a) at 514.5 nm ($P = 60$ μW, $t = 2 \times 90$ s), where a shift and a broadening of the peaks are observed in the IF structure; (b) at 632.817 nm ($P =$ 50 μW, $t = 2 \times 60$ s). The differentiation of the two structures is easy, particularly with the asymmetrical peak at 460 cm^{-1}

of these two peaks in the 2H-MoS_2 and the IF-MoS_2 structures varies, which allows a differentiation between the fullerene and the lamellar form.

New Raman analyses were carried out at 514.5 nm (Figure 2.7(a)), which revealed a light displacement and a significant widening of the peaks in the IF structure. The use of an He/Ne laser with a wavelength of 632.817 nm (Figure 2.7(b)) enabled these two structures to be distinguished because of the resonant Raman conditions described above.

A comparative analysis of the two structures (fullerenes and lamellar) by EXAFS (extended X-ray absorption fine structure) does not show a notable difference in the radial distributions of the first two layers of neighbours around molybdenum (Mo-S distance of 0.2404 nm and Mo-Mo of 0.3158 nm). Only a reduction in the intensity of the peak corresponding to the distance Mo-Mo is observed in the spectra. Cizaire *et al.* attributed this reduction to a greater disorder in the IF structure or the grain boundaries [29].

X-ray absorption near edge structure (XANES) spectroscopy (in the total electron yield (TEY) mode) is well adapted for the study of chemical bonds and electronic structures of atoms located at the extreme surface of solids. A XANES analysis on the threshold of absorption K of sulfur (2480 eV) shows the very weak oxidation of these IFs (Figure 2.8(a)). Another XANES analysis performed on the L threshold of sulfur (175 eV) shows the presence of sulfate bonds (Figure 2.8(b)) [19]. Sulfate bonds are not present on the surface of a 2H-MoS_2 freshly cleaved in vacuum. Comparison of these two analyses leads to the conclusion of a sulfatation on the extreme surface of IF. The depth of analysis is 50 nm for the K threshold but only a few nanometres for the L threshold, so this oxidation concerns only the extreme surface (first sheets). An XPS analysis confirmed the great chemical purity of these compounds, the quantity of oxygen being lower than 1 % in atomic concentration. This analysis also showed the absence of carbon of contamination, which shows the chemical inertness of the surface of these nanoparticles. An analysis performed by electron energy loss spectroscopy (EELS) on

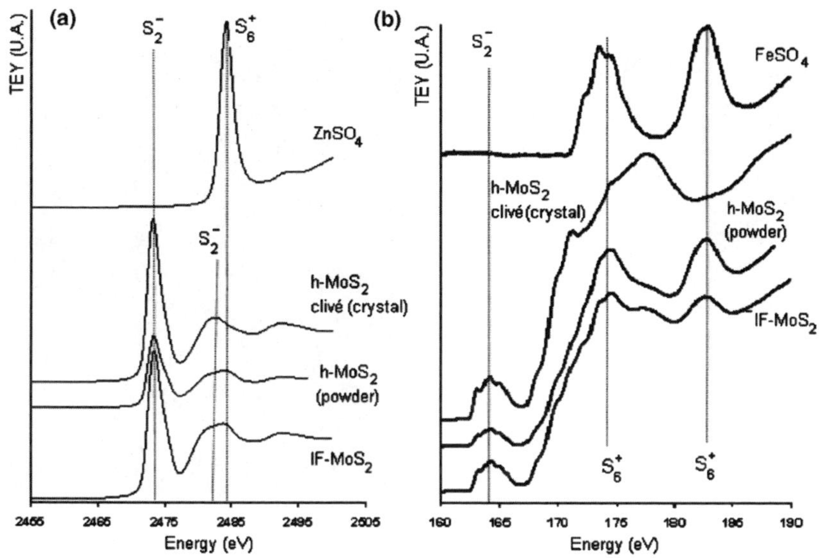

Figure 2.8 (a) S K-edge XANES analysis; (b) S L-edge analysis showing a significant amount of sulfate on the extreme surface of IF-MoS_2 [19]

one IF-MoS_2 (size probe of 1 nm) shows only the presence of Mo and S with a ratio Mo/S equal to 0.47 [29].

Observed by TEM, IF-WS_2 has a mean diameter of about 140 nm, higher than the IF-MoS_2. They are also more faceted (Figure 2.9), due to the higher number of walls. Srolovitz *et al.* calculated that from a certain number of layers fullerenes can no longer be spherical [30], but are faceted. They also showed that dislocations can tend to gather to form grain boundaries

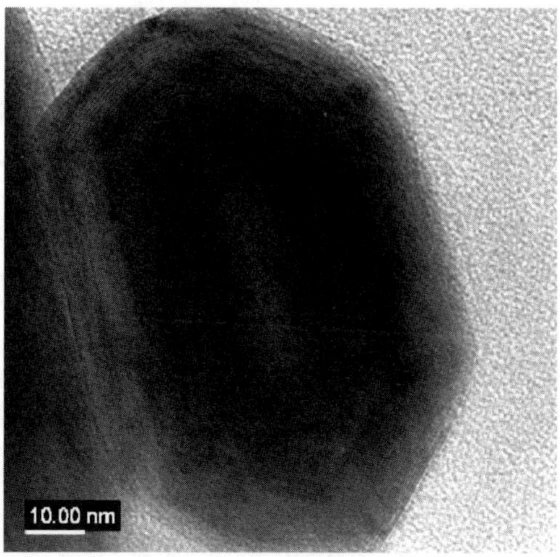

Figure 2.9 HRTEM (high-resolution TEM) micrographs of IF-WS_2 showing the multiwall structure. They have a more faceted shape. Due to their larger diameter, their multiwall structure is more difficult to observe

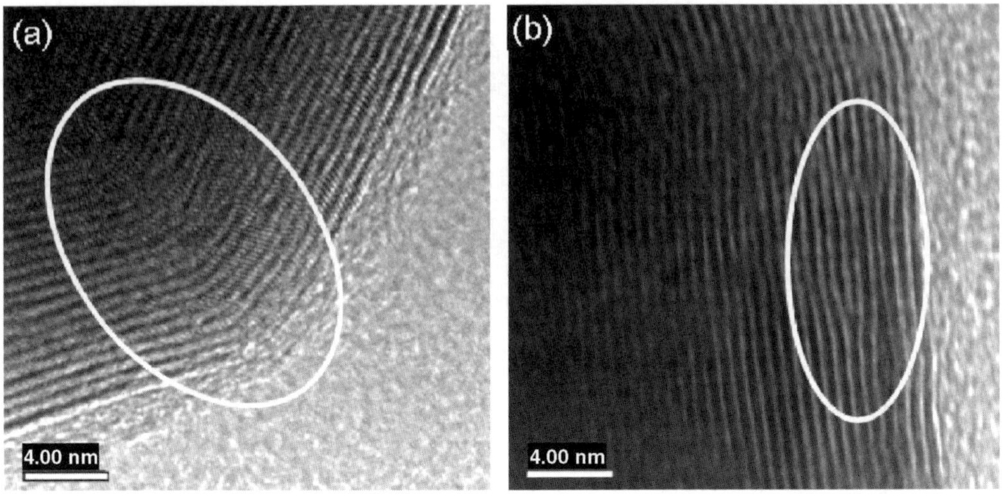

Figure 2.10 HRTEM micrographs showing (a) the curvature between two crystals and (b) stacking faults

[31]. Therefore fullerenes can be described as an assembly of WS_2 crystals. Figure 2.10(a) shows the curvature of layers inside a fullerene. Edge dislocations can also be observed in the structure (Figure 2.10(b)). These dislocations lead to a reduction of mechanical strains inside the structure [31].

Some external layers are sometimes not closed. An edge is thus observed (Figure 2.11(a)). Several layers composing fullerenes can be observed in Figure 2.11. This makes it possible to estimate the interlayer spacing. From Figure 2.11(c), it is estimated to be 0.72 nm. On some

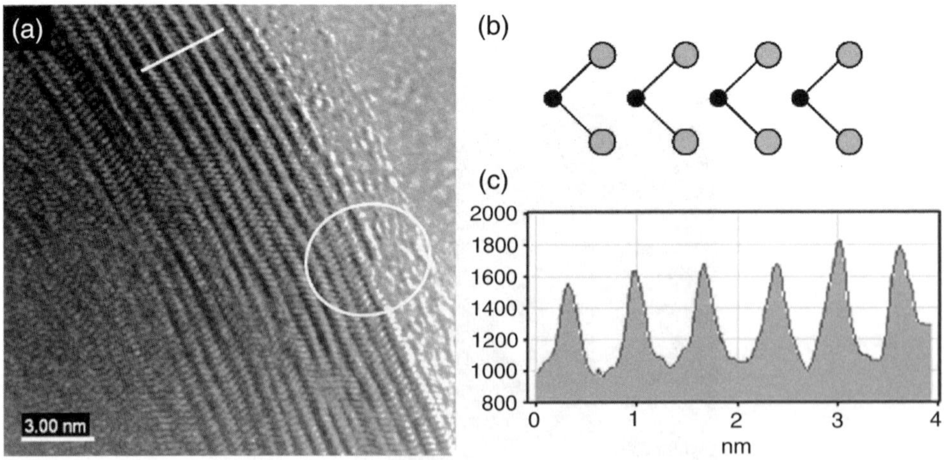

Figure 2.11 (a) HRTEM image of an IF which permits the different sheets and the presence of an unclosed sheet to be distinguished; (b) schematic representation of a WS_2 sheet observed in a [2110] orientation; (c) a grey-level profile on a line perpendicular to the sheets which allows the distance between the sheets to be measured

layers, the characteristic structure of the sandwich S-W-S can also be observed. Isshiki *et al.* showed that transmission electron microscopy is a very useful technique to observe atoms of different layers and possible defects with a [2110] orientation of the layers [32].

Analyses performed by nuclear magnetic resonance (NMR) show the presence of residual water in the fullerenes sample [33]. The presence of water is directly related to the reaction used for the synthesis:

$$MO_3 + H_2 + 2H_2S \rightarrow IF - MS_2 + 3H_2O \quad (1)$$

From the NMR results, the authors conclude that the water is present in the defects inside the structure (as shown in Figure 2.10) or at the surface. Annealing of the powder after reaction leads to a reduction of the H_1 signal, which is equal to one-quarter of the signal of the as-prepared sample. It can therefore be assumed that the water at the surface can be removed by annealing of the IF sample. Nevertheless, no oxygen was detected by the EELS analysis performed by Cizaire *et al.* [29]. Furthermore, the interlayer separation determined by X-ray diffractometry (XRD) (see below) is not sufficient to allow the presence of water molecules. The presence of water was also used to explain the results obtained by scanning tunneling microscopy (STM) by Azulay *et al.* [13]. Furthermore, the authors suggested that the results may be explained by the presence of water inside the fullerenes rather than at their surface. The presence of water inside fullerenes could be confirmed by performing a friction test on fullerene coatings in the ultrahigh vacuum (UHV) environment and by using a gas detector to monitor the presence of water.

In XRD patterns of lamellar compounds, (hk0) peaks bring information on the atom arrangement inside a basal plane while (00l) peaks are representative of the structural organization along the *c* axis. A comparison between patterns shows that IFs adopt the same structure as $2H\text{-}WS_2$ (Figure 2.12). A broadening and a shift of the (002) peak is observed for the IF structure. The broadening can be explained by the nanometre size of the crystals inside the

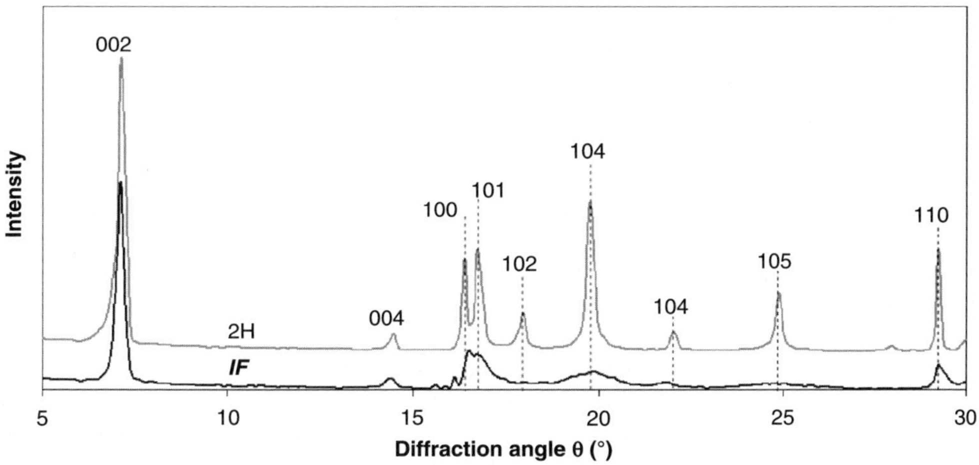

Figure 2.12 Comparison of the XRD patterns of $IF\text{-}WS_2$ and $2H\text{-}WS_2$ (Guinier powder diffractometer in the transmission position, Cu Kα1 primary beam, $\lambda = 0.15406$ nm). The (002) broadened peak in the $IF\text{-}WS_2$ spectrum indicates an expansion of 0.2 % of the *c* parameter. The (h0l) peaks are broadened in the IF, corresponding to stacking disorder

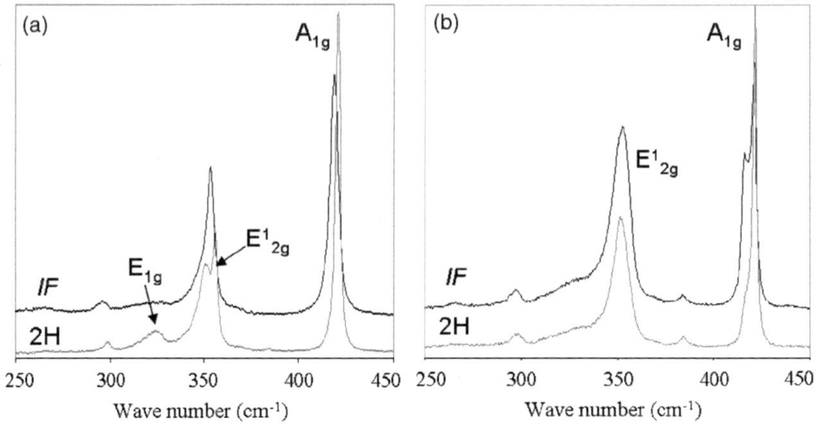

Figure 2.13 Comparison of Raman spectra of IF-WS_2 and 2H-WS_2: (a) at 514.5 nm showing a shift of the peaks and (b) at 632.817 nm showing a weak difference between IF and 2H. A wavelength of 514.5 nm permits easier differentiation

IF structure. The shift corresponds to an expansion between two adjacent WS_2 layers estimated to 0.2 %. Such expansion of the layer spacing in crystals with weak bonding between the layers is very common if the perfection of the layer stacking is lost. The (h0l) peaks are very broad in the IF structure. This proves the presence of disorder inside basal planes.

Raman spectra of the two structures (obtained with a laser of 632.817 nm wavelength) are very similar (Figure 2.13(b)). A more marked contribution at a wave number slightly lower than the peak A_{1g} is nevertheless observed for the IF structure. The A_{1g} peak appears broadened for the IF structure. A shoulder is also observed and its contribution can be attributed to a second-order Raman effect [34]. Analyses performed with a laser of 514.5 nm lead to a better differentiation of the two structures (Figure 2.13(a)). In the IF spectrum, two characteristic peaks (modes E^1_{2g} and A_{1g}) are significantly shifted. This shift can be attributed to residual strains in the fullerene structure.

A chemical analysis of IF-WS_2 determined by X-ray photoelectron spectroscopy (XPS) analysis shows a low content of carbon of contamination but the presence of oxygen (Figure 2.14(a)). For this analysis, fullerenes were deposited on a substrate covered by a thin gold layer. Decomposition of the tungsten 4f peak (W4f7 at 32.69 eV) indicates that tungsten is only in the sulfide form (Figure 14(b)). However, decomposition of the sulfur 2p peak (Figure 14(c)) shows two contributions for sulfur: WS_2 (162.37 eV) and the sulfate form (168.9 eV). The presence of sulfate in these IFs is confirmed by the oxygen peak present only in sulfate form.

Time-of-flight secondary ion mass spectrometry (TOF-SIMS) analyses were performed from the two structures (on two different points for each sample) by using a TOF-SIMS IV (ION-TOF with a bismuth gun). Several oxides and sulfates were studied to characterize as precisely as possible the chemical composition of the surface of the two structures. Table 2.2 summarizes standardized intensities of the signal of these species compared to the signal of S_2 species. For fullerenes, analyses indicate the presence of sulfate only for position 2. Position 1 corresponds to a very pure part of the sample. A comparison of results obtained with IF and

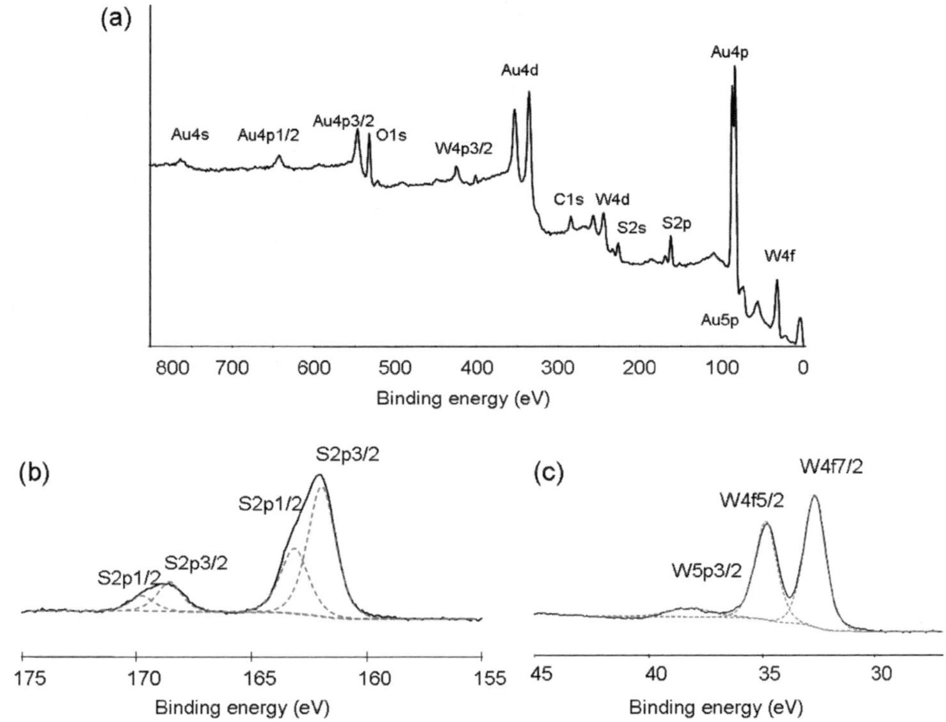

Figure 2.14 XPS analysis of IF-WS_2: (a) general spectrum; (b) the fit of the S2p peak shows that the sulfur is sulfide as well as sulfate (168.9 eV); (c) the fit of the W4f indicates that the tungsten is only sulfide

2H shows that more oxides and sulfates are present on the 2H surface. This confirms results obtained by Cizaire [19] for IF-MoS_2 and 2H-MoS_2. The closed structure of fullerenes limits oxidation, due to the absence of edges.

To resume, IF-MoS_2 and IF-WS_2 used in this study can be described as quasi-spherical and multiwall structures. Several analyses (EELS, TOF-SIMS) show their purity. Closed structures present the advantage to be chemically inert. Few sulfate forms are detected by TOF-SIMS or XANES.

Table 2.2 Intensities of signals of species observed during TOF-SIMS (in negative polarity) of 2H-WS_2 and IF-WS_2 (two analyses for each sample). Intensities are standardized compared to the one of S_2 signals. It can be observed that there is a greater proportion of sulfide species compared to oxide and sulfate species for the IF structure, particularly for the first analysis

	S	S_2	O	HSO_4	WS_3	WO_3	$W(SO_4)_2$	WS_3/WO_4
IF-WS_2	718.0	100.0	418.0	1775.0	101.0	33.0		3.1
IF-WS_2	689.0	100.0	727.0	3708.0	114.0	63.1	2.1	1.8
2H-WS_2	1240.0	100.0	1089.0	5006.0	126.0	78.5	2.0	1.6
2H-WS_2	982.0	100.0	846.0	5537.0	163.0	87.3	3.8	1.9

2.3 IF-MoS_2 and IF-WS_2 as Additives in Boundary Lubrication

The C_{60} fullerene structure has interested tribologists for a long time. Because their structure looks like a ball, C_{60} was supposed to have interesting tribological properties. It was supposed that fullerenes could 'roll' in the contact, the system being like a ball bearing on a nanometre scale [35]. In fact, results obtained are contradictory: some studies report low friction with C_{60} films [35], but other authors found that it is higher than the coefficient of friction for amorphous carbon or diamond [36]. A study using atomic force microscopy (AFM) also showed that the rotation of C_{60} does not reduce the friction [37]. Miura and Kamiya studied the behaviour of a single layer of C_{60} between two layers of graphite [38]. This system can be compared to a ball bearing. A stick-slip motion of C_{60} is observed. A low friction is measured, which suggests a future use of this system could be in micromachines.

As in the case of C_{60}, the quasi-spherical closed structure of the IF presents several advantages:

- the absence of edges;
- the size of these objects which is at least twenty times smaller than the 2H-MS_2 powder.

Lamellar structures are often used but adhere easily to surfaces by their flexibility, leading them to conform easily to the topography of asperities. IF structures, due to the two properties indicated previously, do not adhere easily to surfaces [11]. Moreover, their hollow and multiwall structure gives them great elasticity. These particles can also be used as separators between two surfaces, thus preventing direct contact between them [11].

The studies carried out on the tribological properties of the IF-MS_2 have to date been focused on the IF-WS_2. Added to oils and greases or impregnated into a porous matrix, IF-WS_2 presents better friction-reducing and antiwear properties than the corresponding lamellar structure [39]. A comparison between tribological properties of IF-WS_2, 2H-WS_2 and 2H-MoS_2 shows clearly that the IF structure presents better friction-reducing properties [40]. Two reasons explain these results: first, fullerenes avoid contact between asperities [39]; second, the absence of dangling bonds make the fullerenes much more chemically stable.

The problem of lamellar structures like 2H-MoS_2 and 2H-WS_2 is the easy oxidation at the edges of crystals. These structures easily adhere to surfaces, while fullerenes do not, as Rapoport *et al.* showed using surface force microscopy (SFM). IFs are swept by the tip in the contact image mode, which proves their weak adhesion on a surface [41]. In the case of porous matrices, platelets are perpendicular to the surface involving their rapid deterioration during friction, whereas the fullerenes enter into the asperities, which act as reservoirs [42]. The spherical structure of the IF avoids problems of orientation encountered with 2H platelets [43].

Greenberg *et al.* compared the effects of IF addition in oil under three lubrication regimes: hydrodynamics, mixed and limit [44]. They based the lubrication mechanism of IF on a film formation on surfaces and showed that the IF are most effective at mixed lubrication because all conditions for a good efficiency of IF are combined. In the hydrodynamic regime, fullerenes do not have interactions with surfaces. In a boundary lubrication regime, film formed on surfaces is quickly removed, due to contact severity.

Drummond *et al.* studied properties of transfer films formed during friction tests of IF-WS_2 or 2H-WS_2 dispersions in tetradecane between two mica surfaces. First, they proved that transfer film is formed under shear, but not under hydrostatic pressures [45]. Transfer films obtained from fullerenes are made of small and very thin islands (1.2 to 2.4 nm) distributed in a homogeneous way on surfaces. Tribofilm obtained with 2H is rough and disorganized and is composed of particles from 15 to 40 nm in height and 0.1 to 0.8 μm broad.

Recently, Rapoport *et al.* studied the effect of IF-WS_2 rubbed on alumina surface [46]. They demonstrated that IF-WS_2 is preserved in the pores of the sintered alumina. It is peeled off and supplied to the solid lubricant during a long-term test. This mechanism will be developed in Section 2.8.

The Nanomaterials Society launched the marketing of 'Nanolub', a lubricant made up of a lubricating base (oil or grease) and IF-WS_2. A production of 2000 tons/year is considered. Industrial tests proved that Nanolub had better lubricating properties compared with grease or oil alone, or oils containing 2H-MoS_2 platelets [47]. Nevertheless, precise action mechanisms of these additives are still unknown.

Chhowalla and Amaratunga studied the tribological properties of IF-MoS_2 coatings in the presence of moisture and compared the results with those obtained from 2H-MoS_2 coatings [48]. With IF coatings, lower friction coefficients are obtained. These best performances can be attributed to the absence of edges due to the closed structure of the IF.

Recently, Katz *et al.* proposed the use of IF-WS_2 nanoparticles as lubricant for orthodontic wires [49]. Ni-P coatings containing IF-WS_2 are very useful in reducing the friction between the wires and the brackets glued on teeth. Zou *et al.* found interesting results with Ni-P-(IF-MoS_2) coatings tested in air and vacuum, confirming the improvement of Ni-P coatings with the addition of MoS_2 fullerenes [50]. The results obtained by Katz are very promising and this example demonstrates the wide range of applications possible for these nanoparticles. Furthermore, tests of toxicology of these nanoparticles have demonstrated that they are not toxic in oral administration or upon exposure to skin. Thus, these nanoparticles seem very promising to reduce both friction and wear in many systems.

2.3.1 IF-MoS_2

Cizaire *et al.* studied IF-MoS_2 as lubrication additives in boundary lubrication [29]. Added at 1 wt % to a poly-alpha-olefin (PAO) synthetic base oil, IF-MoS_2 led to a very low friction coefficient of about 0.06. Nevertheless, no clear lubrication mechanism was proposed. In this section, works related to the study of the effect of the concentration and pressure on the tribological properties of IF-MoS_2 are reported. Tests were carried out on the environmental pin-on-flat tribometer (described in the Appendix in Section A.1).

A decrease in the friction coefficient is observed when the IF concentration is increased up to 1 wt %. This percentage can be considered as the optimal concentration (Figure 2.15) since a concentration of 2 wt % does not improve results and leads to problems of dispersion and stability of the suspension.

Figure 2.16 presents results obtained for several contact pressures: 0.66, 0.83, 1.12 and 1.42 GPa (pressures calculated using Hertzian theory) with an optimal concentration of 1 wt %. A decrease in the friction coefficient is observed when the contact pressure is increased. Moreover, the low friction coefficient is obtained from the very beginning of the tests. This characteristic is very interesting for potential industrial applications since it is known that traditional

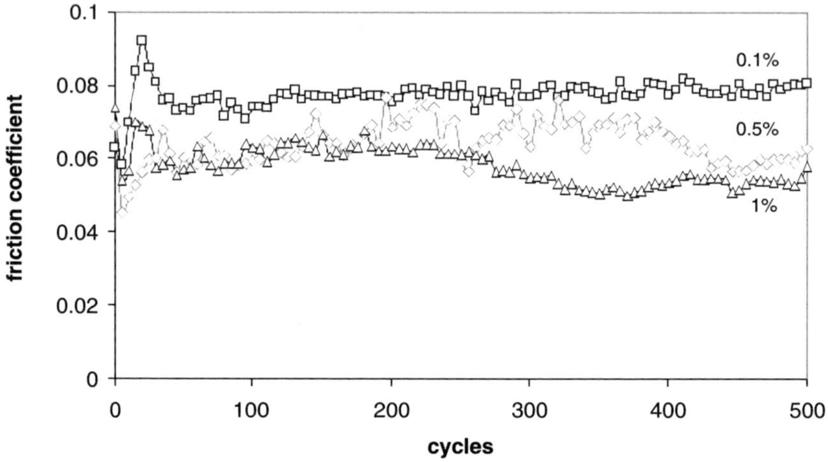

Figure 2.15 Boundary lubrication with IF-MoS_2 dispersed in PAO ($P = 1.12$ GPa, $v = 2.5$ mm/s, $T = 20\,°C$). Effect of concentration on the friction coefficient

additives currently used have to be thermally activated to be efficient. These materials could solve the problem of the cold start of thermal engines.

A comparison of the results obtained with the PAO and the PAO+IF shows that the addition of IF leads also to a significant reduction in wear (Table 2.3). Nevertheless, wear scars have a larger diameter than the Hertz calculated diameter, in particular at low contact pressure. This indicates a slight wear phenomenon of the pin. With IF, a thin tribofilm is observed

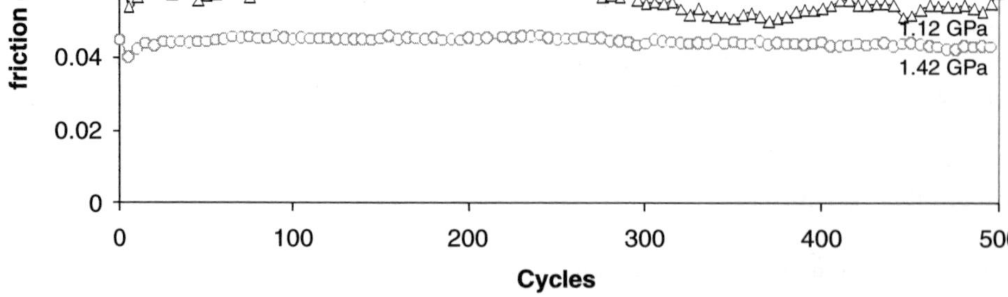

Figure 2.16 Effect of the contact pressure on the friction coefficient with IF-MoS_2 dispersed at 1 wt % in PAO ($v = 2.5$ mm/s, $T = 20\,°C$)

Table 2.3 Diameter (in μm) of wear scars observed on the pin after friction tests compared with the Hertz calculated diameter

	PAO	1 % IF-MoS_2	Hertz calculated diameter
0.66 GPa	115	110	54
0.83 GPa	170	120	68
1.12 GPa	175	125	92
1.42 GPa	180	140	116

on the pin. Cizaire estimated its thickness to 30 nm by Auger analysis coupled with ionic etching [19].

Raman analyses were performed on standards of molybdenum oxide and sulfate to determine the characteristic lines of these compounds and thus to detect their possible presence on surfaces during friction and after friction (Figure 2.17). MoO_3 presents two peaks at 820 and 996 cm^{-1}, which is in agreement with the literature [51] and WO_3 two peaks at 717 and 807 cm^{-1} [52]. Analyses on the wear scar after a test of 500 cycles at a contact pressure of 1.12 GPa were carried out (Figure 2.18). They clearly show the absence of oxide in this tribofilm and in particular the absence of MoO_3. The spectrum obtained can be simulated by a linear combination of the IF-MoS_2 spectrum (70 %) and the 2H-MoS_2 spectrum (30 %) (Figure 2.18). This analysis clearly shows the formation of 2H-MoS_2 on the surface during friction.

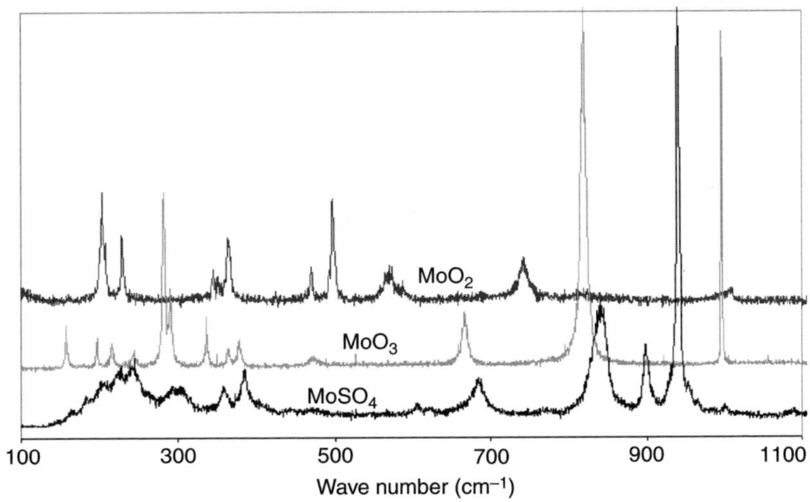

Figure 2.17 Raman spectra of MoO_2, MoO_3 and $MoSO_4$ obtained using a red laser ($\lambda = 632.817$ nm)

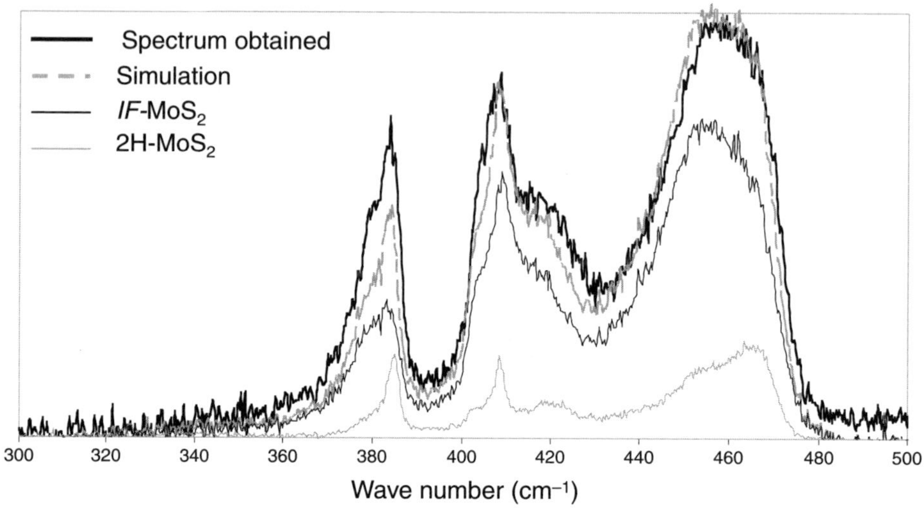

Figure 2.18 Raman spectrum of the film present on the pin compared with a simulation corresponding to a mixture of 70 % IF and 30 % 2H

2.3.1.1 Friction Tests at High Temperature (70 °C)

In the case of the IF-MoS_2 tested at high temperature, the friction coefficient is fairly high and not very stable (Figure 2.19). These results are explained by a problem of dispersion of the IF-MoS_2 in PAO. The higher the temperature, the more difficult is the dispersion of the fullerenes in PAO. This proves the necessity to use a dispersant at high temperature.

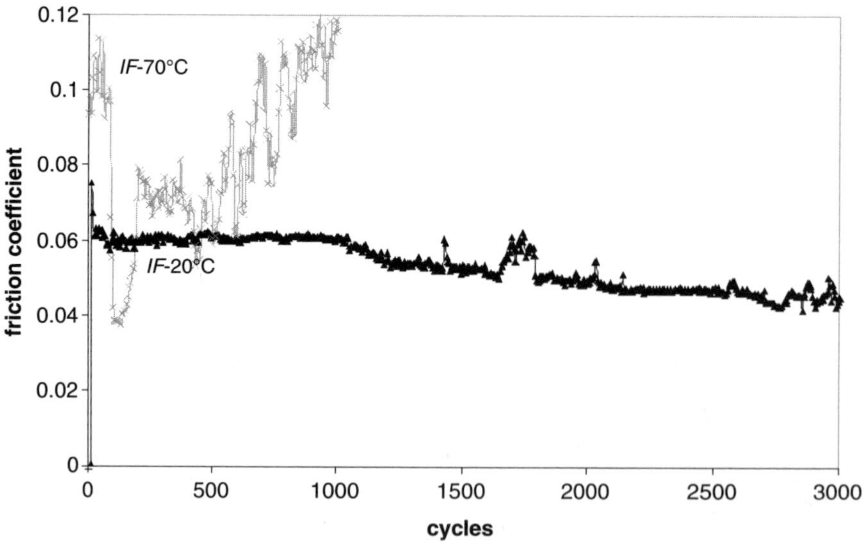

Figure 2.19 Temperature effect on friction tests with IF-MoS_2 (P = 1.12 GPa, v = 2.5 mm/s)

2.3.2 IF-WS_2

In a similar way the same kind of study was performed with IF-WS_2. The optimal concentration of IF-WS_2 in PAO was first determined. The same tendency as with IF-MoS_2 is observed. The friction coefficient decreases when the IF-WS_2 concentration in oil increases. The lowest friction is obtained with a concentration of 1 wt %, for which a value of 0.04 is reached (Figure 2.20). Whatever the concentrations, the friction coefficient is low and stable from the first cycles [53]. Pure PAO presents antifriction properties after 200 cycles, which is still not explained.

Friction tests carried out at several contact pressures, from 0.33 to 1.72 GPa, show a diminution of the friction coefficient when the contact pressure increases (Table 2.4). A critical contact pressure, from which the friction coefficient is very low (0.04), can be determined and is equal to 0.83 GPa. Whatever the load, the addition of IF-WS_2 in PAO leads to a significant reduction of wear compared to that observed with pure PAO (Table 2.4). The wear observed is very low since the wear scar diameter is very similar to the Hertz calculated diameter.

The feature of the wear scar observed on the pin is particular because it presents a crown around the contact area (Figure 2.21(a)). This particular feature is observed whatever the contact pressure tested. A study using scanning electron microscopy (SEM) highlights the different characteristic areas of the wear scar (Figure 2.21(b)); a thick and smooth film is observed in the middle of the scar (Figure 2.21(d)), while a rough film is observed in the crown (Figure 2.21(c)). It seems that the film in the crown contains intact fullerenes, which would be imprisoned around the zone of contact. These fullerenes could lead to a bearing effect, which will decrease the contact area pressure since the contact area is more important.

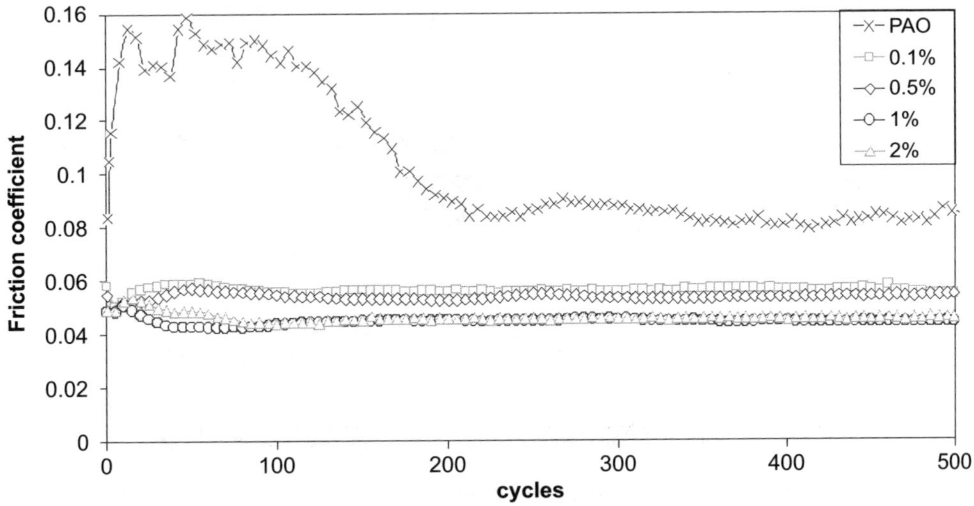

Figure 2.20 Effect of the concentration of IF-WS_2 in PAO on the friction coefficient ($P = 1.12$ GPa, $v = 2.5$ mm/s, $T = 20\,°C$). An optimal concentration of 1 wt % leads to a very low friction coefficient (0.04)

Table 2.4 Values of friction coefficient and wear scar diameters obtained at different contact pressures (lubricants: PAO or PAO + 1 % IF-WS_2, $v = 2.5$ mm/s, $T = 20\,^\circ$C). The pressure leads to a decrease of the friction coefficient. A plateau is observed from 0.83 GPa. The wear observed with IF-WS_2 is very weak (diameter of the wear scar is equal to the Hertz calculated diameter)

			PAO		IF-WS_2 1 % PAO	
Pression (GPa)	Ball diameter (mm)	Hertz calculated diameter (μm)	Friction coefficient	Wear scar diameter on the pin (μm)	Friction coefficient	Wear scar diameter on the pin (μm)
0.33	12	54	0.26	85	0.06	65
0.42	12	68	0.26	115	0.05	80
0.52	12	86	0.16	150	0.05	95
0.66	6	54	0.16	115	0.05	75
0.83	6	68	0.25	170	0.04	86
1.12	6	92	0.16/0.09	175	0.04	100
1.42	6	116	0.08	180	0.04	120
1.72	4.5	106	0.07	150	0.04	115

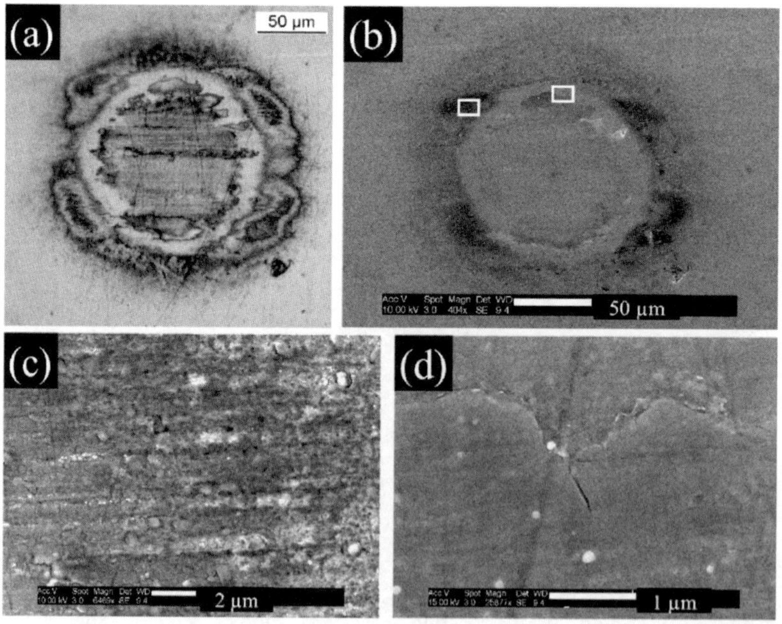

Figure 2.21 (a) Optical image of the wear scar diameter observed on the pin; (b) SEM image of the same scar; (c) detail of the film observed in the crown; (d) detail of the film in the centre of the scar. Friction test at 1.72 GPa, $v = 2.5$ mm/s, $T = 20\,^\circ$C

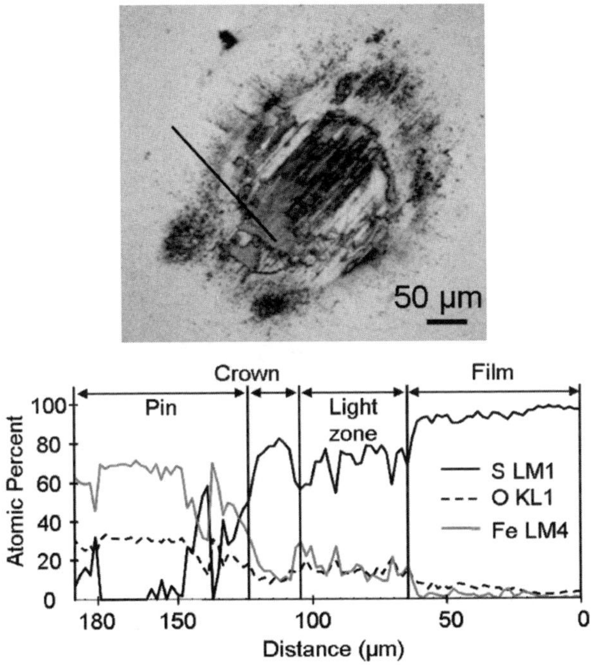

Figure 2.22 An AES line scan on the friction wear scar: (a) optical image showing the line analysed; (b) profile of different elements (S, O, Fe). Friction test at 0.83 GPa, $v = 2.5$ mm/s, $T = 20\,^{\circ}$C

To determine a possible difference in chemical composition between the film and the crown, an Auger profile was performed on the various parts of the wear scar: the film and the crown (see line on Figure 2.22). Sulfur, iron and oxygen were followed. Tungsten is very difficult to detect by Auger. In the central part of the film, only sulfur is detected, indicating the presence of a thick film made of tungsten disulfide. A slight decrease in sulfur is then detected with an increase in iron and oxygen between the thick film and the crown. This indicates the presence of a thin film not observed by optical and scanning electron microscopy. Inside the crown, intensities of the peaks of iron and oxygen decrease. The crown is then composed of WS_2 and does not contain tungsten oxides (oxygen observed is due to the substrate). Absence of tungsten oxides was confirmed by Raman spectroscopy. Characteristic peaks of WO_3 and WO_2 were not observed on the spectra (Figure 2.23). Combining these results with SEM observations, it can be deduced that this crown consists of IF agglomerations that have adhered to the pin.

2.3.2.1 Study of the Bearing Effect

To highlight the phenomenon of the bearing effect, two methods are developed. The first one consists in varying the convergent angle. The second consists in varying the size of the IF-WS_2 particles and testing them with the same experimental conditions.

By using balls of different diameters (6 and 12 mm) the convergent angle of the ball on the flat is modified (see Figure 2.24). Comparison of the experiments carried out with these

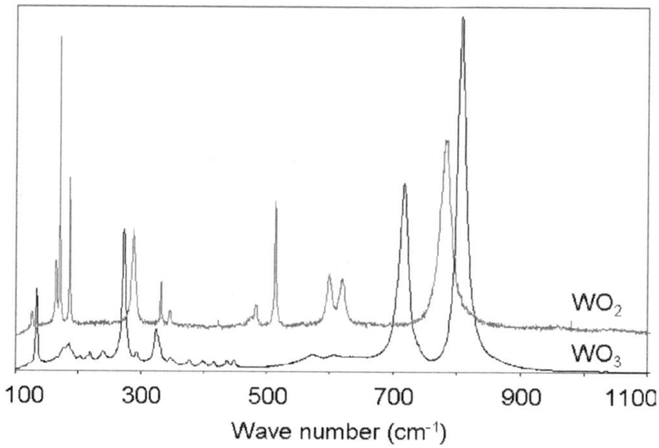

Figure 2.23 Raman spectra of WO_2 and WO_3 obtained with a green laser ($\lambda = 514.5$ nm)

two balls at the same contact pressure shows the effect of the convergent angle (Figure 2.25). The friction coefficient is lower with a ball of large diameter. This is in agreement with the hypothesis of the phenomenon of the bearing effect. It is more important in the case of a ball of large diameter (see Figure 2.25).

Observations of wear scars on pins and flats clearly show this phenomenon of the bearing effect (Figure 2.26), in particular on the flat on which a central zone and two lateral zones can be observed (Figure 2.26(d)). Moreover, the wear scar observed on the pin has a lower diameter (120 μm) than the corresponding Hertz diameter (136 μm), indicating that the pressure inside the contact area is lower than the theoretical calculated pressure.

Friction tests performed with the same contact pressure but with an increasing number of cycles show the formation and the evolution of the film and the crown (Figure 2.27). After 20 cycles (Figure 2.27(a)), a film is already observed and its thickness increases during the friction test (Figure 2.27(b) and (c)). In the same way, the crown around the contact area darkens throughout the test, indicating that more and more particles adhere on the pin. The formation of film from the very beginning of the test can explain the low friction coefficient observed when the test starts.

The friction coefficient obtained with IF-WS_2 of 90 nm diameter is similar to the one obtained with the 140 nm diameter IF-WS_2 at the end of the test. The friction coefficient is nevertheless higher at the beginning of the test before decreasing progressively (Figure 2.28). The lowest friction coefficient observed with IF-WS_2 of large diameter can be explained by

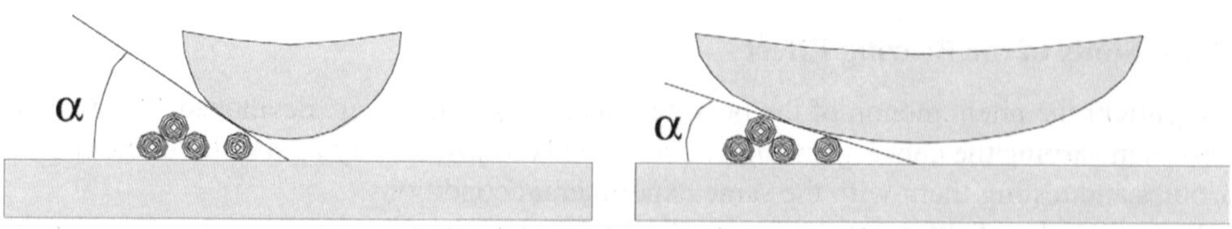

Figure 2.24 Schematic representation of the convergent angle with the two different pins used

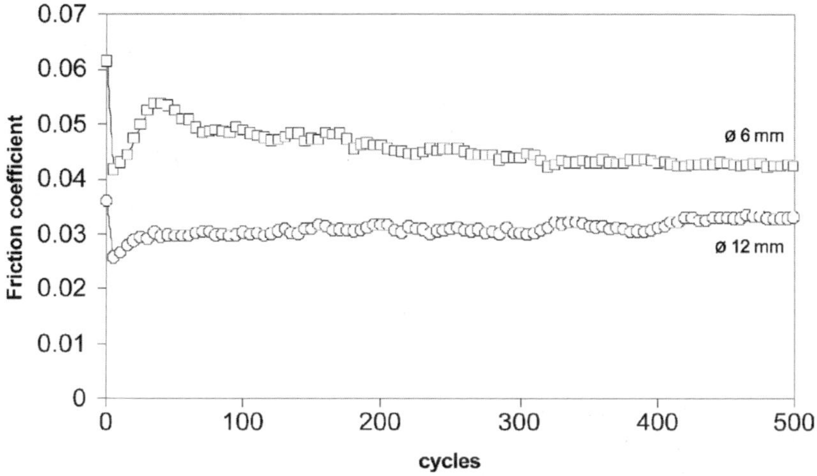

Figure 2.25 Friction coefficient evolution function of the pin diameter ($P = 0.83$ GPa, $v = 2.5$ mm/s, $T = 20\,^{\circ}$C). A higher diameter pin leads to obtain a lower friction coefficient, indicating an effect of the convergent angle

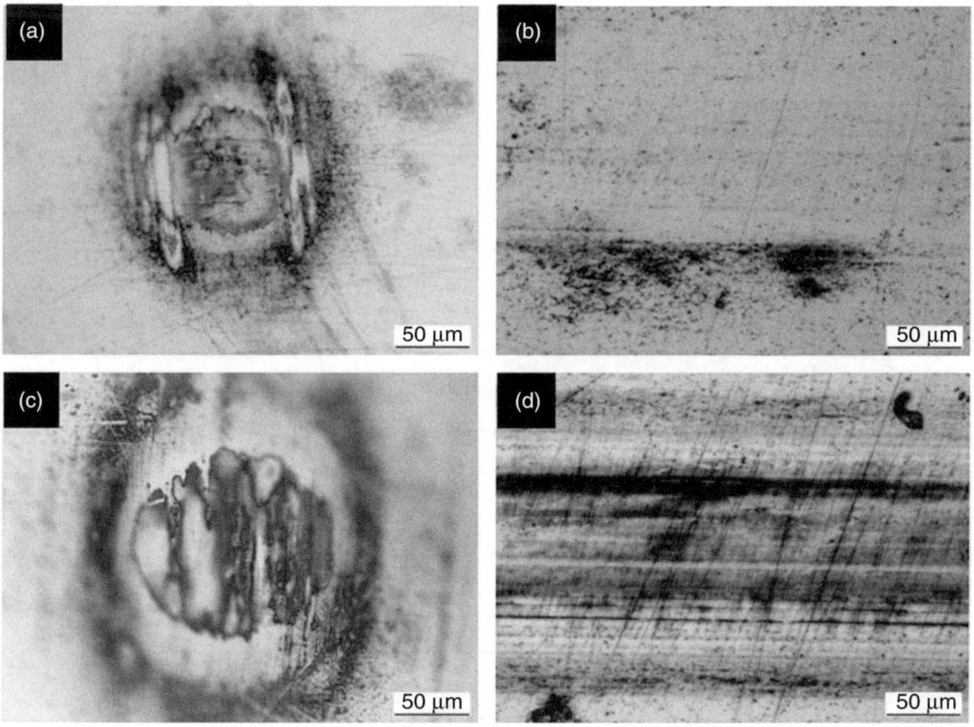

Figure 2.26 Wear scars observed on pins and flats: with a 6 mm diameter pin (a and b) and with a 12 mm diameter pin (c and d)

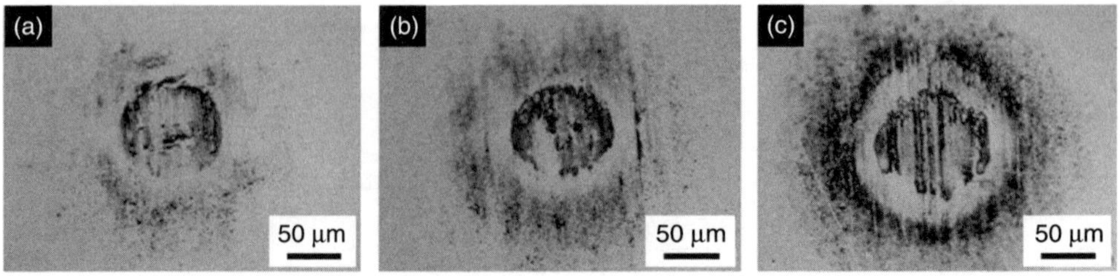

Figure 2.27 Wear scars observed on pins after tests at different numbers of cycles (P = 1.12 GPa, v = 2.5 mm/s, T = 20 °C): (a) 20 cycles, (b) 50 cycles and (c) 500 cycles

the presence of the crown around the contact area, which is observed from the beginning of the test (Figure 2.27) and which leads to a decrease of the pressure inside the contact area. This crown is not observed with IF-WS_2 of 90 nm diameter.

The sliding contact used here is composed of a sphere on a plane. Under normal loading, the solids are reversibly deformed (purely elastic and isotropic behaviour of the substrates). Thanks to Hertzian theory, the contact area is determined to be circular with a radius equal to a (Figure 2.29). The calculation of the distance between the ball and the flat where the crown appears would lead to a better understanding of the formation mechanism of the crown. The formula of the deformation of the ball determined by Cameron gives the distance between the pin and flat at a given distance (r) [54]:

$$h = \frac{a^2}{\pi R}\left[\sqrt{\frac{r^2}{a^2 - 1}} - \left(2 - \frac{r^2}{a^2}\right)\cos^{-1}\left(\frac{a}{r}\right)\right] \tag{2}$$

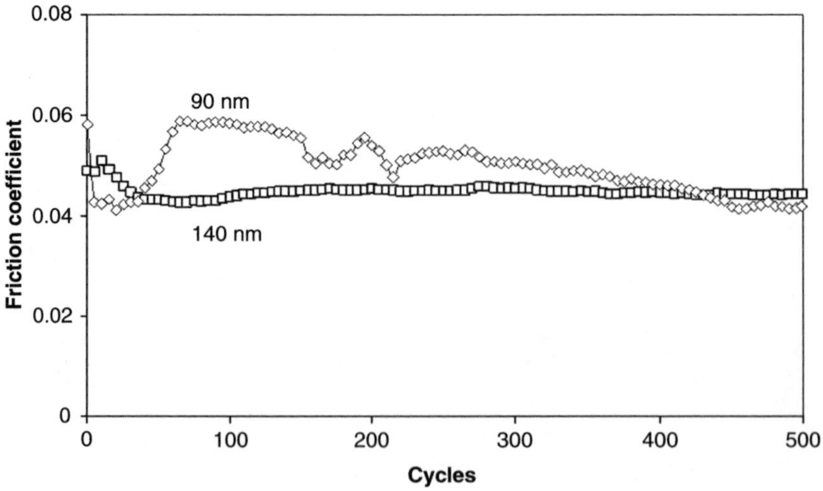

Figure 2.28 Comparison between two different size samples (P = 1.12 GPa, v = 2.5 mm/s, T = 20 °C). With smaller IF, the friction coefficient is higher at the beginning of the test but decreases regularly to obtain the same value than the other sample at 400 cycles

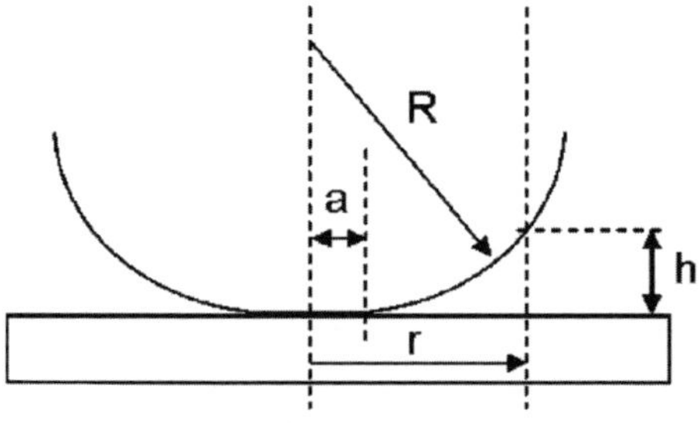

$$h = \frac{a^2}{\Pi R}\left[\sqrt{\frac{r^2}{a^2 - 1}} - \left(2 - \frac{r^2}{a^2}\right)\cos^{-1}\left(\frac{a}{r}\right)\right]$$

Figure 2.29 Geometry of the elastically deformed pin and the expression of the height (h) between the pin and flat. It depends on the pin radius R, on the pressure inside the contact and on materials of the two antagonist surfaces

Using distances r measured on optical pictures of wear scar obtained at different contact pressures and with balls of different diameters, results show the same distance of 500 nm whatever the contact pressure (Table 2.5). This height corresponds to the main size of fullerene aggregates. Since the fullerenes of large diameters interact more easily than those of small diameters, IF-WS_2 of diameter 140 nm will form aggregates more easily.

As previously mentioned, the dispersion of these fullerenes in oil is quite difficult and some of them remain in the form of aggregates. Their presence seems not to be negligible and brings a contribution to the tribological properties of these systems. It is nevertheless necessary that some fullerenes should not be aggregated to supply the contact area. However, the role of the crown still remains a mystery. Moshkovith *et al.* studied the effect of aggregate size on the tribological properties of IF-WS_2 dispersed in a paraffin base oil. By increasing the time of mixing, the size of aggregates decreases and the probability of the penetration of IF-WS_2 in the contact area is increased [55]. Furthermore, reproducibility of friction experiments is increased. A new kind of experiment developed by Perfiliev *et al.* brought important clues to solve this problem [56]. They investigated the role of vibrations of the tribological systems on the tribological properties of nanoparticles and particularly on their penetration in the contact area. By applying forced mechanical excitations to the tribological systems, they obtained a very low friction coefficient (Figure 2.30). They explained these results by a better penetration of nanoparticles and aggregates inside the contact area. Nanoparticles enter the contact area easily because the mechanical excitation leads to a variation of the gap between rubbed surfaces. The variation of the amplitude of excitation seems to have an influence on the results.

Dispersions of IF-WS_2 can also be studied by the 'wet scanning transmission electron microscopy (STEM)' technique, which allows the observation of nano-objects dispersed in a

Table 2.5 Calculation of the distances between the pin and flat

Ball (mm)	6	6	3	3	3	2,5
P (GPa)	0.33	0.42	0.66	0.83	1.42	1.72
a (μm)	27	34	27	34	58	53
Distance r (μm)	80	95	65	65	80	85
Height (μm)	0.43	0.59	0.50	0.41	0.54	0.63

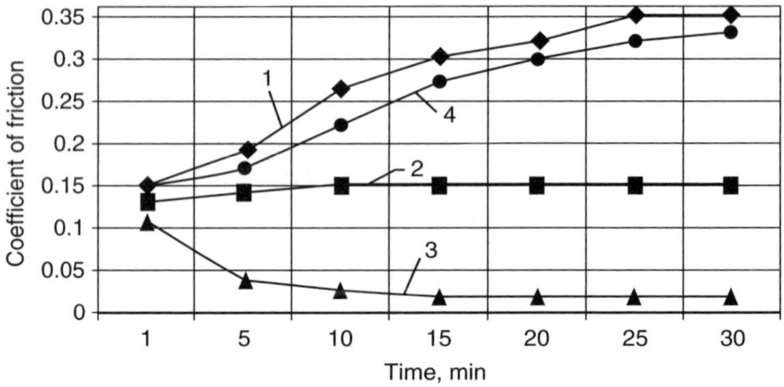

Figure 2.30 Friction coefficient of pure paraffin oil without mechanical excitations of the system (1) and with excitations (4), and paraffin oil containing IF-WS2 without (2) and with excitations (3). With kind permission from Springer Science and Business Media from Perfiliev *et al.* [56], Figure 2, © (2006) Springer Science and Business Media

liquid phase. This technique was developed by Gilbert Thollet and Agnès Bogner at MATEIS Laboratory at INSA de Lyon [57]. A droplet of dispersion is deposited on a TEM grid and observed in an environmental scanning electron microscope in a transmitting mode (the detector is placed below the TEM grid). The only requirement is to prepare an oil film thin enough to be 'transparent' to electrons. This technique is very useful to characterize dispersions in the liquid phase, such as a colloidal solution of nanoparticles or natural rubber latex. In the case of IF-WS_2 in oil, fullerene aggregates of several sizes were observed in oil (Figure 2.31). Owing to the field emission gun of the microscope used, a very good resolution was obtained. Thus, isolated fullerenes were observed (Figure 2.31(b) and (c)) and the faceted structure of some fullerenes can also be distinguished.

2.3.2.2 Comparison with 2H-WS_2

A comparison between IF-WS_2 and the corresponding lamellar structure indicates that fullerenes present better tribological properties than the lamellar structure at two different contact pressures (Figure 2.32). This confirms the advantage of using the fullerene structure instead of the corresponding lamellar structure.

2.3.2.3 IF-WS_2 Coatings: Friction under Ultrahigh Vacuum

Experiments under ultrahigh vacuum allow work to take place in a perfectly controlled environment, free from any contamination, and to study the interactions between IF-WS_2 and the pin (film formation). Grossiord showed that friction under ultrahigh vacuum simulates an 'instantaneous' lubricated contact. Under boundary lubrication regime no external molecule can enter [58]. In practice, friction under ultrahigh vacuum also simulates the conditions in the space environment. Grossiord [58] performed the experiments with an ultrahigh vacuum tribometer (see the Appendix, Section A.3) coupled with *in situ* surface analysis techniques (XPS and Auger electron microscopy (AES)). Due to this machine, an analysis of the surface

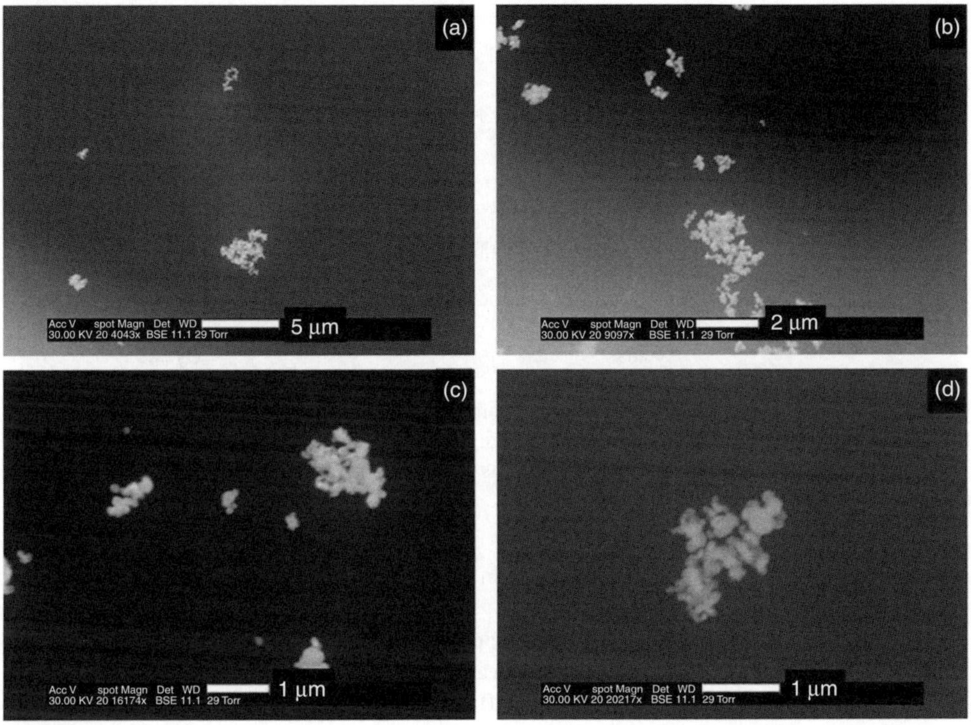

Figure 2.31 Characterization of the IF-WS_2 dispersion in oil by the 'wet STEM' technique. This technique allows to observe fullerenes in oil

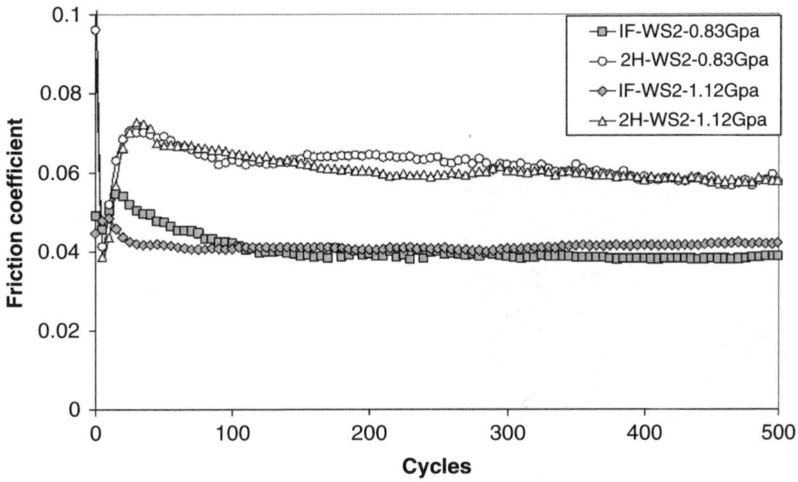

Figure 2.32 Friction coefficient evolution obtained with IF-WS_2 and 2H-WS_2 at 1 wt % in PAO at two different contact pressures (0.83 and 1.12 GPa). A lower friction coefficient is obtained with IF-WS_2 at the two studied contact pressures

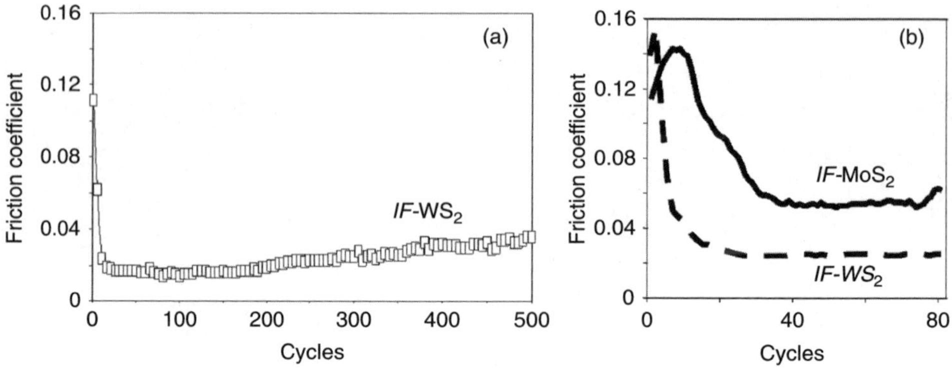

Figure 2.33 Friction coefficient obtained for a contact between a steel pin and IF-WS_2 coating under ultrahigh vacuum: (a) $P = 500$ MPa, $v = 0.05$ mm/s, $T = 20\,°C$, (b) $P = 500$ MPa, $v = 0.01$ mm/s, $T = 20\,°C$. A very low friction coefficient (0.02) is obtained, lower that those obtained with IF-MoS_2 coating

can immediately be performed after the friction test, without exposing surfaces to ambient air, thus avoiding any contamination.

A first deposit was carried out by toluene evaporation on a flat made of AISI 52100 polished mirror. The flat was heated to allow a faster evaporation of solvent.

Friction coefficients obtained are very low (0.02 to 0.04) (Figure 2.33) and lower than those measured with an IF-MoS_2 deposit tested under the same conditions [19]. Auger analyses carried out on wear scar of the pin confirm the formation of a film on the pin during the friction test. This film is composed only of W and S (Figure 2.34).

A second deposit made by burnishing was tested. The deposit by burnishing consists in rubbing IF-WS_2 on the surface with a clean rag to create a very thin deposit (Figure 2.35(a)). The layer obtained by this method is not observed optically. Thanks to an in-depth profile of

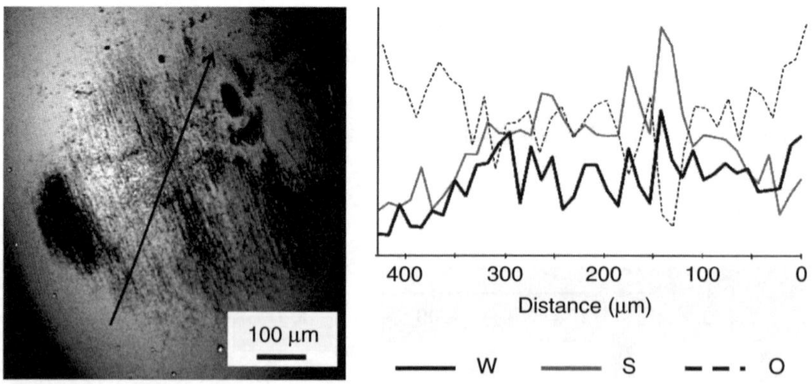

Figure 2.34 Optical image of the pin ($P = 500$ MPa, $v = 0.01$ mm/s, $T = 20\,°C$, 80 cycles) with a line scan of W, S and O along the line indicated on the image. This analysis clearly shows the formation of a WS_2 film on the pin

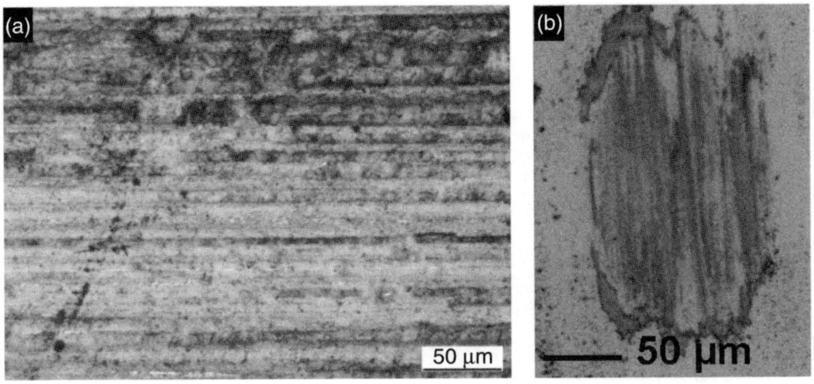

Figure 2.35 IF-WS_2 coating obtained by burnishing. (b) Wear scar observed on the pin after the friction test

distribution of the elements (W, S, Fe, O and Cr), performed by dynamic secondary ion mass spectrometry (SIMS), the thickness of this layer can be estimated to 200 nm.

For the two temperatures tested, the friction coefficient always reaches a very low value close to 0.007 during the first thirty cycles (Figure 2.36). At 25 °C, the friction coefficient then increases regularly to reach a value of 0.02. In spite of the increase in the friction coefficient, the value remains nevertheless lower than the one obtained with the first deposit. This shows the performances of coatings obtained by burnishing. At the end of the friction test, a transfer film is observed on the pin (Figure 2.35(b)), but no wear scar is observed on the flat (Figure 2.35(a)). This shows the good resistance of these coatings during friction tests and can explain

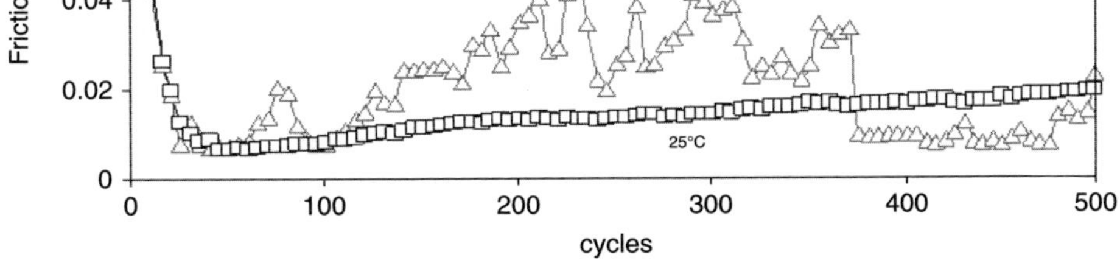

Figure 2.36 Friction coefficients obtained with a IF-WS_2 coating deposited by burnishing. This coating was studied at 25 and −135 °C ($v = 0.05$ mm/s, $P = 500$ MPa). In the frame, the friction coefficient obtained during cycle 50 at 25 °C shows the very low value measured. The friction coefficient increases with the change in direction of the friction (1800 μm)

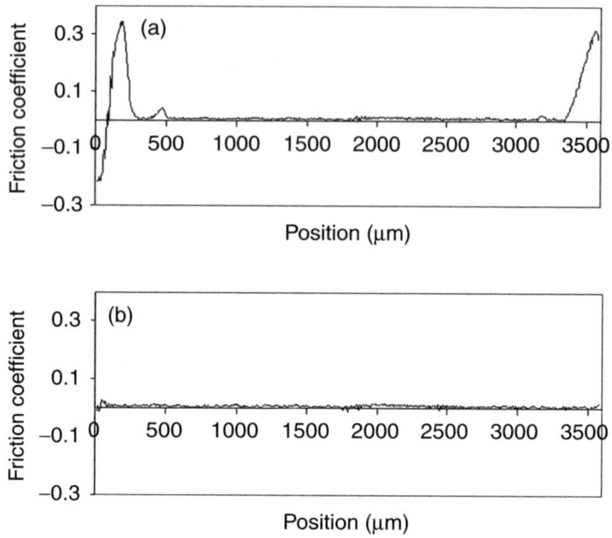

Figure 2.37 Value of the friction coefficient during a stroke at −135 °C: (a) cycle 141 and (b) cycle 44. During cycle 141, high values are observed at the ends of the scar. This can explain the increase in the average value of the friction coefficient observed during the test (Figure 2.36)

the differences observed between the two types of coatings. In the case of deposits carried out by solvent evaporation, they do not adhere perfectly to the surface and are swept (the snow-plough effect).

With friction tests performed at −135 °C, the friction coefficient is much less stable. To understand this phenomenon, it is necessary to follow the evolution of the friction coefficient during one cycle (Figure 2.37). When the friction coefficient is very low and stable, as in the case of cycle 44 (Figure 2.37(b)), the value is low and stable for all the positions. On the other hand, when friction is higher (Figure 2.37(a)), the value of friction is locally very high (at the ends of the track). This explains the increase in the coefficient of friction. It is nevertheless important to note that friction is very weak globally and is higher locally.

Analyses by Auger spectroscopy realized on transfer film observed on the pin after friction (Figure 2.38) show differences between the two tests. At −135 °C, a carbon peak certainly responsible for higher friction is observed. This carbon comes from the condensation of residual hydrocarbons on a cold surface but no clear explanation has been proposed to interpret the effect of the temperature.

Chhowalla and Amaratunga obtained very low friction coefficients with IF-MoS_2 coatings (0.01) [48]. Nevertheless, the test carried out by Cizaire was more severe since it was done with a reciprocating movement [19]. However, these results show the tribological properties of IF-MoS_2 coatings under ultrahigh vacuum.

In the case of IF-WS_2 coatings, the second technique of deposit seems to be more appropriate since coatings have a better adhesion on surfaces. Nevertheless, even when the deposit does not adhere correctly to the substrate, friction is low and can be explained by the formation of a transfer film on the pin. Recently, Moshkovich *et al.* studied several burnishing deposition methods of MoS_2 and WS_2 on steel. By using ball burnishing with vibration, they

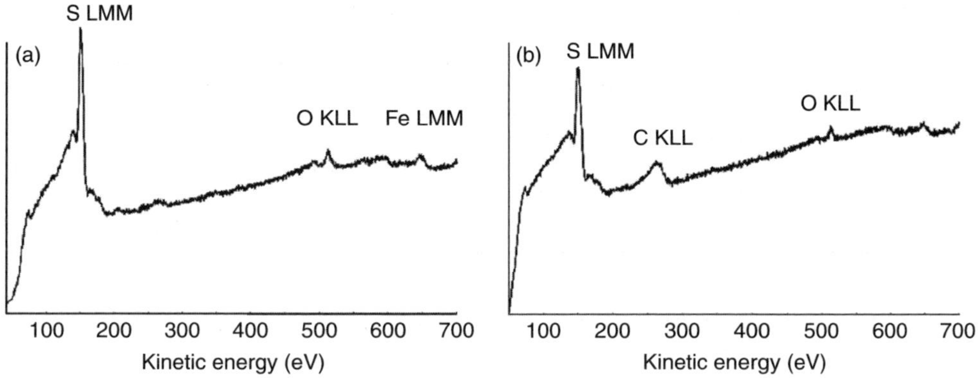

Figure 2.38 Auger analyses of the wear scars on the pin after friction tests: (a) at 25 °C and (b) at −135 °C

improved the friction coefficient by 60 % compared to classical burnishing and the life of the coating is 79 times longer [59]. These new deposition methods may prove interesting in order to improve fullerene deposition on steel surfaces.

It is difficult to explain the differences between MoS_2 and WS_2 coatings. WS_2 coatings are little used in contrast to MoS_2 coatings. A recent work by Iwaki *et al.* reveals differences in behaviour between these two types of coatings under ultrahigh vacuum [60]. With MoS_2 coatings, a low friction coefficient is obtained whatever the temperatures tested, but with WS_2 coatings, the low friction coefficient is only obtained at low temperatures (Figure 2.39). This work is still in progress.

2.3.3 Other Fullerenes

Two other types of IF, synthesized at the Weizmann Institute of Science, were studied: IF-NbS_2 and IF-TaS_2. They are synthesized in the gas phase by reaction between a

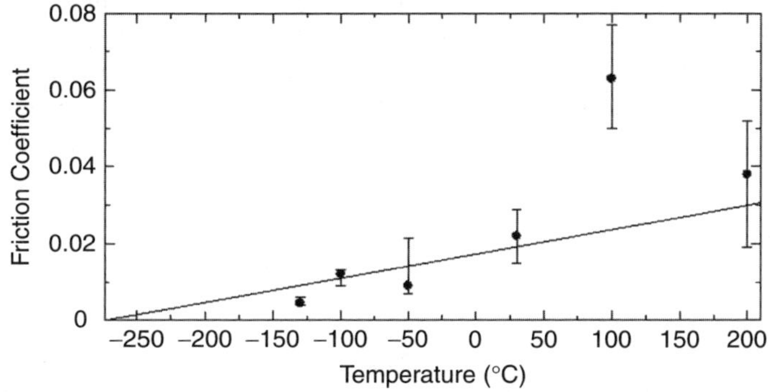

Figure 2.39 Steady-state friction coefficient of the WS_2 coatings as a function of the temperature. Line fitting was made by excluding the irregular data at 100 °C and also by making the friction coefficient 0 at 0 K [60]

Table 2.6 Friction coefficients obtained with dispersions of different IFs at 1 wt % at different contact pressures ($v = 2.5$ mm/s, $T = 20\,^{\circ}$C)

Pressure (GPa)	PAO	IF-WS_2 (140 nm)	IF-MoS_2	IF-NbS_2	IF-TaS_2
0.66	0.16	0.05	0.08	0.09	0.12
0.83	0.25	0.04	0.07	0.09	0.11
1.12	0.16/0.09	0.04	0.06	0.07	0.09
1.42	0.08	0.04	0.05	0.06	

pentachloride ($NbCl_5$ or $TaCl_5$) and H_2S [61, 62]. The mean diameter of IF-NbS_2 is close to 60 nm and close to 50 nm for IF-TaS_2. The presence of tantalum oxide on the surface layer was observed by energy filtered transmission electron microscopy (EFTEM) analyses by Schuffenhauer [62]. The presence of this oxide layer can have a bad effect on the tribological properties of IF-TaS_2 nanoparticles.

Tables 2.6 and 2.7 compare results obtained for different IF dispersed at 1wt % in oil. IF-WS_2 with a diameter of 140 nm present the best reducing properties of friction. Then come IF-MoS_2, IF-NbS_2 and IF-TaS_2. What concern anti-wear properties, the same classification can be made, with a wear which can be considered negligible in the case of IF-WS_2. With IF-TaS_2, the diameter of the wear scar is higher than those obtained with pure PAO. These particles would be more abrasive. This can be explained by the presence of oxygen on the surface of the IF [62].

Among the lamellar structures of metal dichalcogenide type (MX_2), MoS_2, WS_2, $NbSe_2$ and WSe_2 are considered as the best solid lubricants [63]. Metal dichalcogenides do not have the same crystallographic structure, which depends on the group to which the metal belongs [64]. Jamison considered the various structures and connected the tribological properties of the MX_2 to their crystallographic structure. The capacity to form films and adherence of these films are also connected to their structures (see Figure 2.40) [63]. For example, NbS_2 is known not to form films or to form films that do not present interesting tribological properties. On the contrary, MoS_2 and the WS_2 form films on surfaces.

Waghray *et al.* studied films formed by metal dichalcogenides MX_2 (M = Mo, W, Nb, Ta, X = S, Se, Te) deposited by burnishing or sputtering. For films obtained by burnishing, all the metal dichalcogenides except NbS_2 and $TaTe_2$ form adherent films on steel surfaces [65]. This study explains the results observed by SEM on tribofilm obtained with IF-NbS_2. This film is discontinuous and has a multilayer structure (Figure 2.41). On the contrary, in the case of the IF-WS_2, a continuous film is observed on the surface. This difference in aspect of films

Table 2.7 Wear scar diameters (in μm) observed on pins for friction tests mentioned in Table 2.6

Pressure (GPa)	Hertz calculated diameter	PAO	IF-WS_2 (140 nm)	IF-MoS_2	IF-NbS_2	IF-TaS_2
0.66	54	115	75	110	120	130
0.83	68	170	86	120	140	150
1.12	92	175	100	120	155	190
1.42	116	180	120	140	180	

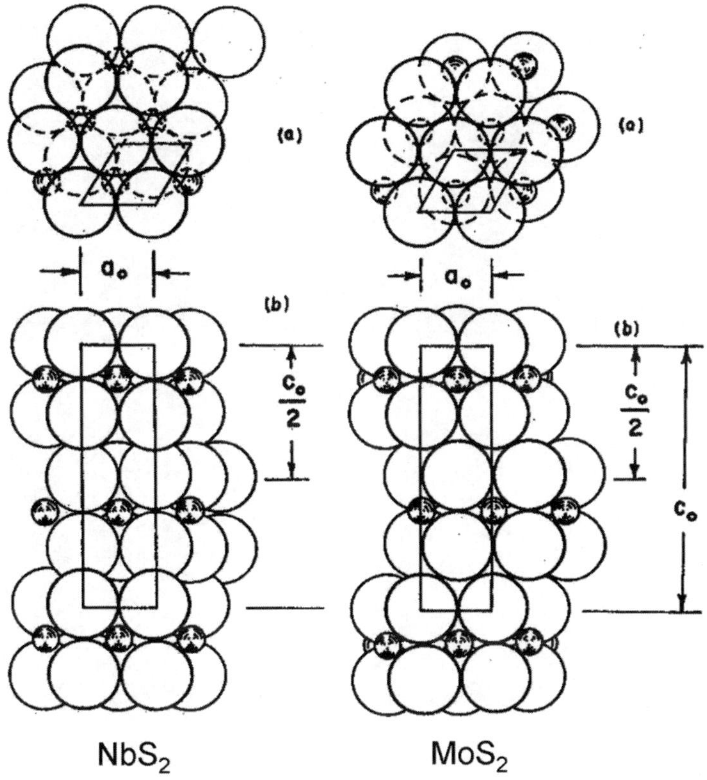

Figure 2.40 Crystalline structures of 2H-NbS_2 and 2H-MoS_2. These schematic representations show the difference between the two structures. Taken from Jamison [63] by courtesy of ASLE, © (1972) American Society of Lubrication Engineers

can explain the worse performances observed in the case of NbS_2. In the case of IF-TaS_2, no film is observed on surfaces. The results are not easily explainable.

2.4 NT-MoS_2 and NT-WS_2 Nanotubes as Lubricant Additives

Nanotubes of MoS_2 can be obtained by a modification of the IF-MoS_2 synthesis conditions (see Section 2.2) [66]. Other synthesis techniques exist: thermal decomposition of $(NH_4)_2MoS_4$ in dihydrogen atmosphere [67, 68] or by chemical transport [69]. With the synthesis techniques proposed by Feldman and Tenne [66] and Nath *et al.* [67], the nanotubes obtained have a multiwall structure (see Figure 2.42). Annal-Therese *et al.* recently developed a synthesis method for WS_2 nanotubes starting from WO_3 nanorods [70]. Mutiwall nanotubes of NbS_2 (and also of TaS_2, HfS_2 and ZrS_2) were synthesized by reduction of metal trioxide in H_2 atmosphere [71].

Synthesis by chemical transport is commonly used to synthesize lamellar crystals of metal dichlacogenides. Remskar *et al.* synthesized MoS_2 and WS_2 nanotubes by placing MS_2 platelets (M = Mo, W), C_{60} and iodine in an evacuated sealed ampoule [69]. This ampoule is

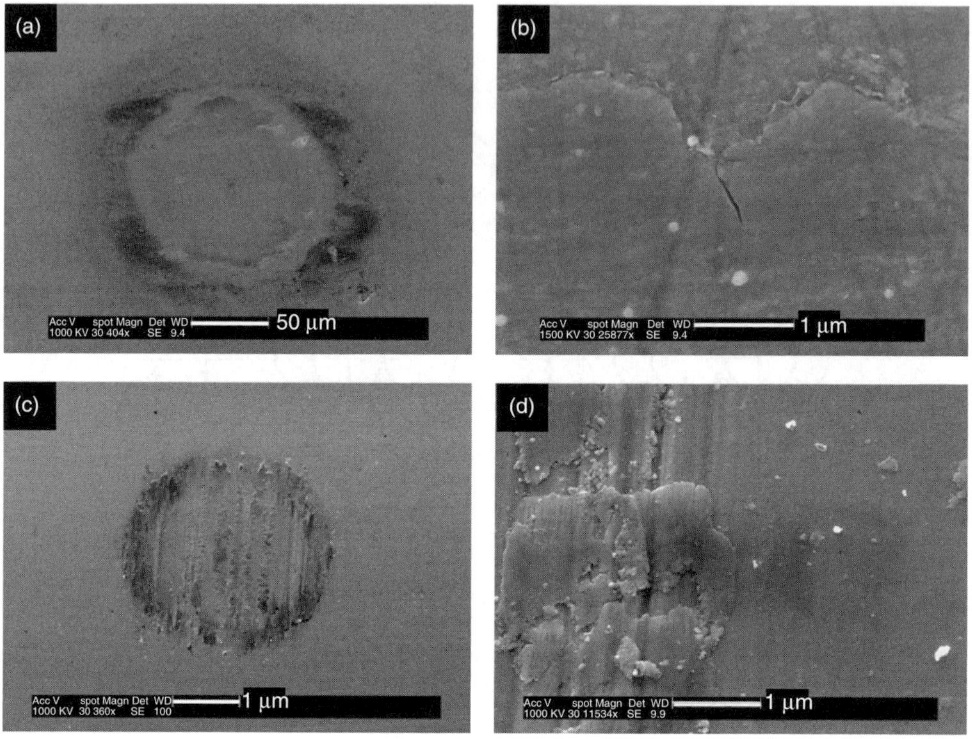

Figure 2.41 SEM images of wear scars observed on the pin, with enlargement showing films aspects on surfaces, after a friction test with IF-WS_2 (a and b) and IF-NbS_2 (c and d). These images show differences between films. In the case of IF-WS_2, the film is smooth and uniform. With IF-NbS_2, the film is discontinuous and has a multilayer structure

then placed in a chamber with a temperature gradient of 6 K/cm. Reactants are in the higher temperature part and the reaction is run for three weeks. At the end of the synthesis time, a 'furry-like' sample is found. It is composed of nanotubes bundles.

The first samples synthesized by Remskar *et al.*, with a chemical formula of $MoS_2I_{1/3}$, were described as MoS_2 single-wall nanotubes weakly bonded in bundles with iodine in the

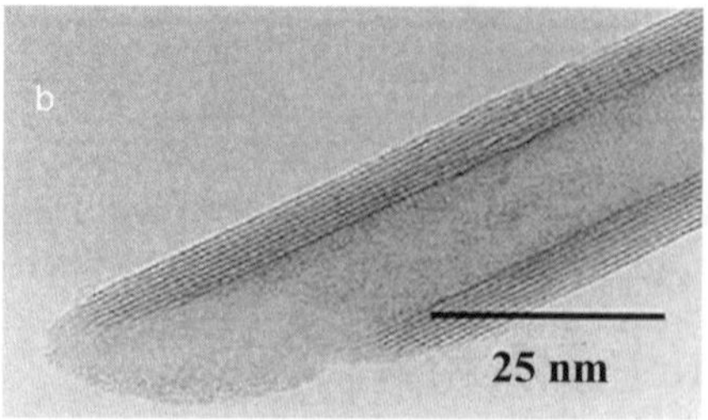

Figure 2.42 TEM micrograph of an MoS_2 nanotube. Several walls can be distinguished. Reproduced by permission of Elsevier from Chen *et al.* [68], Figure 3, © (2003) Elsevier

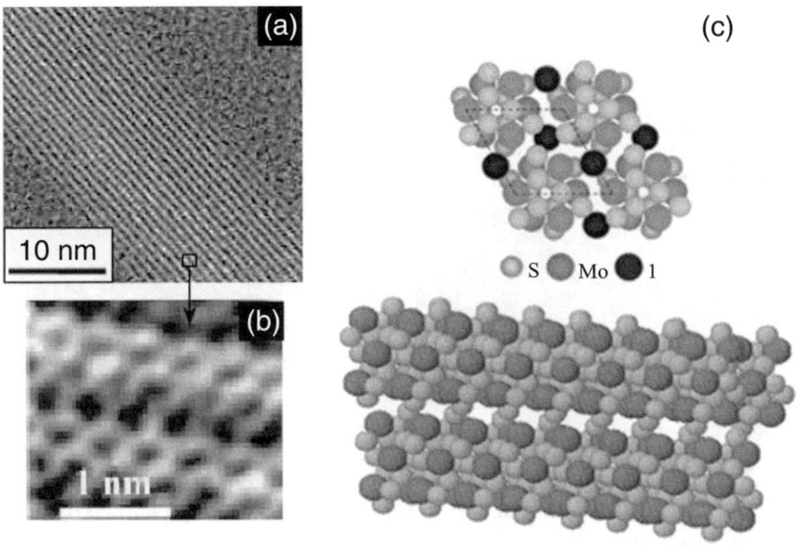

Figure 2.43 $MoS_2I_{1/3}$ nanotubes: (a) TEM micrograph showing the bundle structure; (b) HRTEM micrograph of two nanotubes in a bundle; (c) model of the bundle structure. Reprinted with permission from AAAS from Remskar *et al.* [69], Figures 2c and 3, © (2001) American Association for the Advancement of Science

interstitial position [69]. On the TEM picture of one bundle, several nanotubes can be observed (Figure 2.43(a)). A high-resolution TEM (HRTEM) picture (Figure 2.43(b)) allows the proposed structure (Figure 2.43(c)) to be confirmed, even if no pictures were simulated from this model.

Wu *et al.* recently reported the synthesis of NbS_2 nanofils by a similar synthesis method than that used by Remskar [72]. They described the sample formed by using the term 'nanowires' to describe their long length. This term is perhaps not applicable because they are composed of several NbS_2 sheets, so the term 'multiwall' would be more explicit. The sample is composed of NbS_2 multiwall nanotubes bonded in bundles with iodine in the interstitial position, similar to the structure described by Remskar for $MoS_2I_{1/3}$ nanotubes.

The mechanical properties of WS_2 nanotubes were recently studied by AFM [73]. A nanotube was stuck at the end of an AFM tip and pushed on to a silicone surface. The Young modulus was found to be 170 GPa. Tensile strength tests were performed inside an SEM [74]. A Young's modulus of 152 GPa, a tensile strength of 3.7 to16.3 GPa and an elongation of 5 to 14 % were determined. The value obtained for the Young's modulus is very close to that obtained from theoretical calculations. It was explained by the absence of defects inside the nanotube structures. These properties show that there could be many further applications for this material. Since MoS_2 is isostructural to WS_2, MoS_2 nanotubes may present the same properties. Further studies will focus on the study of the shear modulus or the effect of the chirality of nanotubes on their properties [74].

WS_2 nanotubes tested were synthesized at Johannes Gutenberg Universität of Mainz [70]. These tubes have a multiwall structure and an average diameter of 30 nm for a length of

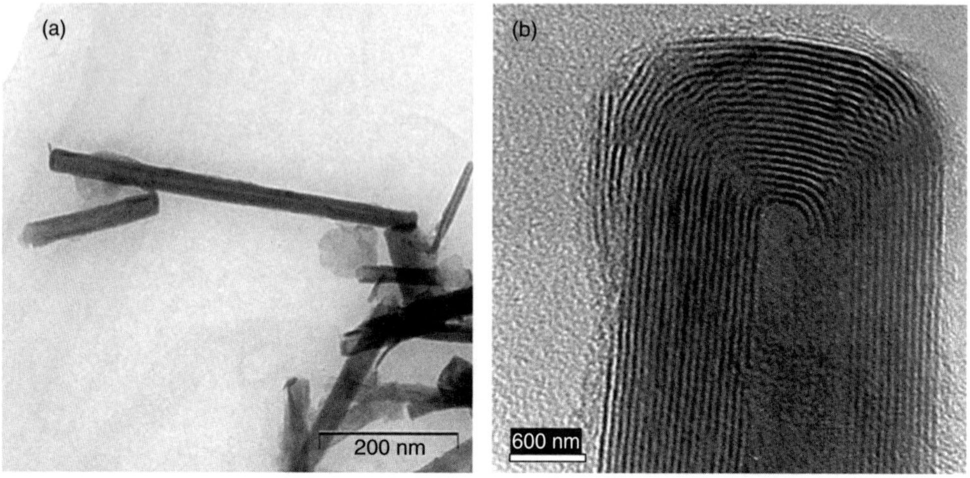

Figure 2.44 (a) TEM image of WS_2 nanotubes; (b) HR-TEM micrograph of the extremity of a nanotube showing the square shape

approximately 500 nm (Figure 2.44(a)). Their ends have a square shape, as shown in Figure 2.44(b), and contain some defects. This is due to the presence of strains only at the ends of the tubes. As in the case of IF-WS_2, Raman spectroscopy with a wavelength of 514.5 nm allows an easy differentiation to be made between 2H-WS_2 and NT-WS_2 (Figure 2.45).

The tribological tests show similar tribological properties to those observed with IF-MoS_2 or small-diameter IF-WS_2. An optimal concentration of 1 wt % leads to a very low friction coefficient (0.06), stable all along the test (Figure 2.46). Contact pressure seems to have no important effect on tribological properties of WS_2 nanotubes, contrary to IF-MS_2 (Figure 2.47). The addition of NT-WS_2 in PAO leads to a reduction in wear (Table 2.8), but is less important than in the case of IF-WS_2 (see Table 2.4).

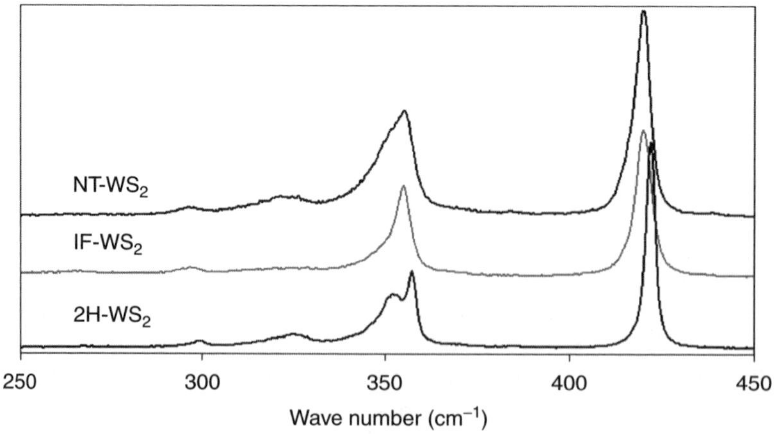

Figure 2.45 Raman spectra of NT-WS_2 compared to those of 2H-WS_2 and IF-WS_2 ($\lambda = 514.5$ nm)

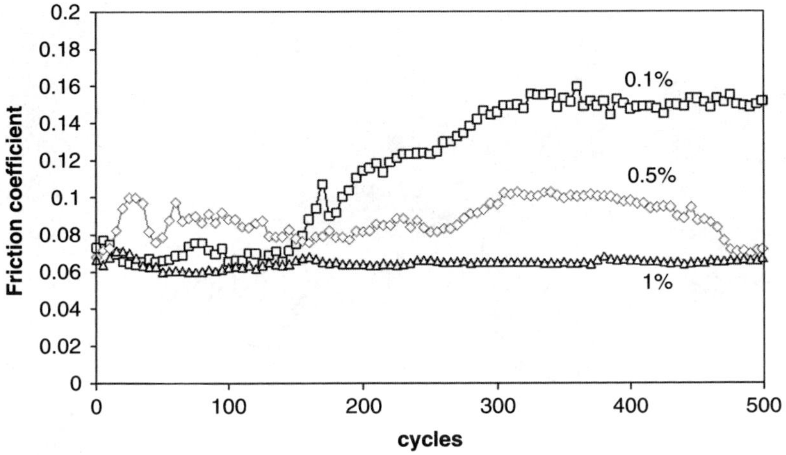

Figure 2.46 Effect of NT-WS_2 concentration in PAO on the friction coefficient (v = 2.5 mm/s, P = 1.12 GPa, T = 20 °C). A 1 wt % concentration leads to a very low friction coefficient (0.06)

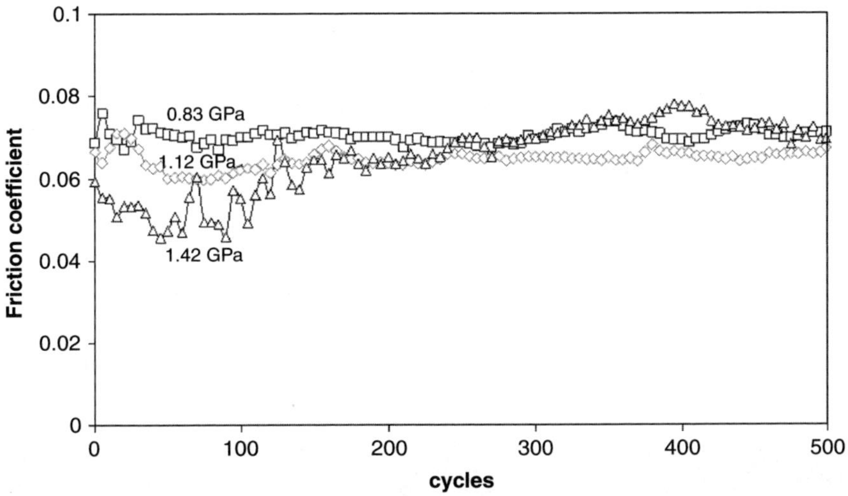

Figure 2.47 Effect of contact pressures on the friction coefficient with a 1 wt % NT-WS_2 dispersion

Table 2.8 Wear scars diameters (in μm) observed on the pins after friction tests with PAO + 1 wt % NT-WS_2

Pressure (GPa)	Wear scar diameter	Hertz diameter	Wear scar diameter obtained with PAO
0.83	110	68	170
1.12	125	92	175
1.42	155	116	180

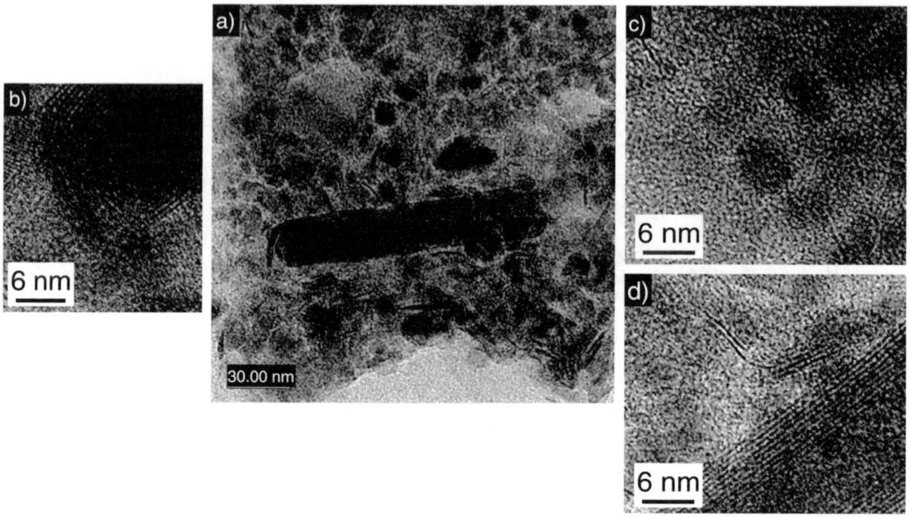

Figure 2.48 HRTEM image of wear particles observed after friction. These particles contain (a) intact nanotubes and (d) sheets. An extremity of the nanotube is damaged (b)

TEM observations of wear particles show the presence of intact nanotubes after friction tests (Figure 2.48(a)). Only the end of the nanotube is damaged (Figure 2.48(b)), which is explained by the presence of defects only at the ends of nanotubes. Wear particles also contain WS_2 sheets (Figure 2.48(d)). These observations indicate that only some layers of the nanotubes are exfoliated. Analyses carried out by energy dispersive X-ray spectroscopy (EDS) on the dark particles indicate that they are iron particles (Figure 2.48(c)).

The film formed on the pin is not homogeneous (Figure 2.49(a)) and intact nanotubes can be observed (Figure 2.49(b)). Thus, nanotubes undergo few structural modifications in spite of observing WS_2 sheets on TEM pictures. Nanotubes, with their cylindrical shape, have less strains in their structure than IF structures and are thus less sensitive to pressure and shear.

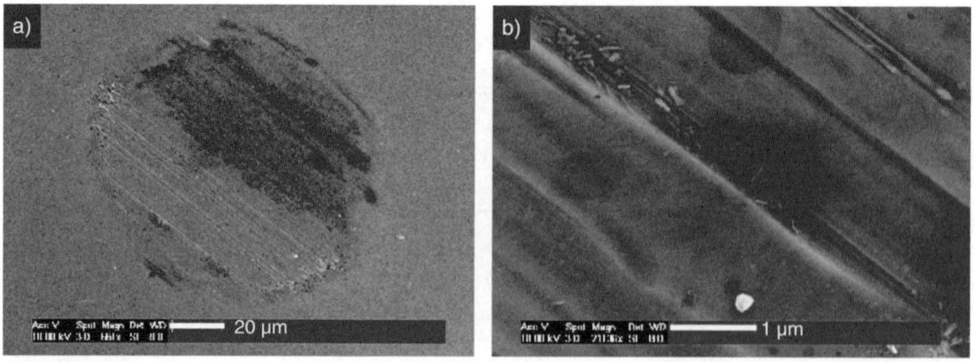

Figure 2.49 SEM images of the wear scar observed on the pin after a friction test with NT-WS_2. The film observed on the scar is not homogeneous (a). Intact nanotubes can be observed in the films (b)

2.5 Lubrication by a Mixture of Fullerenes

Superlubricity (friction coefficient below 0.01) was only observed under UHV or an inert environment. Two different mechanisms can involve this phenomenon:

- sliding between two hard carbon coatings terminated by hydrogen atoms and very smooth, which is the case with diamond-like carbon coatings [75];
- friction between two incommensurate crystal surfaces, as Martin *et al.* observed with the friction of a steel pin on an MoS_2 coating [6, 76].

In the case quoted above, low friction between two incommensurable surfaces is explained by rotational anisotropy. A layer is shifted by a certain angle compared to the other layer, thus allowing an easier slip, with the atoms no longer coinciding (Figure 2.50(b)). The second case to be considered is the friction of two layers with different interatomic distances (Figure 2.50(c)). This condition allows an absence of coincidence between the atoms. In this section, friction experiments of IF mixtures are developed in order to study the possibility of the sliding of sheets with different interatomic distances.

Three types of fullerenes were used: IF-MoS_2, IF-WS_2 and IF-NbS_2. A comparison of parameters of the corresponding lamellar structures shows that they are different for the three IFs. For 2H-MoS_2, $a = 0.3116$ nm and $c = 0.6149$ nm (ASTM 37-1492). For 2H-WS_2, $a = 0.3154$ nm and $c = 0.6181$ nm (ASTM 08-0237). For 3R-NbS_2, $a = 0.3335$ nm and $c = 0.896$ nm (ASTM 38-1367), the structure of NbS_2 in the IF being of the 3R type [61]. If two layers of two different metal dichalcogenides slide on each other, the system will be in conditions of incommensurability and thus low friction could be obtained.

Two types of mixture were tested: a mixture of 50/50 IF-WS_2/IF-MoS_2 and a mixture of 80/20 IF-WS_2/IF-NbS_2. In the case of the IF-WS_2/IF-NbS_2 mixture, the proportions were adapted to the quantity of matter available (the method of synthesis of the IF-NbS_2 does not allow much product to be obtained). Friction coefficients obtained are slightly higher than those obtained with pure IF-WS_2 (Figure 2.51).

Observation by TEM (STEM) of the wear particles collected on the pin after a friction test performed with an IF-WS_2/IF-MoS_2 mixture shows the presence of sheets. By using a high-angle annular dark field (HAADF) detector an image with a contrast depending on the atomic

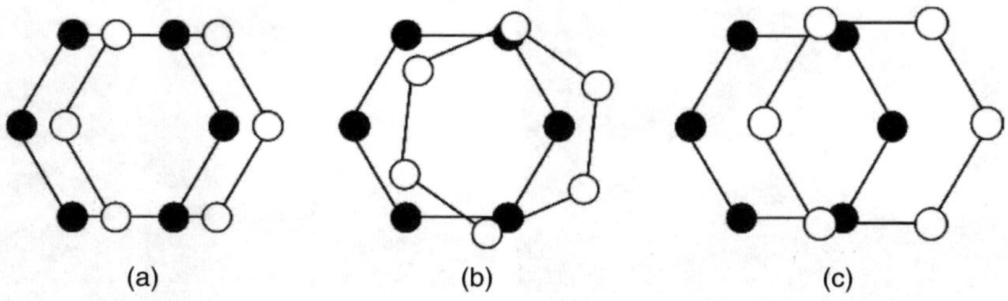

Figure 2.50 Schematic representation of two superimposed atomic layers of an hexagonal structure. In case (a), the atoms coincide perfectly and friction will be high. In case (b), the layer of white atoms has rotated. There is no coincidence between atoms, so the friction will be low. In case (c), the two patterns do not have the same distance, so there is no coincidence between atoms

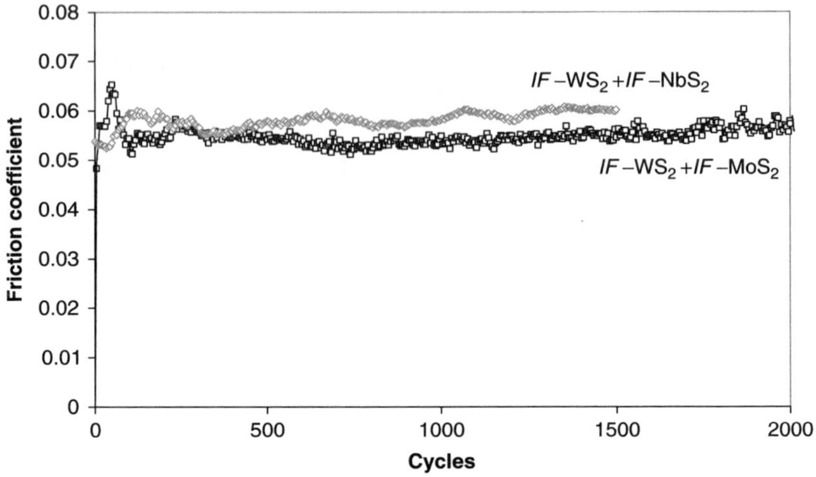

Figure 2.51 Friction coefficients obtained with two IF mixtures: 50/50 IF-WS_2/IF-MoS_2 and 80/20 IF-WS_2/IF-NbS_2 ($P = 1.12$ GPa, $T = 20\,^\circ$C). Friction coefficients are slightly higher than those measured for pure IF-WS_2

number can be obtained. This technique was used to see if fullerenes were well mixed before friction tests. In Figure 2.52, IF-MoS_2 and IF-WS_2 can be distinguished. IF-MoS_2 corresponds to the small grey particles while IF-WS_2 corresponds to the larger white particles.

A chemical cartography of the film formed during the friction test was performed by using a synchrotron beam. With the synchrotron radiation a very good space resolution (90 nm) and also a very good spectral resolution (400 meV) can be obtained. Figure 2.53 presents

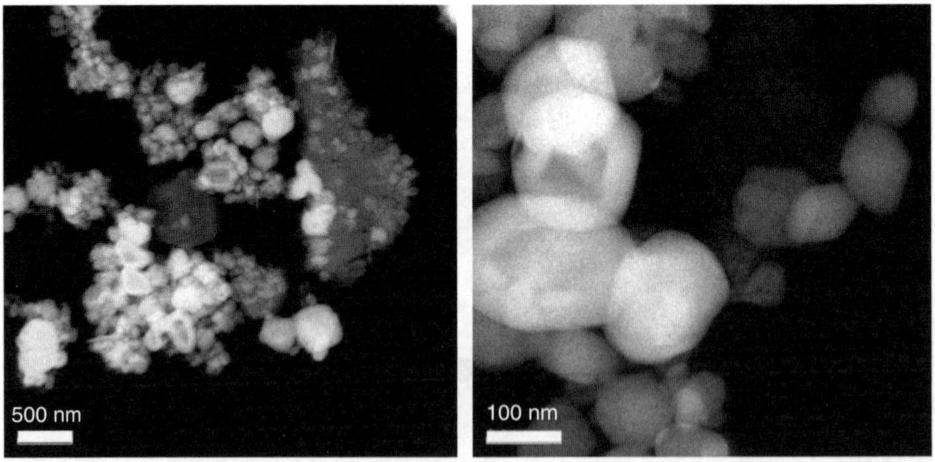

Figure 2.52 TEM images of the IF-WS_2/IF-MoS_2 mixture obtained with a high-angle annular dark field (HAADF) detector. These images have a z-contrast that allows IF-WS_2 and IF-MoS_2 to be distinguished easily (Mo: Z = 42 and W: Z = 74)

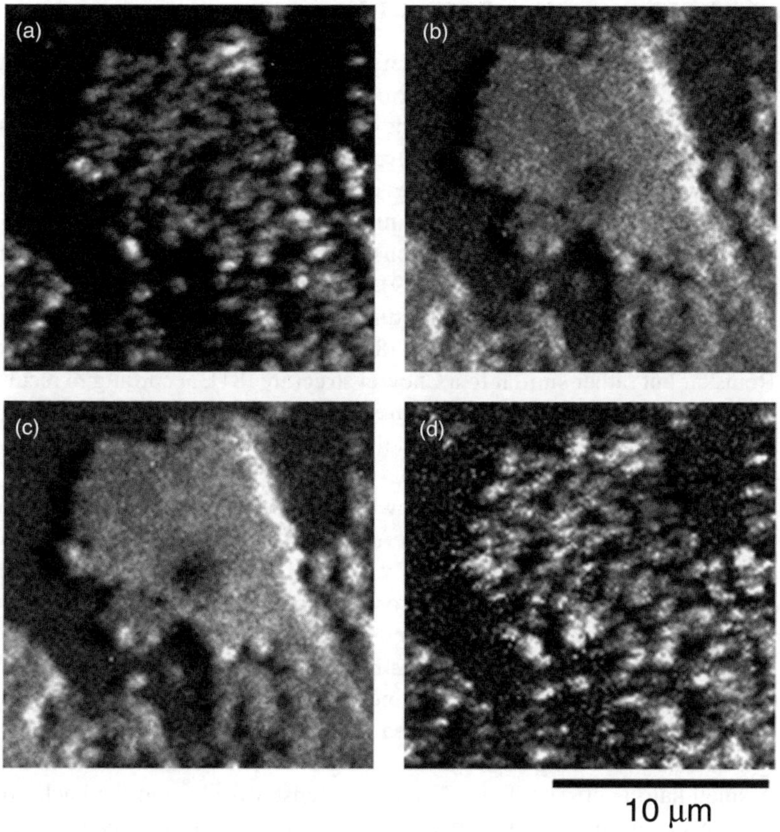

Figure 2.53 Chemical mappings of (a) W, (b) Mo and (c) S realized on the flat after the friction test with the mixture IF-WS_2/IF-MoS_2. These chemical mappings were realized with a synchrotron beam (photoemission electron microscope). Image (d) represents the ratio between the W and the Mo images

the cartography of tungsten, molybdenum and sulfur (the photoemission electron microscopy (PEEM) technique). These cartographies mainly show a topographical contrast. A ratio of these images avoids the problem of topographical contrast. Figure 2.53(d) presents the W/Mo ratio and shows that tungsten is not uniformly distributed compared to molybdenum. From these results, it can be concluded that molybdenum is well distributed on all the surfaces and tungsten forms islands.

Friction coefficients obtained in the case of these mixtures are low but do not allow the awaited suprafriction to be obtained. Although IFs seem to be well mixed (TEM), surface analyses show a bad distribution of tungsten. This bad distribution can explain why suprafriction is not obtained. Another experiment could be to rub a pin coated with a tribofilm made up of MoS_2 with an IF-WS_2 dispersion. A thorough study of this type of system would certainly be interesting to carry out. A similar study to the one performed by Dienwiebel *et al.* in nanotribology in the case of graphite would also be interesting [77]. This study would consist in rubbing an MoS_2 single sheet, deposited on a tip, on a WS_2 film to see whether suprafriction is obtained.

2.6 Tribological Properties of Mo-S-I Nanowires

Mo-S-I nanowires were synthesized at the Jozef Stefan Institute by chemical transport synthesis. The preparation of these products is described precisely in the patent PCT/EP04/001870. Mihailovic chose the name of nanowires to describe the important length of the sample (of the order of a millimetre) and the absence of cavity in the structure of the wire. A company named Mo6 was created to study the properties of this kind of compound and to market them. These nanowires present, for example, field emission properties [78].

A study carried out by *ab initio* calculations showed that the structure is certainly more complex than that described by Remskar [79]. Various stoichiometries of Mo-S-I systems were simulated, but the most probable structure today is not known and all the simulated structures can be synthesized. Vrbanic *et al.* [80] presume that this structure is not that described by Remskar but rather similar to a Chevrel structure [81], according to results obtained by Raman spectroscopy analyses. The structure proposed by Vrbanic is a wire composed of a core made of sulfur atoms, surrounded by molybdenum atoms and finally iodine atoms (see Figure 2.54) [82].

Samples studied here are bundles of nanowires of several millimetres in length (Figure 2.55(a)), of different diameters up to 500 nm (depending of the number of nanowires in the bundle). Figure 2.55(b) presents nanowires of various diameters on a perforated carbon film.

Observations by transmission electron microscopy show the presence of some MoS_2 sheets in the sample. The XRD technique was used in complement to confirm the presence of 2H-MoS_2 in large quantities but not in the pristine material.

A comparison between XRD spectra obtained for 2H-MoS_2 and nanowires shows a big difference between them (Figure 2.56). The peak at 7.2° consistent with that of the (002) peak of the hexagonal MoS_2 (ordering of the layers along the *c* axis) is not visible in the nanowire spectrum. A small halo is observed at 7.2° but its intensity is close to the background noise. Other peaks observed in the nanowire pattern could correspond to reflections characteristic of the ordering in the basal plane of 2H-MoS_2. These peaks are, however, shifted compared to those in hexagonal MoS_2, probably due to the presence of iodine, which could modify the lattice parameters, or to the introduction of strain owing to curvature of the layer.

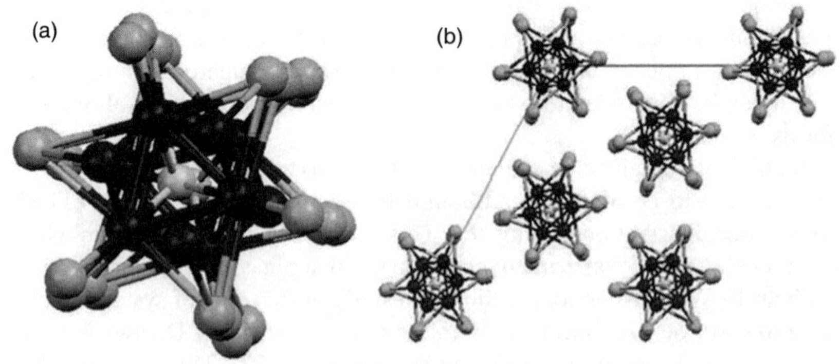

Figure 2.54 Structure model of nanowires from [82]. Sulfur atoms are in the centre and iodine atoms in the outlayer

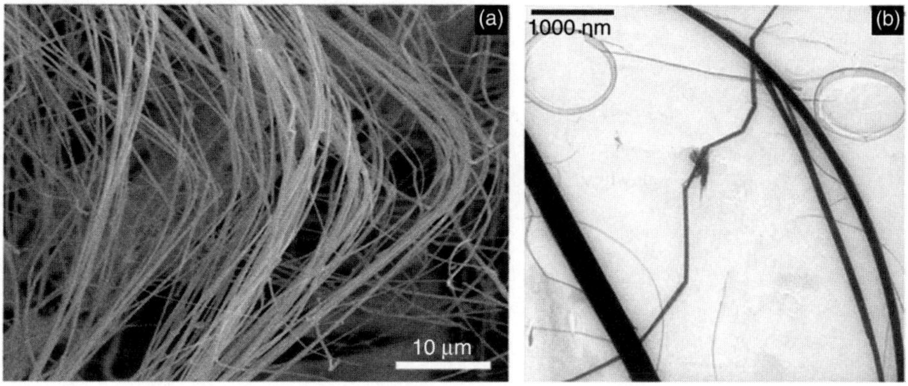

Figure 2.55 (a) SEM image of Mo-S-I nanowires. (b) TEM image showing bundles with different diameters

Raman spectroscopy using an He/Ne laser (632.817 nm) was also used to characterize the sample. Spectra obtained are very similar to those obtained by Vrbanic *et al.* [80]. Three peaks at 220, 285 and 320 cm^{-1}, not characteristics of a 2H-MoS_2 structure, are observed on the Mo-S-I nanowire spectrum (Figure 2.57). A comparison of these two spectra proves the presence of MoS_2 sheets in the sample. Nevertheless, Raman spectroscopy does not allow them to be quantized.

The combination of several analytical techniques proves the presence of some MoS_2 sheets in the nanowire sample. Nevertheless, their concentration is low and can be estimated to be lower than 5 %.

2.6.1 Influence of the Nanowire Concentration in PAO on the Tribological Properties

The influence of the nanowire concentration in PAO on the friction was studied using a pin-on-flat tribometer (contact pressure of 0.83 GPa, sliding velocity of 2.5 mm/s, 20 °C, 500 cycles

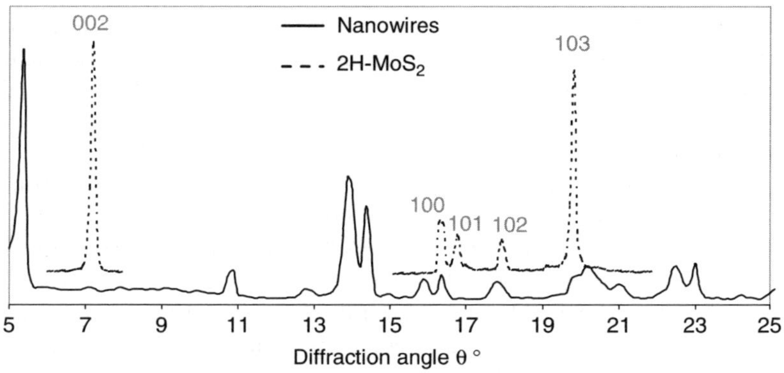

Figure 2.56 XRD patterns of Mo-S-I nanowires compared with 2H-MoS_2

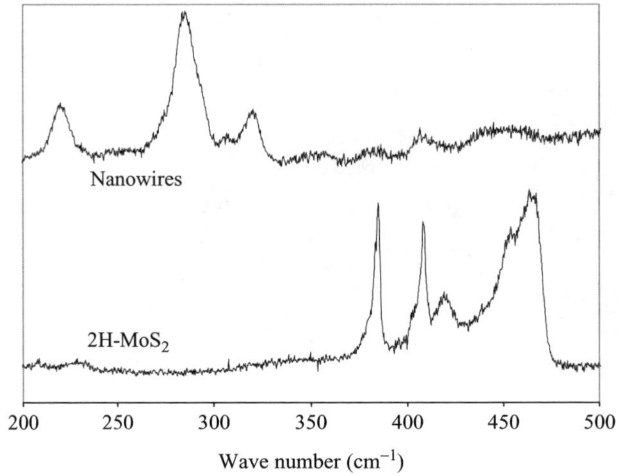

Figure 2.57 Raman spectrum of Mo-S-I nanowires and 2H-MoS_2 (wavelength of 632.817 nm). The spectra are very different but the nanowires are slightly contaminated by 2H-MoS_2

corresponding to a sliding distance of 2.5 m). Nanowires were dispersed in PAO using an ultrasonic bath. No dispersant was used in order to avoid a chemical reaction of the dispersant during friction, thus ensuring that only the effect of the nanowires was tested.

With the experimental conditions indicated above, pure PAO presents a high friction coefficient (0.25). The addition of 0.5 wt % of nanowires led to a reduction of 20 % of this value [83]. With 1 or 2 wt % of nanowires, the results are more spectacular. The friction coefficient is close to 0.04. This value is usually difficult to reach in a boundary lubrication regime.

Figure 2.58 shows the presence of two regimes of lubrication with Mo-S-I nanowires. During the first 200 cycles, the friction coefficients are low and stable. Then, an increase in the friction is observed. A more precise analysis of the curves shows that only a few cycles are necessary to reach the low friction coefficient. This supposes that a structural evolution of nanowires occurs inside the contact area, rather than a chemical reaction for which an induction period, which can be long, is required.

As in the case of IF-MoS_2, Figure 2.59 clearly shows a correlation between the contact pressure and the friction coefficient value, with lower friction coefficient values obtained at high contact pressures. The two lubrication regimes are also observed, the increase of the friction appearing more quickly at high contact pressures: 30 cycles at 1.12 GPa and 15 cycles at 1.42 GPa.

A study of the evolution of the electrical contact resistance (R_c) during friction tests explains the evolution of the friction coefficient (Figure 2.60). A decrease in R_c coincides with an increase in the friction coefficient. This means that the insulating film formed on surfaces has been removed. The triboscopic image confirms the formation of a tribofilm on the pin at the beginning of the test, the value of R_c being equal all along the track, and then its removal at 200 cycles [84]. This explains the increase in the friction coefficient observed after 200 cycles (Figure 2.58), which is caused by a defect in providing nanowires inside the contact so that the tribofilm is not regenerated. This defect can be explained by an insufficient dispersion of nanowires in PAO and the alternative movement of the pin supporting the deposit of the

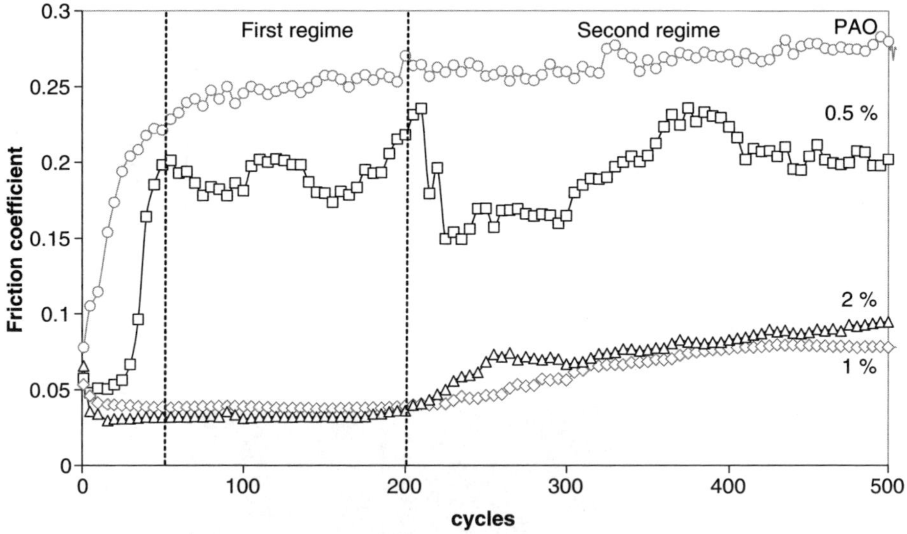

Figure 2.58 Influence of nanowire concentration in PAO on the friction coefficient ($P = 0.83$ GPa, $v = 2.5$ mm/s, $T = 20\,°C$). A very low friction coefficient is observed during 200 cycles

particles at the ends of the track (the snow-plough effect). Figure 2.60 also indicates the formation of a film at the end of the test, but this film does not affect the value of the friction coefficient. It does, however, correspond to an oxide layer. This is proved by the Auger analysis where an oxygen peak can be observed.

Observation of a wear scar on the pin by optical microscopy indicates that the addition of nanowires leads to a reduction in the wear. When the friction is low, the pin and flat are not

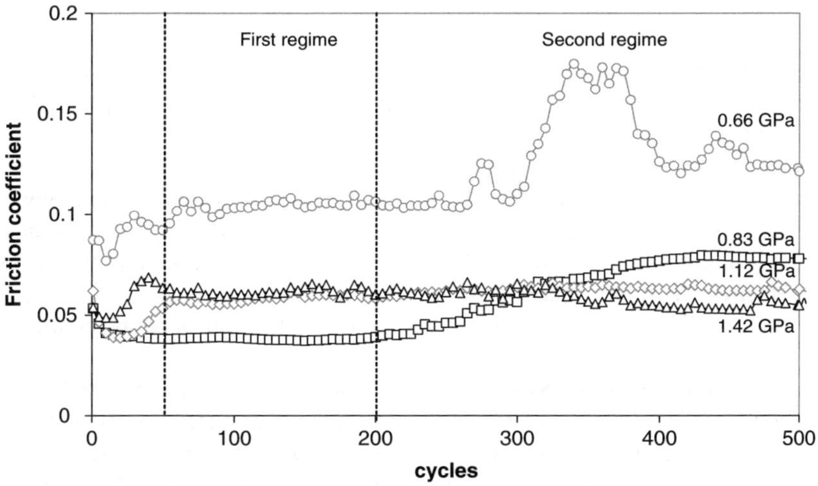

Figure 2.59 Effect of the contact pressure on the friction coefficient with the optimal concentration of 1 wt % ($v = 2.5$ mm/s, $T = 20\,°C$). Pressure improves the efficiency of the Mo-S-I nanowire additive

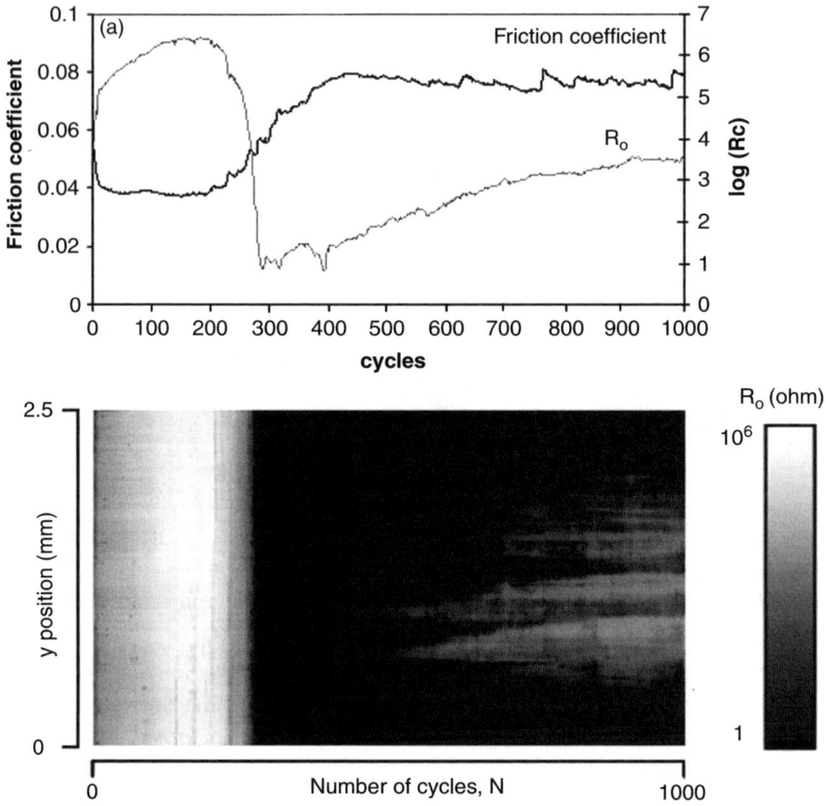

Figure 2.60 Friction coefficient and electrical contact resistance evolutions ($P = 0.83$ GPa, $v = 2.5$ mm/s, $T = 20\,°C$, 1000 cycles), with the triboscopic image of the electrical contact resistance

worn out (Figure 2.61(b) and (d), and a film is observed on the pin. After an increase in the friction coefficient, the wear of the pin is the same as that observed with pure PAO, with an oxide layer on the flat scar.

Characterization by TEM of the wear particles collected on the pin after friction shows the presence of cut and fractured bundles (Figure 2.62(a) and (b)), both with many 2H-MoS_2 sheets. A high-resolution TEM picture and its associated diffractogram confirm that sheets are made of MoS_2 (Figure 2.62(c) and (d)). Distances and angles observed on the diffractogram are in agreement with those obtained for an hexagonal structure observed along a [0001] axis. Nevertheless, TEM does not allow the formation of these sheets to be confirmed during friction.

The massive formation of MoS_2 sheets during friction was clearly pointed out by X-ray diffraction on wear particles collected from the flat after a friction test (500 cycles, 0.83 GPa). The comparison of the spectra obtained before and after friction highlights the appearance of the intense peak at 7.2°, characteristic of the (002) peak in the 2H-MoS_2 structure (Figure 2.63). The formation of MoS_2 sheets inside the contact area during friction can also be concluded.

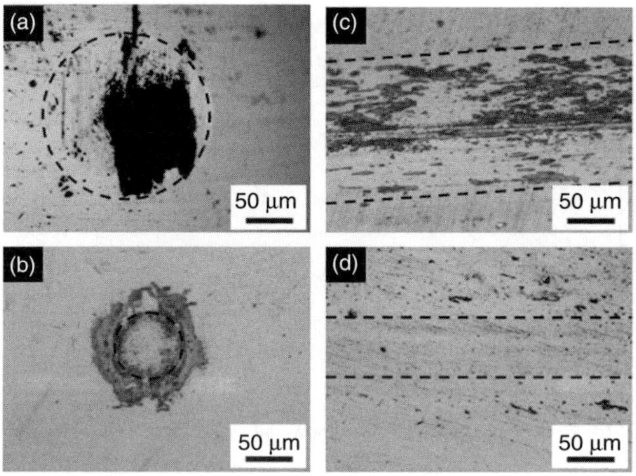

Figure 2.61 Wear scars observed on the pin (left) and on the flat (right) with pure PAO (a and c), with 1 wt % of nanowires after 50 cycles at 0.83 GPa (b and d)

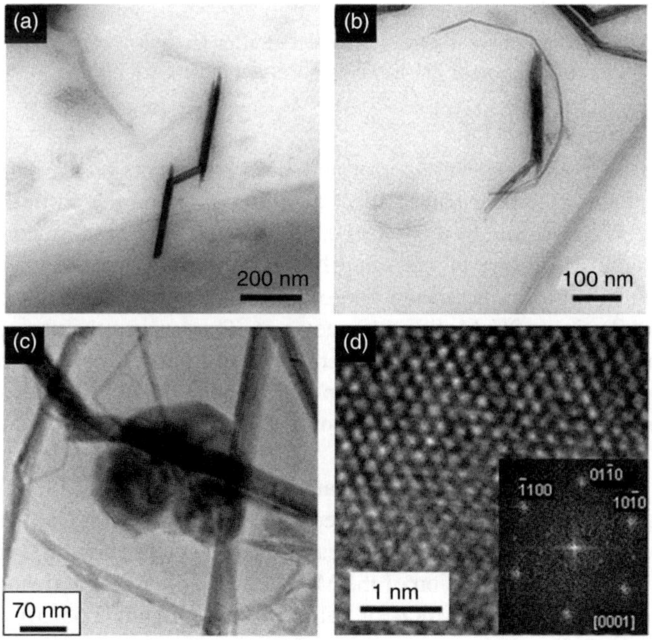

Figure 2.62 (a and b) Bundles cut during friction; (c) MoS_2 sheets observed after friction; (d) HRTEM image of these sheets and its corresponding Fourier transform confirming the hexagonal structure of the sheets

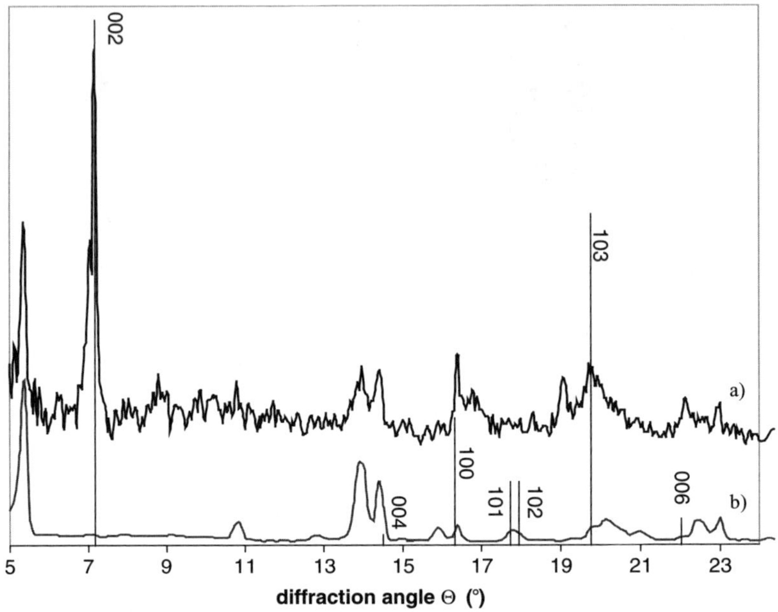

Figure 2.63 (a) XRD pattern of wear particles collected after friction compared to the one of Mo-S-I nanowires; (b) 2H-MoS_2. Formation of 2H-MoS_2 is observed during friction

Raman spectroscopy analyses were performed at several points on the flat after the friction test. Figure 2.64 shows a comparison of spectra obtained: (b) on the particles collected from the flat (the same as those analysed by XRD), (d) on the tribofilm formed in the wear scar after 200 cycles and (c) on one of the small heaps on the wear scar. A spectrum recorded from the tribofilm is very similar to the one of 2H-MoS_2(e). A spectrum of heaps is composed of two contributions: 2H-MoS_2 and nanowires. This proves the presence of 2H-MoS_2 and nanowires inside the wear scar. The spectrum of the particles shows the two contributions, but with a more important contribution of 2H-MoS_2. The comparison of this spectrum with that obtained before friction (a) proves the formation of MoS_2 sheets during the friction test.

An XPS analysis realized on the pristine material seems to confirm the structure proposed by Vrbanic [82], since iodine is mainly detected on the nanowire surface (see the structure in Figure 2.54). The XPS spectrum on the wear scar after 200 cycles shows the presence of MoS_2 sheets (Figure 2.65(a)). Deconvolution of the Mo3d peak indicates that the molybdenum is in the MoS_2 form as well as in the MoO_3 or sulfate forms (Figure 2.65(b)). The presence of MoO_3 could explain the important wear observed on the two antagonist parts.

The formation mechanism of MoS_2 sheets during friction tests can be visualized starting from SEM images of the flat after friction (Figure 2.66). An observation along all the track indicates the presence of bundles with a low length lying randomly along the track. More precise observations showed two morphologies for these bundles. Some of them were dismantled into ones of smaller diameters, while other bundles had platelet structures on their surfaces and extremities (Figure 2.66(c)). These platelets are MoS_2 sheets.

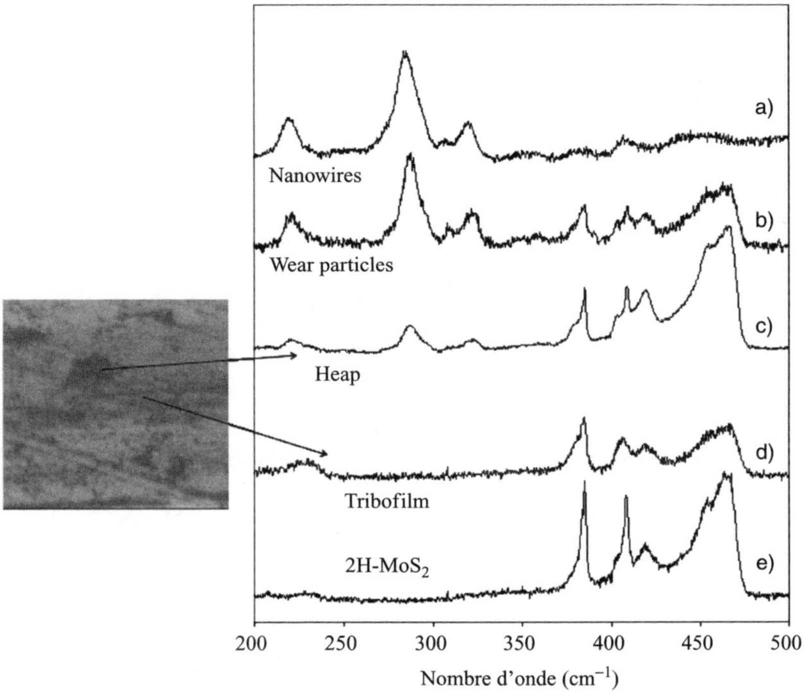

Figure 2.64 Raman spectra of (b) wear particles, (c) particles on the flat and (d) tribofilm compared to Raman spectra of (a) nanowires and (e) 2H-MoS_2 (friction test at 0.83 GPa, $v = 2.5$ mm/s, $T = 20\,^{\circ}C$, 200 cycles). These analyses show the formation of 2H-MoS_2 and the presence of nanowires on the flat after 200 cycles

To summarize this section, the lubrication mechanism of these nanowires is not yet known in detail. It is certainly based on the formation of MoS_2 sheets in the contact. This was determined by XRD and Raman analyses and observed by HRTEM. Two mechanisms can be distinguished: a first phase where the friction coefficient and wear are low with the presence of cut nanowires on the flat (Figure 2.66) and the formation of MoS_2 sheets and a second phase during which the friction coefficient increases. The increase in the friction coefficient coincides with the disappearance of the film (R_c), which can be explained by a supply failure of nanowires inside the contact area. Indeed, the dispersion of nanowires in PAO is not very stable. Furthermore, due to the alternative movement of the pin, nanowires tend to agglomerate at the end of the track by a 'snow-plough' effect. In this study, it has been shown that nanowires present good tribological properties and are potential additives for oil. Further investigations have nevertheless to be done to improve their dispersion in oil. The cutting of the bundles or the use of dispersant can be envisaged. Recently, a study of the dispersion of nanowires in several common solvents was done by Nicolosi *et al.* [85]. By studying the dispersion of nanowires in solvent, the authors found a good way to purify nanowire samples. Dispersion of these purified samples could be easier than the sample studied here. Furthermore, this study can lead to new studies of dispersion capabilities of this material in oil.

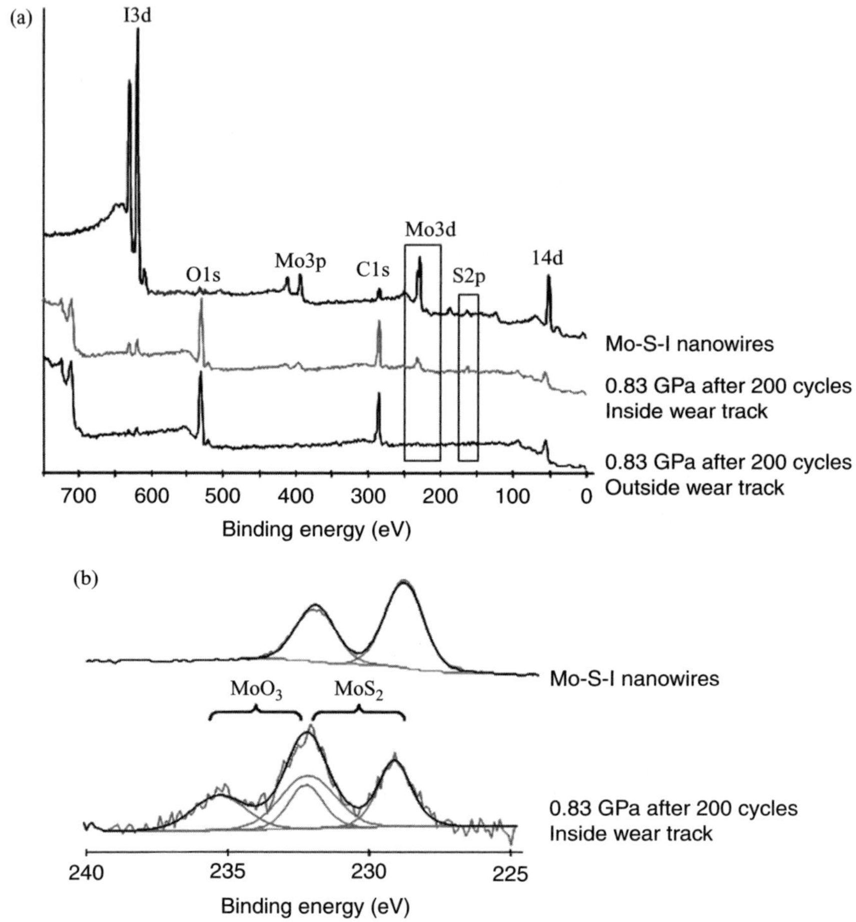

Figure 2.65 (a) Comparison of the XPS spectrum of the pristine material with those obtained on the flat after the friction test of 200 cycles inside and outside the wear track; (b) fit of the Mo3d peak of the pristine material inside the wear track ($P = 0.83$ GPa, $v = 2.5$ mm/s, $T = 20\,°C$)

Figure 2.66 (a) SEM image of a cut bundle; (b) area showing dismantled nanowires, (c) area showing the presence of sheets (friction test: 200 cycles, $P = 0.83$ GPa, $v = 2.5$ mm/s, $T = 20\,°C$)

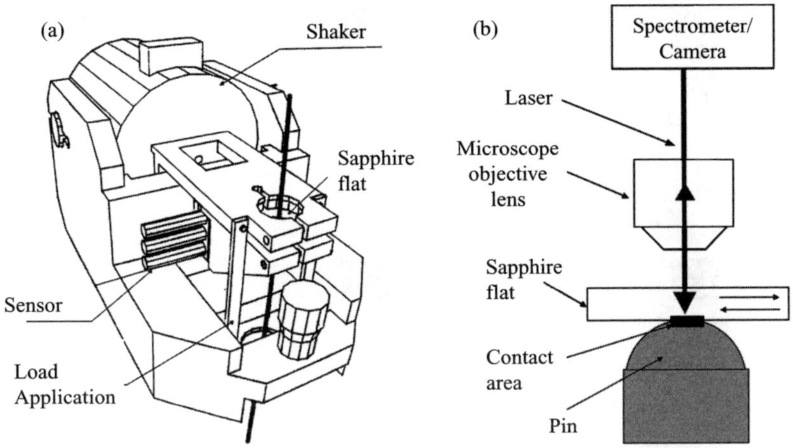

Figure 2.67 (a) Schematic representation of the alternative tribometer using a sapphire flat; (b) principle of the *in situ* video or *in situ* Raman analysis

2.7 Raman Tribometry on IF-MS_2

2.7.1 In situ Observation of the Structures in the Interface

To understand as precisely as possible the lubrication mechanism of the nanoparticles, it is necessary to study their behaviour inside the contact area during a friction test. A tribometer (Figure 2.67) with an alternative movement using a plan made up of transparent sapphire ($e = 1.5$ mm) makes it possible to have two types of analyses during a friction test (without opening the contact area):

- *in situ* optical video to see the global movement and the flow of the particles during friction tests, and the formation of films on surfaces;
- *in situ* Raman spectroscopy to observe possible structural and chemical modifications of nanoparticles during friction tests.

Used for the first time in 1977 by Sliney to study solid lubricants (graphite, MoS_2), *in situ* video during a friction test is a useful technique to understand matter movements inside the contact area [86]. It was used to explain the difference in wear observed after a friction test lubricated by MoS_2 or graphite. MoS_2 flows more easily inside the contact than graphite and forms a more continuous film. Scharf and Singer used *in situ* video to study the role of the third body with diamond-like carbon (DLC) coatings [87] and correlated the instability of the friction coefficient measured with the pulling up and the reformation of transfer films on surfaces. This technique was planned to study the rheology of transfer film formed during a friction test between a transparent surface and a layer of $MoS_{1.6}$ [88]. As now, *in situ* video experiments were carried out under dry conditions. Here, experiments with a lubricated contact are presented. Two aspects are to be considered: the first is the evolution of transfer film and the second is the movement of the nanoparticles inside the contact area and the way in which they supply it.

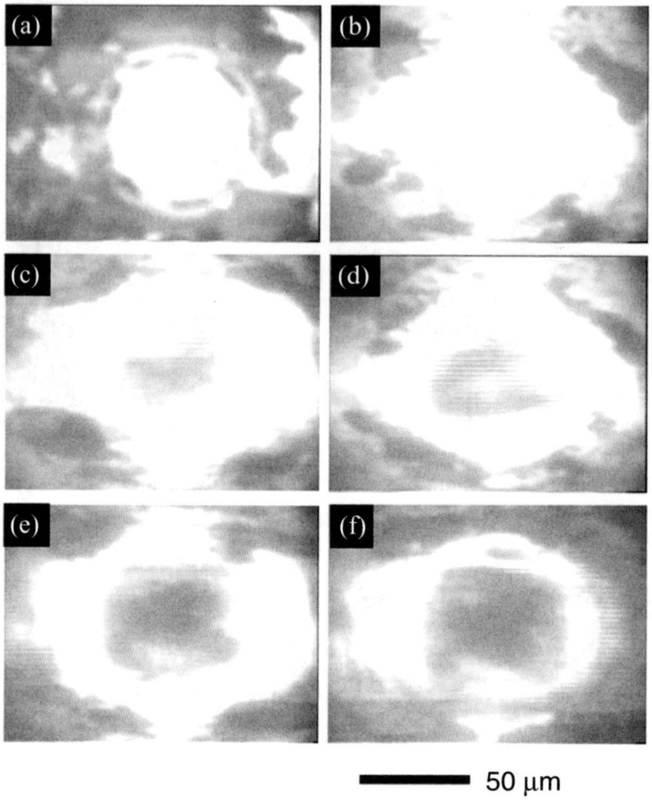

Figure 2.68 Image of the contact area during the friction test with IF-WS_2 at different cycle numbers: (a) 0, (b) 6, (c) 12, (d) 18, (e) 33 and (f) 48. These images show the film formation inside the contact area at the beginning of the test

In situ video during friction tests with IF-WS_2 dispersed in oil showed the formation of a film in the contact area from the first cycles of the test (Figure 2.68). This experiment was carried out with a colour camera that gives a good contrast of the contact area. Thus, the formation of a black tribofilm on the centre of the contact area can be followed continuously during the friction test. From the twelfth cycle, a film is visible on the centre of the contact area (Figure 2.68(c)). This experiment complements observations made on the pins after various numbers of cycles (Figure 2.27) and contributes to a better understanding of the transfer film formation and its evolution during the friction test.

Other experiments carried out with a black and white camera allow the global evolution of the nanoparticles to be followed inside the contact. A comparison of *in situ* video realized with three different IF diameters, IF-MoS_2 (40 nm), IF-NbS_2 (60 nm) and IF-WS_2 (140 nm), reveal a different behaviour for each IF. IF-MoS_2 easily enters the contact area (Figure 2.69(a)). With IF-NbS_2, a crown can be observed inside the contact area (arrow, Figure 2.69(b)). These observations should be connected to those made on wear scars after friction, where the formation of a film, mainly on the circumference of the wear scar, is observed. In the same way,

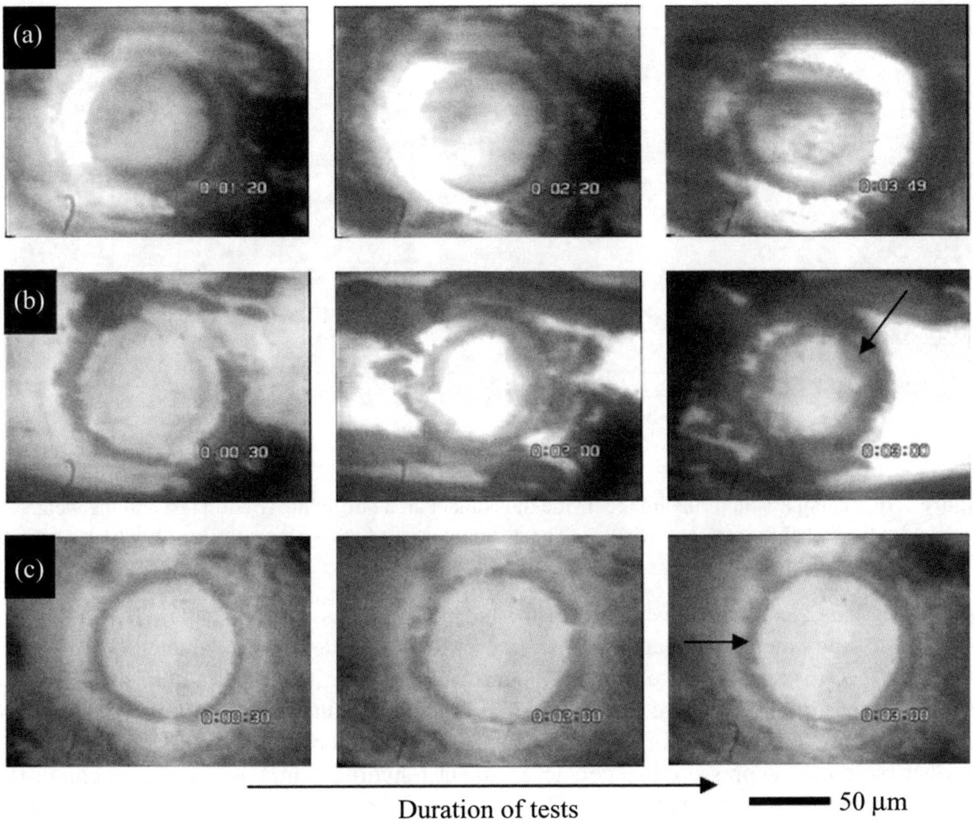

Figure 2.69 Images of the contact area for three IFs: (a) IF-MoS_2, (b) IF-NbS_2 and (c) IF-WS_2 at three times of friction: 30 s, 2 min and 3 min. A different behaviour can be observed for each IF tested

in the case of IF-WS_2, an agglomeration of IFs at the edge of the contact area (arrow, Figure 2.69(c)) and another IF around the contact area, where the crown is observed, can be distinguished. A picture of the wear scar during the friction test can be correlated to an image of the wear scar after the friction test (Figure 2.70).

In situ video highlights the behaviours of the different IFs in the contact and the importance of IF size. IFs of small size penetrate more easily into the contact. In the case of IF-WS_2, these experiments illustrate the formation of the crown observed on the pins after friction.

2.7.2 Raman Tribometry

2.7.2.1 State of the Art

To go further than a simple visualization from the transfers of matter inside the contact, techniques of analysis using a radiation can be used. EXAFS analysis was used by Martin *et al.* to study the behaviour of MoS_2 during friction [89]. Raman spectroscopy was performed on

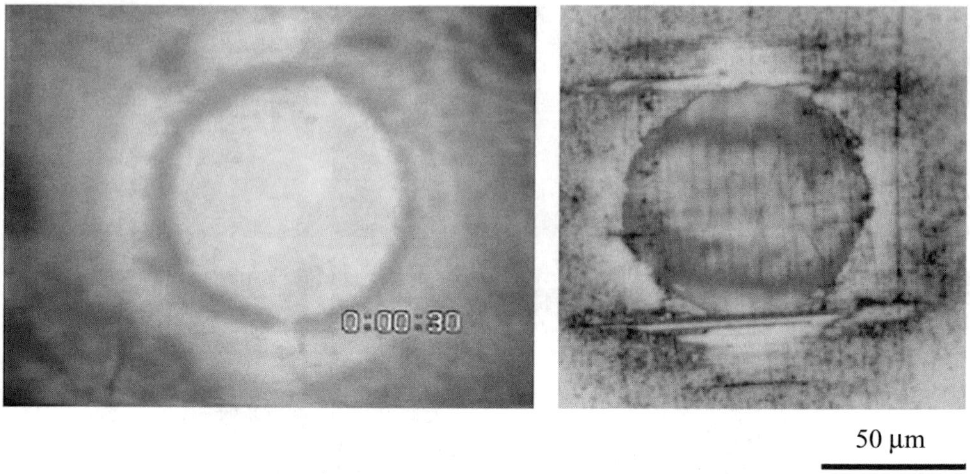

Figure 2.70 Comparison of the image inside the contact area during the friction test and the wear scar observed at the end of the test with IF-WS_2. The different parts of the wear scar (film, crown) can be seen

DLC [87]. This technique is very useful to study the various carbon shapes. It is also very powerful to study metal dichalcogenide structures. It will be demonstrated later.

Raman spectroscopy is very efficient in measuring pressure inside the contact area. Mansot used it to measure the pressure in a sphere/flat contact containing a thin polymer film [90]. Pressure curves obtained are in agreement with Hertzian theory. Similar works determined the distribution of the pressure in a contact in an elastohydrodynamic lubrication regime [91].

In the case, the experiments of *in situ* Raman were planned to understand the mechanism of exfoliation of the nanoparticles in the contact. They were performed in collaboration with the Laboratoire des Sciences de la Terre of Ecole Normale Supérieure de Lyon. The tribometer using a flat made of sapphire was placed directly under the microscope of a Raman spectrometer. The originality of the experimental device used involved the presence of a diamond anvil cell to carry out tests with very high hydrostatic pressures (35 GPa) (Figure 2.71).

The resistance of NT-WS_2 had already been studied using very strong and very short compression [92]. It resists to a pressure of 21 GPa, whereas at 35 GPa nanotubes are no longer observed. The sample is only composed of nanofibres coming from an important structural modification of the nanotubes under the effect of pressure. According to Zhu *et al.*, IFs are less resistant than nanotubes since the IF structure is subjected to more tensions and strains. IFs are more faceted and contain more defects and thus are less resistant. A similar experiment carried out recently on the 'Nanolub' lubricant showed that the IFs are intact after a pressure of 25 GPa [93]. Even if the authors assimilate the pressure exerted to a pressure of shearing, this pressure is different from that exerted in friction tests. A theoretical study of the deformation of fullerenes inside the contact and in particular the role of adhesion and pressure on the exfoliation of fullerenes showed that direct compression is not sufficient to involve the formation of sheets [94].

To understand precisely the exfoliation mechanism of the nanoparticles three types of experiments were planned. First of all, a study of the resistance of nanoparticles to strong hydrostatic pressures by using a diamond anvil cell was performed. Second, a study of the behaviour

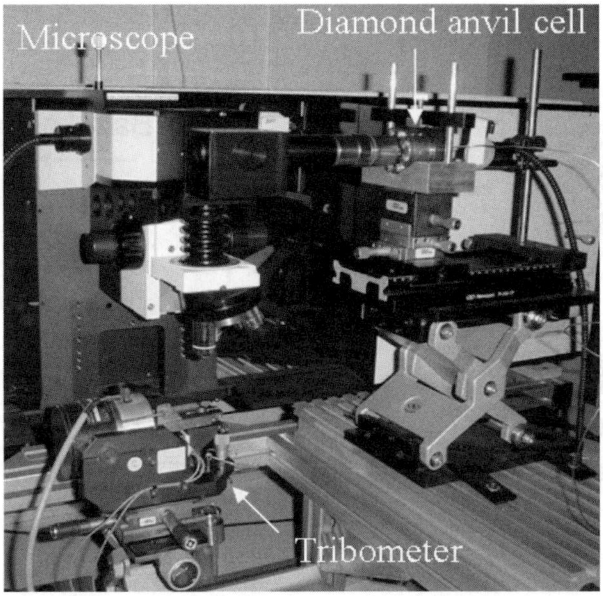

Figure 2.71 Experimental device used for the tribo-Raman study. This device is composed of a pin-on-flat tribometer with a sapphire flat, a microscope of the Raman spectrometer and also a diamond anvil cell

of nanoparticles inside a static Hertzian contact was carried out in order to understand better the effect of pressure on the fullerenes structure inside a ball/flat contact. In the end, analyses were carried out during a friction test to follow the structural evolution of nanoparticles.

2.7.2.2 Standard Spectra

Two modes are characteristics of metal dichalcogenides: E^1_{2g} and A_{1g}. The E^1_{2g} mode corresponds to the movement of atoms in basal planes and the A_{1g} mode to the movement of sulfur atoms along the *c* axis (Figure 2.72).

Raman spectra of the various forms of WS_2 studied (2H, IF and NT) present significant differences (Figure 2.73). Two differences can be pointed out: a shift of the E^1_{2g} peak, leading to a broadening of the peak at 350 cm^{-1} (which is composed of two components, the E^1_{2g} mode and 2 LA), and especially a shift of the A_{1g} peak. These differences are particularly interesting because they permit a structural evolution of the nanoparticles to be followed during friction tests.

Differences are also observed in the case of 2H-MoS_2 and IF-MoS_2, more particularly with a laser wavelength of 632.817 nm (Figure 2.74). The asymmetrical peak allows easy differentiation of the two structures [28].

2.7.2.3 Preliminary Tests under Hydrostatic Pressure

For tests carried out in a diamond anvil cell (DAC), IF-WS_2 was subjected to a maximum pressure of 35 GPa. With IF-MoS_2, the diamond anvil cell experiment was carried out only

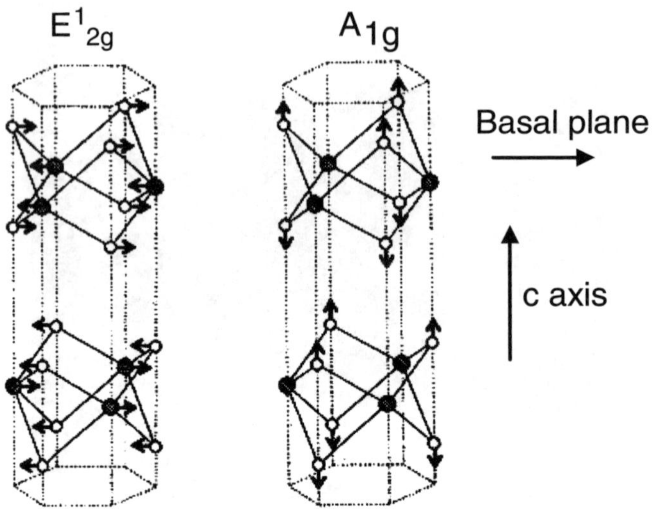

Figure 2.72 E^1_{2g} and A_{1g} modes of metallic dichalcogenides

up to 18 GPa because the signal became too weak. Samples are placed in the cell with PAO as a pressure transmitting medium to obtain a hydrostatic pressure in the cell, as long as the transmitting medium allows it. PAO was chosen as the pressure transmitting medium in order to work under the same conditions as during the friction test. Another point to mention is that PAO does not present characteristic lines between 250 and 550 cm^{-1}, the range of wave numbers corresponding to the studied products (Figure 2.75).

Evolution of Raman spectra with the pressure in the DAC for IF-MS_2 and the corresponding lamellar structure is presented in Figure 2.76 (MoS_2) and Figure 2.77 (WS_2). The comparison

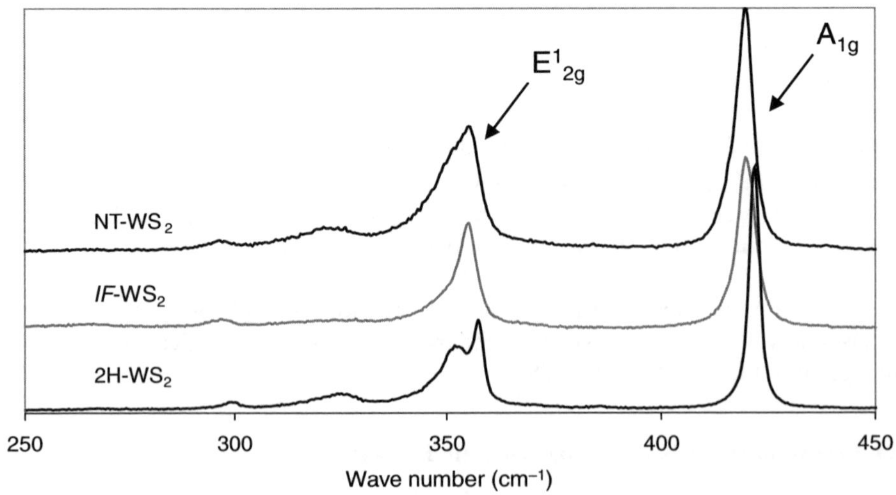

Figure 2.73 Standard spectra of 2H-WS_2, IF-WS_2 and NT-WS_2 ($\lambda = 514.5$ nm)

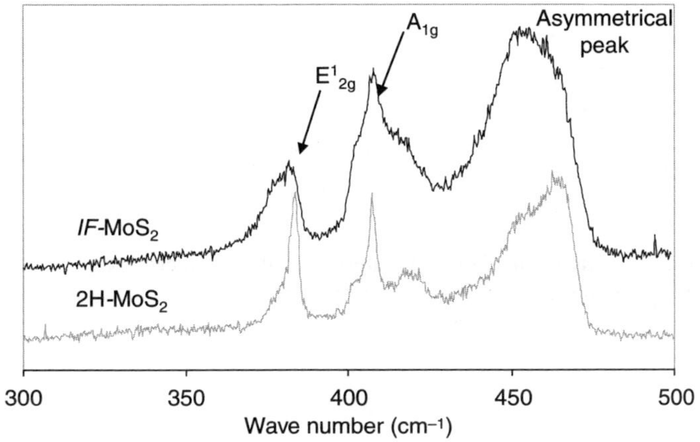

Figure 2.74 Standard spectra of 2H-MoS_2 and IF-MoS_2 (λ = 632.817 nm)

between the pristine products spectra and those recorded after the unloading shows a broadening and a light shift of the characteristic peaks. This tendency is much more marked in the case of IF. This indicates the creation of disorder in the structures. In the case of IF, it could correspond to the formation of some layers. Observations by TEM of the particles, collected after the diamond anvil cell test performed with IF-WS_2, show the formation of some layers at the edge of IF (Figure 2.78). The formation of these layers is, however, not significant and can come from peripheral zones where the pressure would not be hydrostatic [95].

Thus a high hydrostatic pressure is not sufficient to induce an important structural modification of fullerenes. These results confirm simulations made by Schwarz *et al.* which indicate

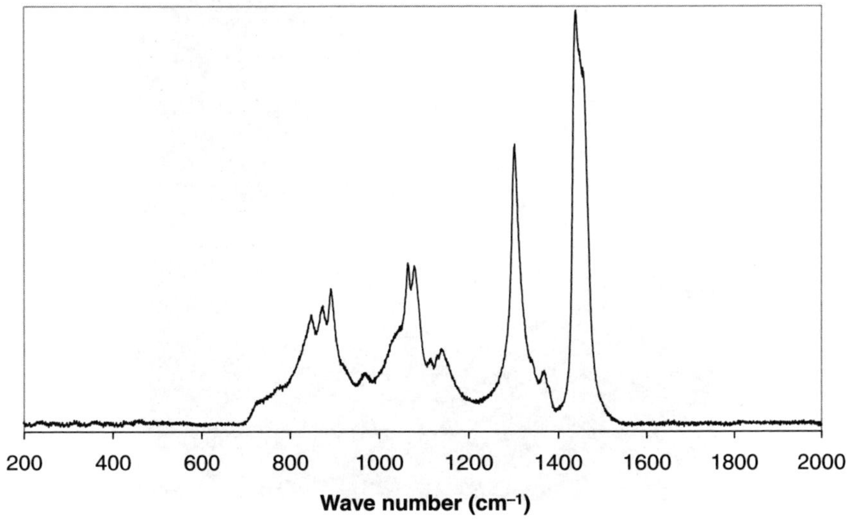

Figure 2.75 Raman spectrum of PAO (λ = 514.5 nm)

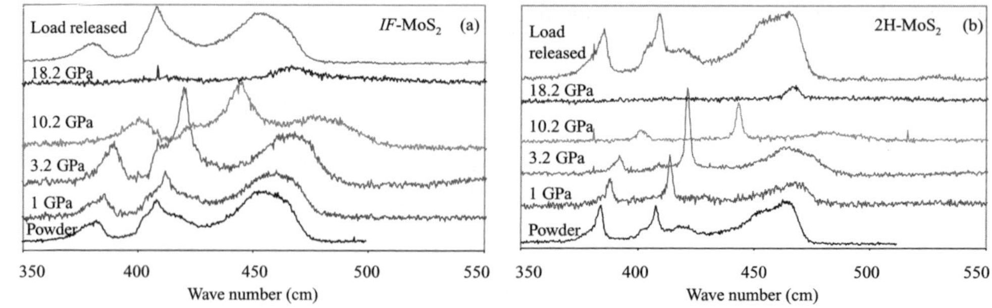

Figure 2.76 Raman spectra obtained during diamond anvil cell experiments with (a) IF-MoS_2 and (b) 2H-MoS_2. For the two samples, the spectrum obtained after the release of the load is similar to the one obtained before the experiment. Nevertheless, a broadening of the peaks is observed

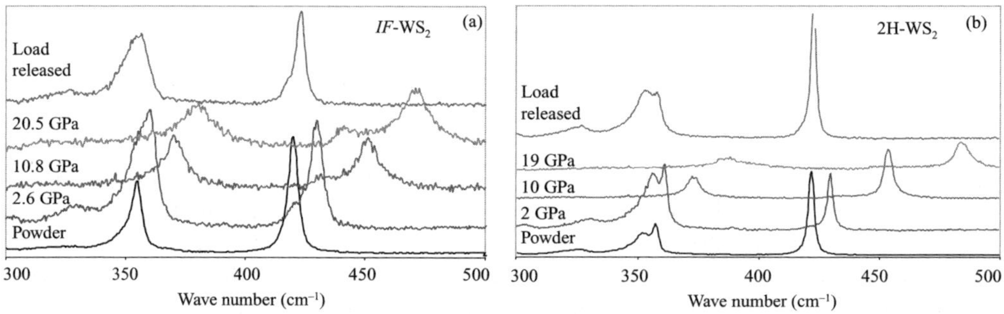

Figure 2.77 Raman spectra obtained during diamond anvil cell experiments with (a) IF-WS_2 and (b) 2H-WS_2. The same conclusion as the one for MoS_2 samples can be made

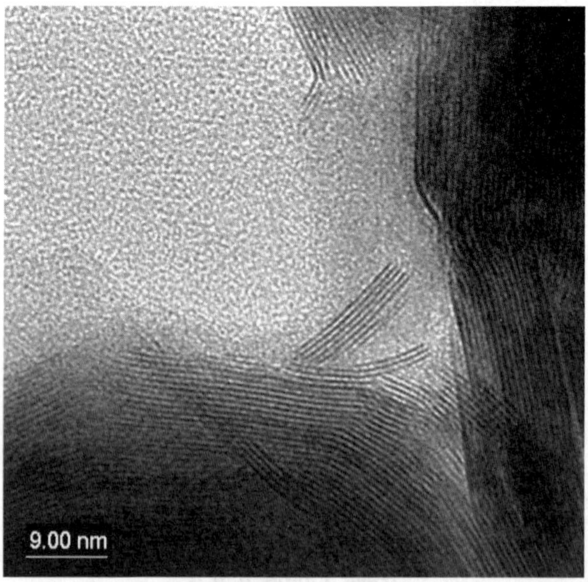

Figure 2.78 Nanoparticles collected after a diamond anvil cell with IF-WS_2. Some sheets are observed on the edge of the IF

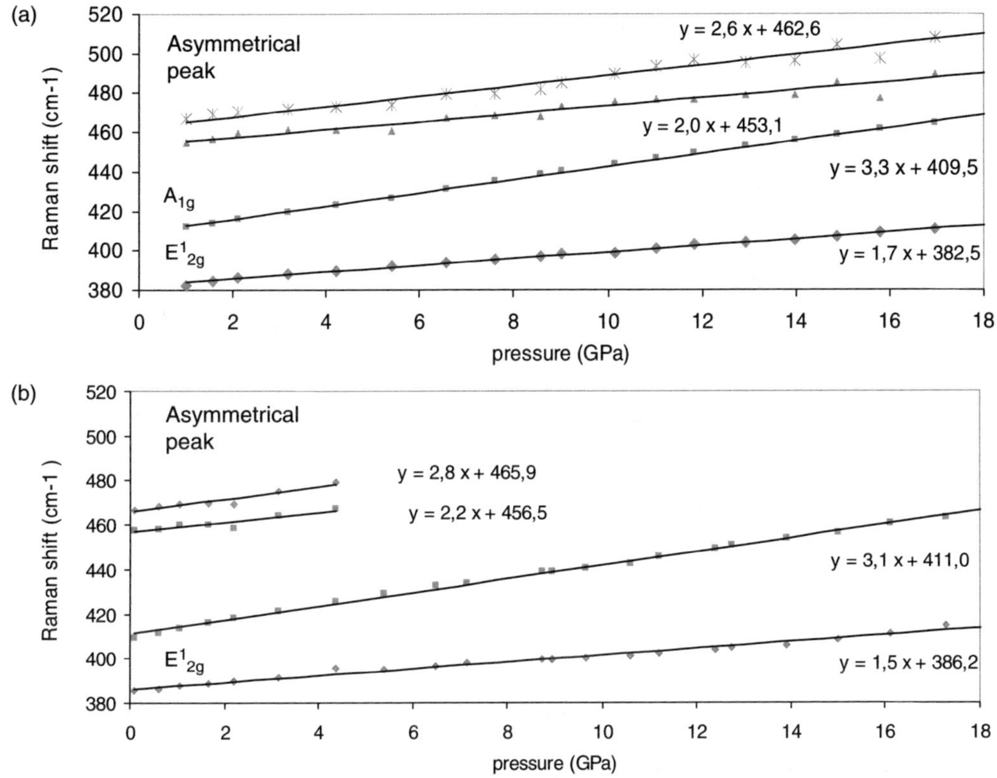

Figure 2.79 Raman shift of E^1_{2g}, A_{1g} modes and the asymmetrical peak for (a) IF-MoS_2 and (b) 2H-MoS_2

that direct compression is not at the origin of the structural modifications of IF inside the contact [94]. Another important point to note is that these experiments can simulate the conditions of an elastohydrodynamic contact. In this case, the thickness of oil is 1 μm and IF is thus subjected to an hydrostatic pressure. Similarly to the diamond anvil cell experiments, the oil film becomes solid when the pressures become high. During the hydrodynamic lubrication regime, IF remains intact and is not active in the contact, thus confirming the results obtained by Greenberg *et al.* [44].

The analysis of all the spectra obtained at known pressures indicates a displacement of the peaks characteristic of the structures (Figures 2.79 and 2.81). The displacement of the bands with the pressure is drawn. Then, the determination of the pressure inside the contact area during a static test or during a friction test is possible.

In the case of 2H-MoS_2, Sugai and Ueda carried out a diamond anvil cell experiment with a different pressure transmitting medium [96]. It used an ethanol–methanol mixture, known to be perfectly hydrostatic up to 10.4 GPa. The evolution of the shift of the peaks with the pressure can be described by a polynomial law (Figure 2.80). Sugai and Ueda expressed the displacement of bands according to shearing coefficients and compression coefficients between chalcogenes planes and between Mo atoms and chalcogenes atoms. This showed that

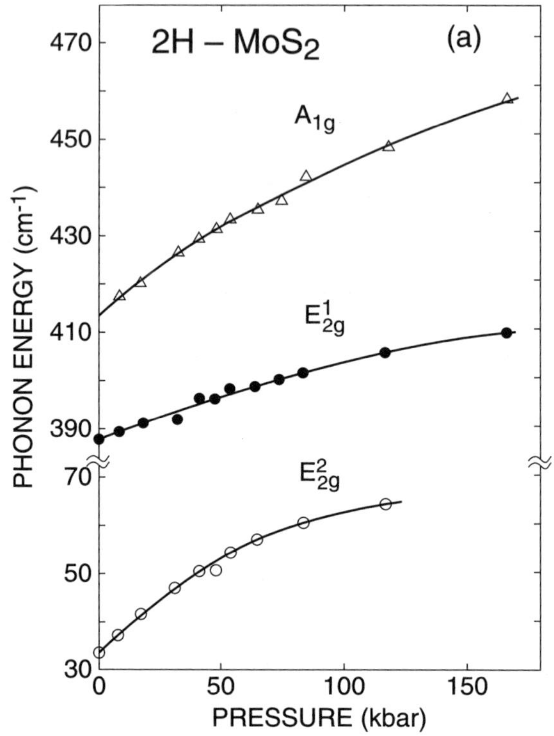

Figure 2.80 Raman shift of the E^1_{2g} and A_{1g} modes for 2H-MoS_2. This experiment was performed using a methanol/ethanol transmission media, which is different from the one used for the present study (PAO). Reprinted with permission of American Physical Society from Sugai and Ueda [96], Figure 2, © (1982) American Physical Society

these coefficients do not vary linearly with the pressure, but increase quickly up to 4 GPa. A comparison of the results obtained by Sugai and Ueda and the present results shows that they are similar. It will now be supposed that the PAO base oil presents rather similar properties of pressure transmission to the ethanol–methanol mixture. Here, the displacement of the peaks with the pressure will be expressed using a linear law. The evolution is thus well represented. Another important point of the Sugai and Ueda study is that it did not observe any important structural modification. This is consistent with the present study.

The displacements of the peaks in IF and 2H-MoS_2 structures are similar (Figure 2.79). For WS_2, the displacement of the E^1_{2g} peak is the same, but the A_{1g} mode presents a break of slope in the case of IF-WS_2 (Figure 2.81).

2.7.2.4 Test under Static Pressure in a Ball-on-Flat Hertzian Contact

Before carrying out friction tests, a study of the effect of the pressure on IF in a static contact using the tribometer without tangential displacement was performed. The Raman spectrum of

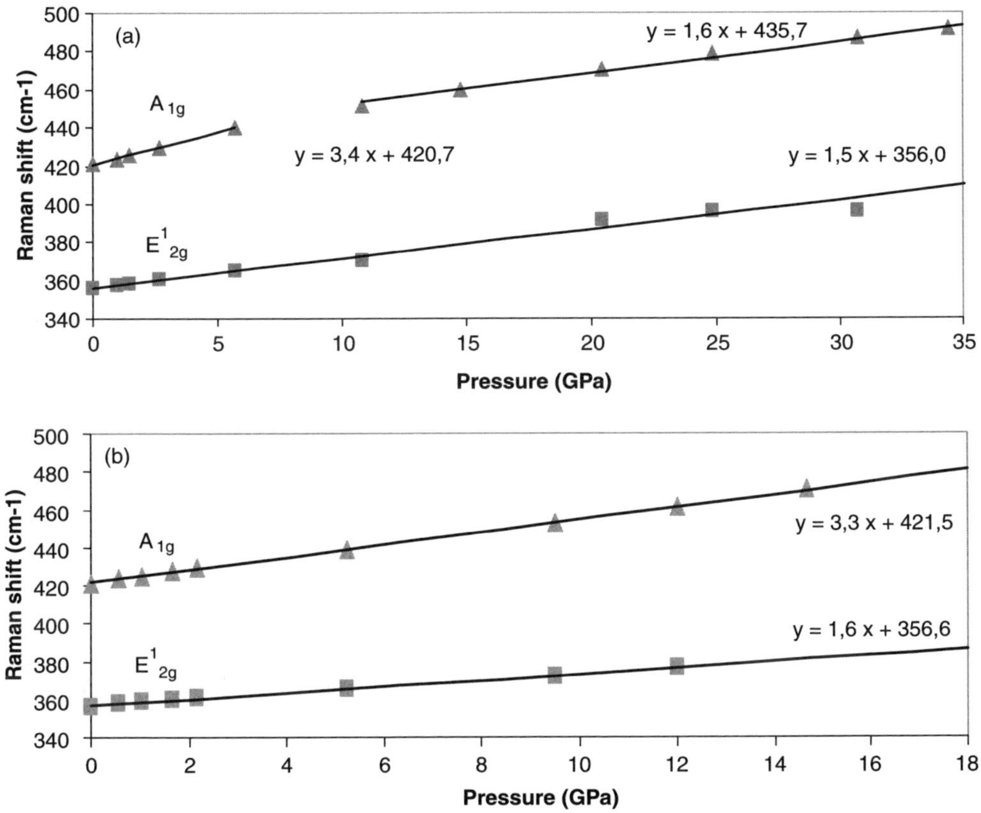

Figure 2.81 Raman shift of the E^1_{2g} and A_{1g} modes for (a) IF-WS_2 and (b) 2H-WS_2

the sapphire window used for these tests is reported in Figure 2.82. This spectrum is mainly composed of three peaks, 418, 577 and 752 cm^{-1}, which interfere with the peaks of IF.

Figures 2.83 and 2.84 present spectra obtained with 2H-WS_2 and IF-WS_2 dispersed in PAO and subjected to a process of load and unload successively. The process of loading consists in recording the spectrum of the pristine product put on the ball, closing the contact and then applying loads from 1 to 10 N. The unloading corresponds to the opposite process. The diameter of the zone analysed inside the contact area is approximately of 5 μm. After closing the contact, the appearance of a peak to 418 cm^{-1} can be noticed. This peak is due to the sapphire flat and its intensity depends on focusing the beam during Raman spectrum acquisition. For the two samples, a displacement of this peak due to the pressure is observed, which shows an elastic deformation of sapphire during the test.

In the case of 2H-WS_2 (Figure 2.83), a widening of the peak to 350 cm^{-1} is observed. This indicates the creation of disorder in the 2H structure. The spectrum after unloading is different from the spectrum of the pristine material product. This confirms a modification of the 2H structure inside a static contact. A comparison of the spectra obtained in the diamond anvil cell (Figure 2.77(b)) and in the static contact (Figure 2.83) shows a change in the form of the peak only inside the static contact. This change can be attributed to an orientation of

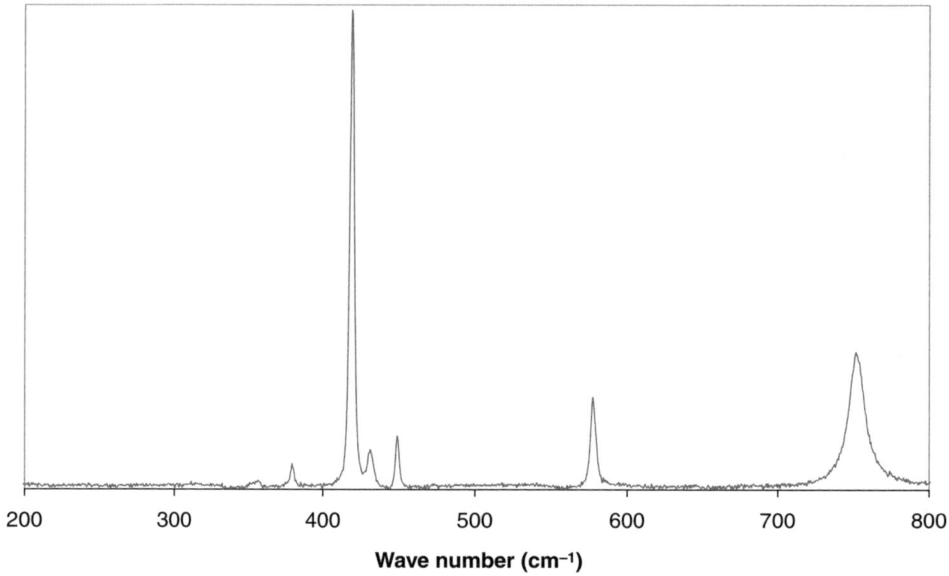

Figure 2.82 Raman spectrum of the sapphire flat ($\lambda = 514.5$ nm)

2H-WS_2 under nonhydrostatic pressure. This orientation would then explain the difference between peaks obtained before and after the loading–unloading test inside the static contact. 2H-WS_2 particles are also crushed into smaller ones during the friction test, which explains a more important disorder.

With IF-WS_2 (Figure 2.84), a broadening of the peak at 350 cm^{-1} is observed just after closing the contact. The difference between spectra of a pristine material and after the

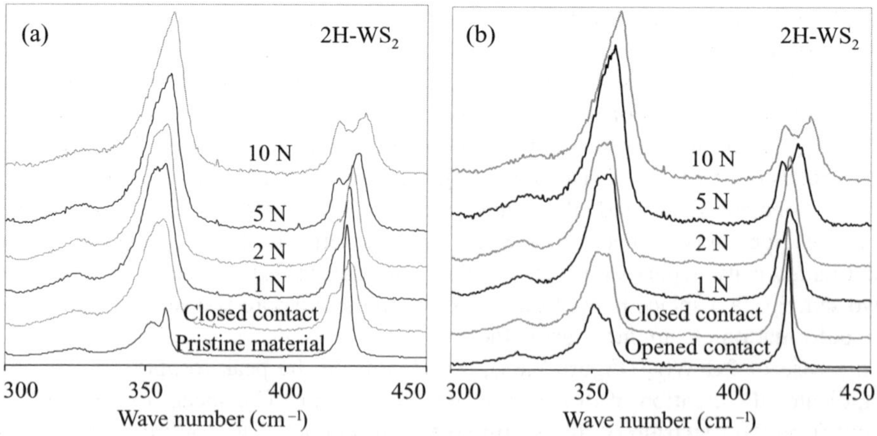

Figure 2.83 Raman spectra obtained during (a) load and (b) unload inside a static contact with 2H-WS_2 dispersed in PAO. A difference between spectra obtained before load and after unload can be observed

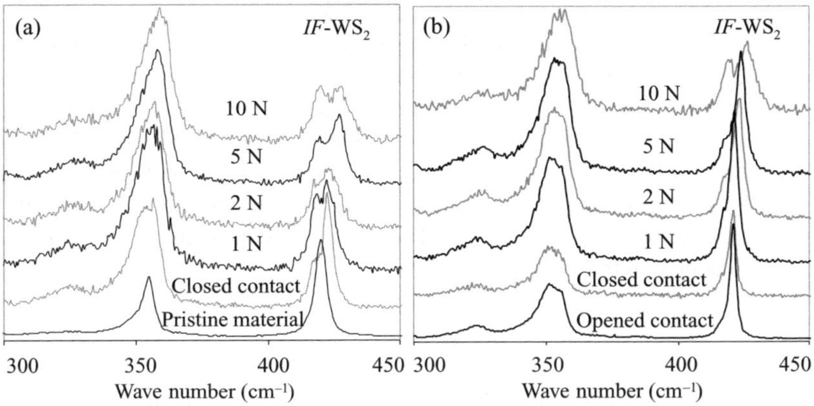

Figure 2.84 Raman spectra obtained during (a) load and (b) unload inside a static contact with IF-WS_2 dispersed in PAO. As in the case of 2H-WS_2, spectra before load and after unload are different

loading–unloading test is more important with IF than with 2H. The spectrum after unloading is similar with that of 2H-WS_2 after unloading. This means that a structural modification of IF in 2H occurred. After diamond anvil cell experiments, a broadening of the peak was observed (Figure 2.77(a)), indicating the appearance of disorder. After this loading–unloading test, the peak is also broadened but changes in form to look like the peak of 2H after the loading–unloading test (Figure 2.84(b)). A strong structural modification of IF thus occurs inside a static contact. This can be explained by the nonhydrostatic pressure exerted on fullerenes inside the static contact (Figure 2.85). Indeed, in the diamond anvil cell, pressure is exerted in an isotropic way, so particles do not undergo important structural modifications. Inside a static contact, pressure is exerted in an anisotropic way. This process can be assimilated in a 'nut-cracker' process.

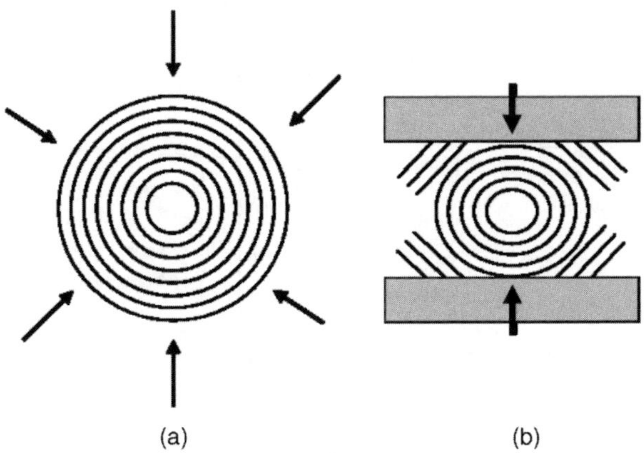

Figure 2.85 Pressure submitted to the IF: (a) inside a diamond anvil cell and (b) inside a static contact. The nonhydrostaticity may explain the structural modification observed only inside the static contact – the nut-cracker effect [95]

In the simple contact configuration used here, the contact pressure can easily be estimated owing to the theory of contact mechanics between two solids [97]. The bases of the calculation are the following: sphere and plane initial geometry, purely static loading, purely elastic and isotropic behaviour of the substrates and infinitely smooth surfaces. Under normal loading, the solids are reversibly deformed with a circular contact zone (see Figure 2.29). The normal pressure generated on the surface is maximum at the centre of the contact zone. The lateral distribution of the contact pressure within the contact zone is quadratic: maximum at the centre with a parabolic decrease towards the edges. The value of the maximum contact pressure P_{max} is nonlinear. It depends on the normal load F_n:

$$P_{max} = kF_n^{2/3} \tag{3}$$

Taking into account the mechanical characteristics of the plane (sapphire) and the sphere (AISI 52100 steel), with the contact geometry and the normal load applied, the maximum contact pressure can be calculated. This gives $P_{max} = 0.78, 0.98, 1.33$ and 1.68 GPa, respectively, for $F_n = 1.0, 2.0, 5.0$ and 10.0 N.

The position of the peaks on spectra obtained at various normal loads enables the pressure applied on the particles inside the contact to be determined. Calculations are carried out using the straight line of Raman displacement obtained from hydrostatic pressure experiments (Figure 2.81). Figure 2.86 presents pressures calculated for 2H-WS_2 and IF-WS_2 compared with those determined by Hertzian theory (contact without particles, contact pressures determined as above). Differences between the experimental and theoretical curves can be observed. Several hypotheses can explain these differences. First, it was found that the analysed point does not correspond to the centre of the contact area. This would explain the low values obtained. The most probable hypothesis is that the pressure is not uniformly distributed in the contact but is more important on the fullerene clusters (Figure 2.87).

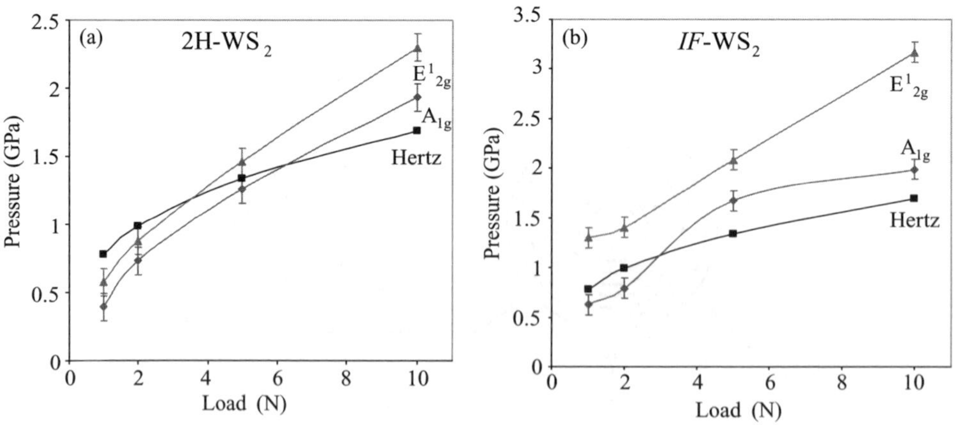

Figure 2.86 Contact pressures determined from Raman shift observed on the Raman spectra obtained at different normal loads with (a) 2H-WS_2 and (b) IF-WS_2. These pressures are compared to those calculated from Hertzian theory

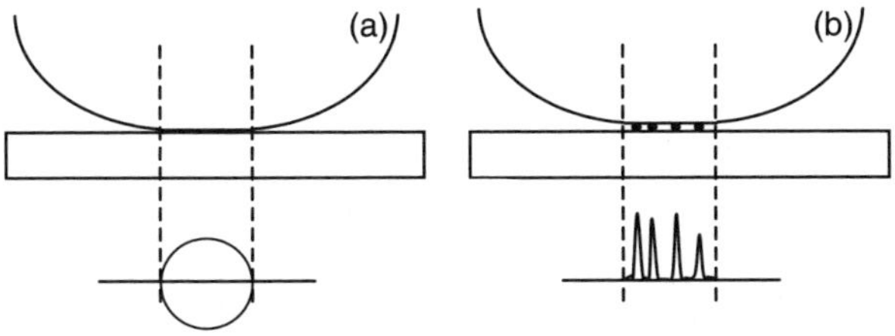

Figure 2.87 Pressure distribution inside (a) a theoretical contact and (b) a contact containing fullerenes

A similar study was carried out with WS_2 nanotubes (Figure 2.88). As in the case of IF, the spectrum after unloading is different from that of the pristine material and looks like the spectrum of 2H-WS_2 after discharge. A comparison of the behaviour of NT-WS_2 and IF-WS_2 under high pressure was made by tracing displacements of the peaks according to the pressure (Figure 2.89). The A_{1g} mode (movement of sulfur atoms along the c axis) shows a similar evolution with the pressure in both structures, whereas the E^1_{2g} mode (vibration of atoms inside a basal plane) seems to be more sensitive to pressure in the nanotube structure than in the fullerene structure. This can be explained by the structure of the two systems. With nanotubes the layers of WS_2 are rolled up, whereas IFs are an assembly of 2H nanocystals. Thus, vibrations in the layers of nanotubes will be more sensitive to the pressure than in IFs. An experiment with the diamond anvil cell could confirm the behaviour of various modes under the effect of hydrostatic pressure.

The behaviour of IF-MoS_2 was studied in the same conditions using a red laser (Figure 2.90). A slight change in form of the asymmetrical peak is observed for IF-MoS_2 at 5 and 10 N. This supposes that the formation of 2H-$MoS_2$is inside the contact area. Nevertheless, the spectrum after the loading–unloading test is very similar to that of the pristine material. Thus, an accurate fit of the spectra would be necessary to analyse the results better and to distinguish the lamellar and the spherical structures. As in the case of IF-WS_2, contact pressures were determined from Raman spectra, and are higher than those calculated by Hertzian theory

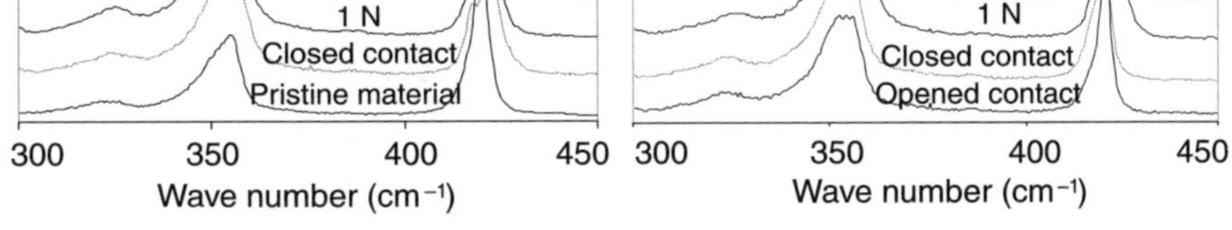

Figure 2.88 Raman spectra obtained during (a) loading and (b) unloading inside a static contact with NT-WS_2 dispersed in PAO

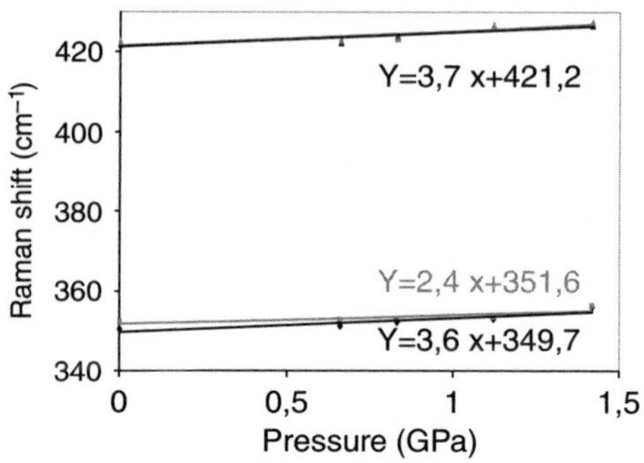

Figure 2.89 Raman shift of E^1_{2g} and A_{1g} modes with NT-WS_2 (black) are compared with those obtained with IF-WS_2 (grey). The A_{1g} mode has the same behaviour in IF or NT structures. A difference is observed for the E^1_{2g} mode

(Figure 2.91). The displacement of the A_{1g} peak is difficult to follow from the presence of other peaks (403 cm^{-1}) and the peak of the sapphire flat.

Raman experiments in the static contact highlighted an important structural modification of IF-WS_2 in the contact area only under static pressure. This structural modification can be explained by the nonhydrostatic pressure subjected to the majority of fullerenes in the contact. The formation of sheets only under static pressure explains the effectiveness of IF from the first cycles of a friction test. The closing of the contact is already sufficient to induce the formation. In the case of 2H-WS_2, large particles are crushed to form smaller particles.

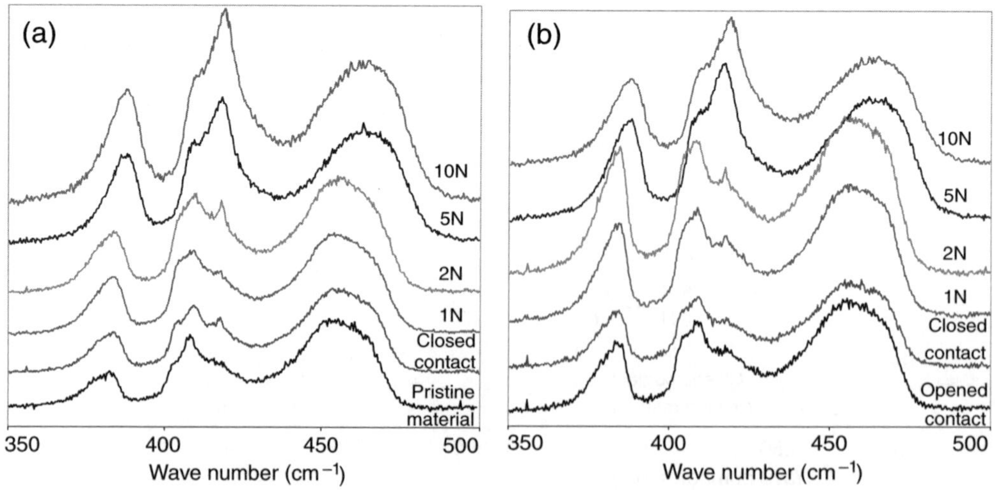

Figure 2.90 Raman spectra obtained during (a) loading and (b) unloading inside a static contact with IF-MoS_2 dispersed in PAO. A difference between spectra of the pristine material and after the test can be observed

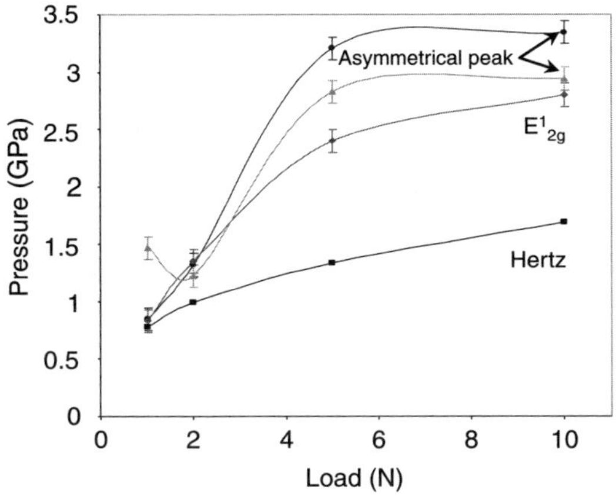

Figure 2.91 Contact pressures determined from Raman shift observed on Raman spectra obtained at different normal loads with IF-MoS_2. These pressures are compared to those calculated from Hertzian theory

2.7.2.5 *In situ* Raman Analyses during Friction Tests with IF-WS_2 and NT-WS_2

Raman spectra of IF-WS_2 and 2H-WS_2 recorded at 514.5 nm are different. Thus, analyses during friction tests were carried out to follow the structural evolution of IFs and in particular the formation kinetics of 2H-WS_2 sheets. Conditions for friction tests are similar to those used for a steel/steel contact: a frequency of 0.5 Hz and a temperature of 20 °C. It is important to note that the tangential movement is stopped during the acquisition of the spectrum (approximately 2 minutes). Figure 2.92 compares the spectra obtained in the contact area after 1 and 4 minutes of friction with those of the initial products. These analyses confirm the massive formation of 2H-WS_2. The A_{1g} peak is shifted to 422 cm^{-1}, which corresponds to the peak

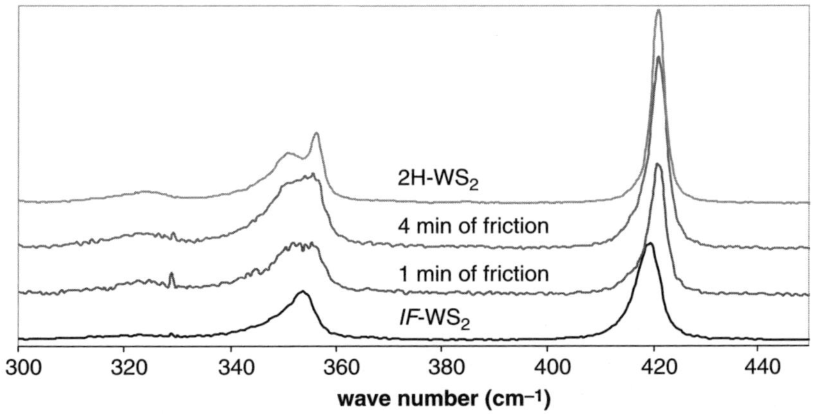

Figure 2.92 Raman spectra realized inside the contact area after 1 and 4 minutes of friction compared with the spectra of IF-WS_2 and 2H-WS_2. These analyses clearly show the formation of 2H-WS_2 just after 1 minute of friction (30 cycles)

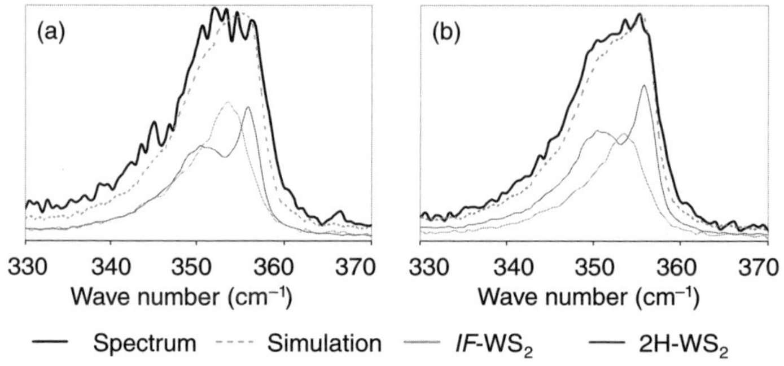

Figure 2.93 Raman spectra obtained during friction compared with simulations of a mixture of IF-WS_2 and 2H-WS_2: (a) after 1 min with 62 % IF-WS_2 and 38 % 2H-WS_2; (b) after 4 min with 50 % IF-WS_2 and 50 % 2H-WS_2

of 2H. A broadening of the peak is observed at 350 cm^{-1}. After 1 minute of friction only, 2H is present in the contact. The proportion of 2H formed can be estimated by simulating the shape of the peak by a linear combination of IF and 2H spectra (Figure 2.93). After 1 minute of friction, the proportion of 2H is about 38 %. The presence of 2H in the contact from the first minute can explain the low friction observed from the very beginning of the test. After 4 minutes, the proportion of 2H is about 50 %. Thus the regular formation of 2H in the contact area is observed.

The evolution of the proportion of 2H in the tribological contact can be traced (Figure 2.94). Values for the static contact were obtained from the spectrum during the load–unloading experiment for a load of 2 N, friction tests being realized at this load. This chart shows the formation of 2H in the contact area during friction and the presence of 2H in the contact before the beginning of friction only after applying the load.

In situ Raman analyses performed during a friction test confirms the continuous delivery of sheets from IF particles during friction (Figure 2.94). In the case of nanotubes, a similar

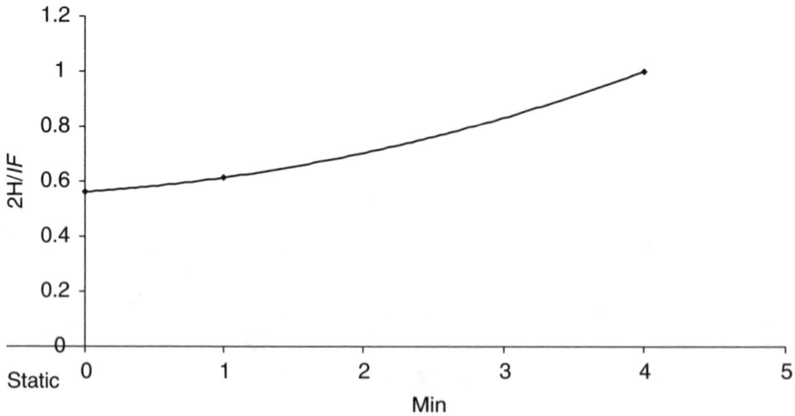

Figure 2.94 Evolution of the 2H ratio inside the contact area during a static experiment or during the friction test with IF-WS_2

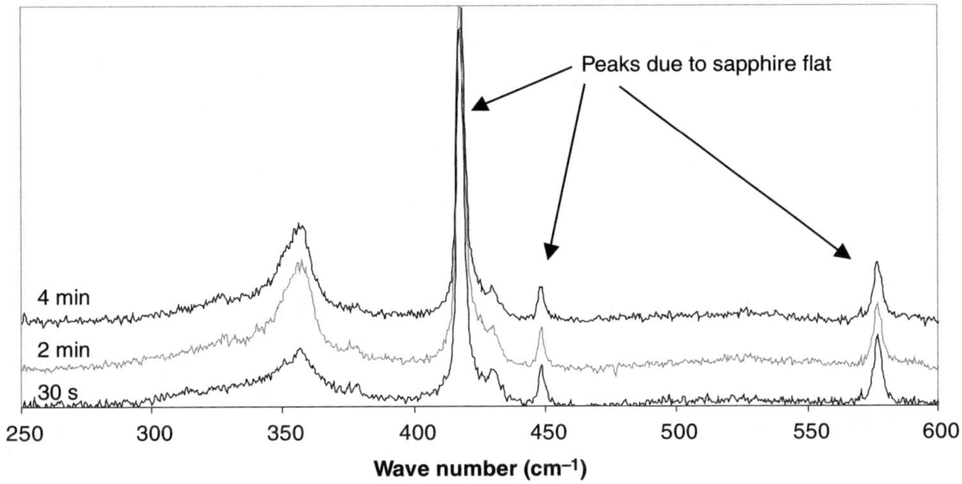

Figure 2.95 Spectra obtained inside the contact area after friction (30 s, 2 min and 4 min) with NT-WS_2. A significant contribution of the sapphire flat can be observed

experiment was carried out. Unfortunately, the Raman signal obtained is very weak and there is an important contribution of the signal from the sapphire flat (Figure 2.95). The formation of WS_2 was not clearly observed.

To highlight the Raman signal due to pure WS_2, the spectrum of the sapphire flat was withdrawn from the experimental spectrum. The spectra obtained are very similar to the spectrum of pristine NT-WS_2 (Figure 2.96). Nevertheless, due to the weakness of the signal, no conclusion about the nature of WS_2 present in the contact area could be made.

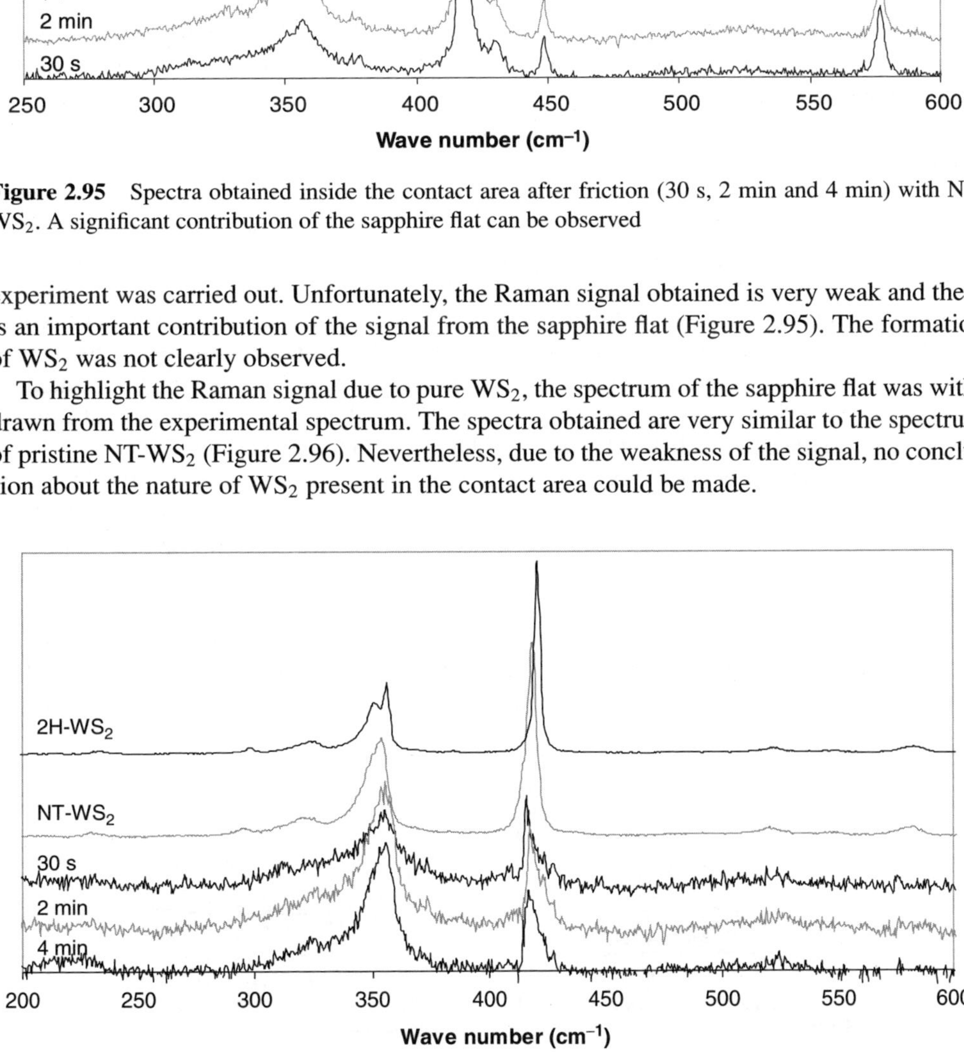

Figure 2.96 Spectra obtained inside the contact area after friction with NT-WS_2 and after substraction of the sapphire flat contribution. These spectra are compared to those of pristine NT-WS_2 and 2H-WS_2

To continue, Raman spectroscopy is very useful for obtaining an understanding of the physical mechanism of lubrication of IF-MS_2. As shown by calculations of Schwarz *et al.*, exfoliation of IFs is due to pressure, but occurs only under adhesion conditions [94]. In this case, pressure subjected to fullerene is no longer hydrostatic, which explains the structural modifications of IF. When subjected to hydrostatic pressure IFs do not undergo important structural modifications. This was proved by the diamond anvil cell experiments. With Raman spectroscopy the structural evolution of IF into lamellar compounds in the contact during a friction test is followed.

2.8 Lubrication Mechanism of IF-MS_2: 'A Drug Delivery' Model

Cizaire *et al.* attributed the friction-reducing properties of IF to a flattening of the particles and an exfoliation of their external layers, releasing MoS_2 sheets inside the contact area [29]. Observations by HRTEM showed the presence of flattened IF and MoS_2 sheets. Low friction can be attributed either to a slip between flattened IF (Figure 2.97(a)) or a slip between MoS_2 sheets (Figure 2.97(b)).

Formation of MoS_2 single sheets inside the contact area has already explained friction-reducing properties of several additives: the molybdenum dithiocarbamate (MoDTC) and the molybdenum dithiophosphate (MoDTP) [58]. Grossiord *et al.* showed the presence of MoS_2 sheets in a carbon matrix or a phosphate matrix, these sheets coming from the chemical decomposition of the molecule of additive. Only a few sheets are necessary to obtain low friction. With these two additives, a low coefficient of friction is obtained after only several minutes of test, which corresponds to the activation time of the additive [98]. Moreover, these additives are efficient only if they are thermally activated (70 °C), whereas IFs are efficient at ambient temperature. In the case of IF, low friction is obtained from the first cycles.

The authors carried out new observations of wear particles and confirmed the formation of MoS_2 sheets during the tests. Some of these sheets are in incommensurate conditions (Figure 2.98). The rotation angle between two sheets, determined from the optical diffractogram, is equal to 30°.

The action mode of IF-MoS_2 is based on an exfoliation of their external sheets to release single sheets inside the contact area. These sheets lead to a reduction of the friction coefficient. Some of these sheets are in incommensurate conditions. This explains the very low friction coefficients observed. The results show that the action mechanism of IF-WS_2 is identical to

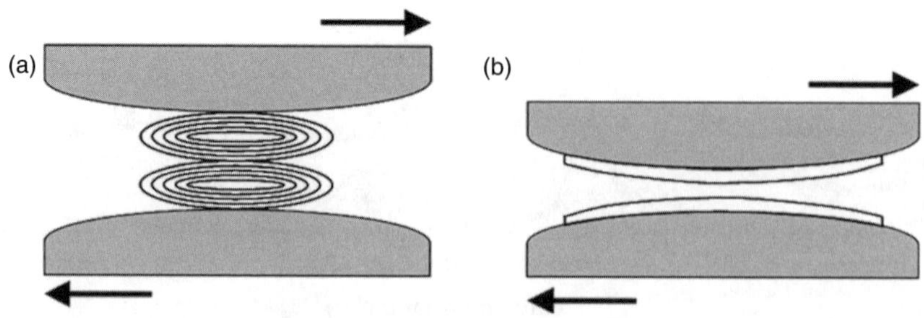

Figure 2.97 Lubrication mechanisms of IF-MoS_2: (a) shear between flattened IF; (b) formation of sheets on the surfaces. The same mechanism is observed with MoDTC

Figure 2.98 HRTEM micrograph of two MoS_2 sheets present in wear particles collected on the pin. These sheets are in incommensurate conditions (angle of 30°)

that of IF-MoS_2, namely an exfoliation of the layers external to the IF and the formation of very thin films on surfaces.

Schwarz *et al.* studied the effect of adhesion and pressure on the tribological properties of fullerenes [94, 99]. Particles adhere to surfaces by van der Waals forces. These forces are proportional to the radius of the particles and independent of the number of layers. Deformations due to adhesion of the particles on surfaces are very weak. Adhesion supports the exfoliation of the particles, but does not trigger it. The exfoliation of fullerenes is due to the pressure exerted on the particles, which destabilizes the structure, a critical pressure being necessary. Exfoliation occurs only under adhesion conditions and concerns only the first layers (1 or 2 first layers). Layers then adhere on surfaces, which minimizes the edge effects and leads to the formation of films on surfaces [99].

Drummond *et al.* studied the properties of films formed on surfaces with IF-WS_2 dispersed in tetradecane [45]. After a friction test, they carried out rinsing of surfaces using solvent to eliminate all IFs present. A new friction test was then performed with pure tetradecane and a friction coefficient similar to that obtained with IF in tetradecane was measured. This experiment illustrates the tribological properties of WS_2 films formed on surfaces. Prasad and Zaenski studied transfer films formed during friction between a steel ball and WS_2 coatings [100]. A very low coefficient of friction (0.04) was obtained in an N_2 atmosphere, with a smooth transfer film made up of very thin WS_2 layers formed on the pin. These thin layers are very flexible and can be curved and thus adhere to surfaces.

TEM observations of the wear particles collected on the pin after the friction test ($P = 1.12$ GPa, $v = 2.5$ mm/s, $T = 20\,°C$) show the presence of many WS_2 single sheets. The diffractogram of Figure 2.99 clearly shows 18 diffraction points, which can be divided into three

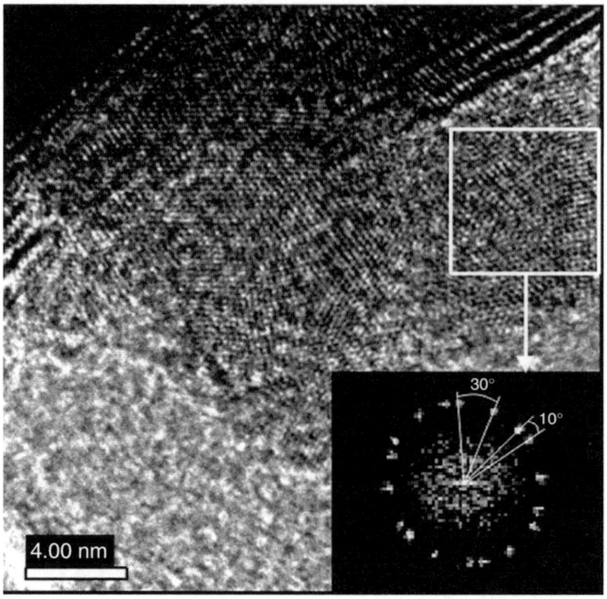

Figure 2.99 HRTEM image of WS_2 sheets in incommensurate conditions. The diffractogram of the framed area shows an angle of 30° or 10° between the three superimposed sheets [53]

times six points. They correspond to the superposition of three figures of diffraction obtained from an hexagonal compact structure directed along the [0001] axis but whose certain planes are shifted an angle of 10° and of 30° compared to a third given orientation. This wear particle is thus composed of at least three superimposed WS_2 sheets with an angle of 30° between two of them and an angle of 10° between two others. With an angle of 10° there is still atomic coincidence, but with an angle of 30° there is no atomic coincidence and the shear is easier. These sheets are in incommensurate conditions and friction can be vanished [76, 101].

The mechanism of action of the nanoparticles is based on the release of sheets in the contact area. Fullerenes, of nanometre size, are able to enter the tribological contact easily, as shown in the *in situ* video experiments, whereas particles of 2H-MS_2 are much larger and circumvent the contact area. These fullerenes are exfoliated to release sheets directly inside the tribological contact. The *in situ* Raman spectroscopy highlights the formation of these layers only in the contact area. The formation of some layers is sufficient to obtain very interesting tribological properties. This formation occurs in the first cycles of the test, which explains the good tribological properties of these additives from the very beginning of the tests. In the case of the diamond anvil cell experiments, simulating an elastohydrodynamic lubrication regime, IFs are little modified and are thus not very active. These additives are thus only active in the boundary lubrication regime, in the critical zones where oil does not have a rheological action.

An analogy between this mechanism and a 'drug delivery mechanism' can be made. The fullerenes are active only in the contact area, by releasing their active substance (individual layers). Figure 2.100 schematically represents the release of the individual layers in the tribological contact.

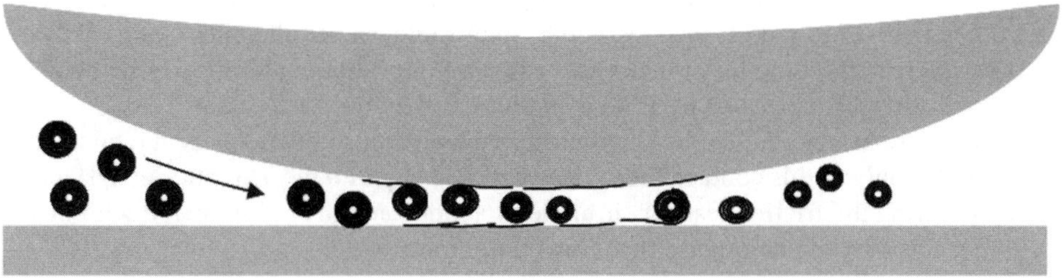

Figure 2.100 Schematic model of the sheet formation inside the tribological contact. The individual sheets are released only in the contact area

This is the same action mechanism as a drug that acts in a very precise part of the body. In medicine, systems of nanometre size (nanospheres, nanocaps, micelles, etc.) are largely used in drugs or as vectors in the field of genetics [102]. For example, micelles are used in drugs against cancer to give a controlled release of the active principles or are applied to the transport of genetic material (Laboratory of Professor Leroux, University of Montréal).

Contrary to drugs that generally require a chemical activation (change of pH, etc.) to be active, IFs undergo a structural modification inside the contact area induced by shearing and pressure. This point is very important because these additives are active at ambient temperature, contrary to MoDTC and ZnDTP, which are active at 100 °C and after an induction period. This type of additive would therefore be a solution to the problem of cold start of engines, the critical phase, in particular in hybrid systems.

The nanometre size of the particles also enables them to penetrate in the asperities and porosities of surfaces. These asperities thus act as reservoirs and can regularly supply the contact area during friction. Rapoport *et al.* have already expressed this hypothesis [42, 46]. The authors have confirmed it using SEM images of surfaces after friction (Figure 2.101). Figure 2.101(a) shows IF-WS_2 in an asperity present on the flat. These IFs can arise to supply the contact area, in the same way as in the case of nanotubes (Figure 2.101(b)). Surface texturing is now widely studied to improve tribological systems. Moshkovith *et al.* recently studied the texturing of a steel surface followed by burnishing the surfaces with MoS_2 particles (2 µm

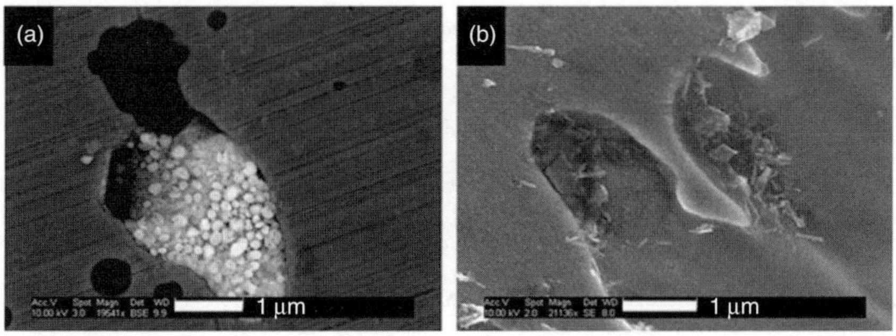

Figure 2.101 SEM images of IF-WS_2 trapped in asperities of the flat surface. Surface asperities can be assimilated to reservoirs, which will release the nanoparticles in the contact area during friction

in size) [103]. The burnishing was performed to fill the asperities with MoS_2 particles. The same work with IF-MoS_2 and IF-WS_2 would be interesting, since fullerenes could easily enter the asperities created by the texturing. Otherwise, friction tests could directly be performed with textured surfaces lubricated by PAO containing fullerenes.

The nanometre size of the studied particles enables them to enter easily the contact area and to be active only in this contact zone. They can also slip into the asperities of surfaces and be released during the friction test. Their mode of action can thus be compared with that of a drug, which acts only at one precise place and continuously.

2.9 Conclusion

Nanoparticles made of metal dichalcogenides present very interesting tribological properties when they are used as lubricant additives or as coatings. They combine friction-reducing and antiwear properties, so can be envisaged as substitutes for the additives actually used. Great progress has been made on their synthesis and in the near future they will be produced in large quantities. Furthermore, progress in the synthesis process has led to the formation of new fullerenes of metal dichalcogenides, which could also present tribological properties. Recently, Margolin *et al.* synthesized IF-TiS_2 nanoparticles [104]. The first tribological results indicate that they have the same properties as IF-MS_2. Jamison classified metals that are able to form lamellar structures and those that present interesting tribological properties [63]. Mo, W and Nb are known to form lamellar structures with good tribological properties, especially MoS_2, WS_2, $NbSe_2$ and WSe_2 [64]. It would therefore be interesting to test IF-$NbSe_2$ and IF-WSe_2 as nanolubricants, or other fullerenes of lamellar structures like IF-TiS_2. IF-MX_2 ($X = S$, Se) can be considered as a big 'family' of potential nanolubricants and further works are necessary to evaluate all of these fullerene-like materials.

Acknowledgements

The authors would like to thank Dr Niles Fleischer of Nanomaterials Ltd and Professor Reshef Tenne of the Weizman Institute of Science (Israel) for providing the fullerenes samples and for their helpful discussions. They would also like to thank Dr Bruno Reynard, Director of Laboratoire des Sciences de la Terre at Ecole Normale Supérieure de Lyon, for his collaboration for the Raman study, and Gilles Montagnac, for his support during the Raman measurements and helpful discussions.

Lucile Joly-Pottuz would like to thank Gilbert Thollet of INSA de Lyon for his help with the 'wet STEM' experiments.

References

1. Winer, W. O., Molybdenum disulphide as a lubricant: a review of the fundamental knowledge, *Wear*, **10**, 1967, 422–452.
2. Sokoloff, J. B., Theory of energy dissipation in sliding crystal surfaces, *Physical Review B*, **42** (1), 1990, 760–765.
3. Hirano, M., Shinjo, K., Kaneko, R. and Murata, Y., Anisotropy of frictional forces in muscovite mica, *Physical Review Letters*, **67** (19), 1991, 2642–2645.
4. Hilton, M. R. and Fleischauer, P. D., TEM lattice imaging of the nanostructure of early-growth sputter-deposited MoS_2 solid lubricant films, *Journal of Materials Research*, **5**, 1990, 406–420.
5. Fleischauer, P. D. and Lince, J. R., A comparison of oxidation and oxygen substitution in MoS_2 solid film lubricant, *Tribology International*, **32**, 1999, 627–636.

6. Martin, J. M., Donnet, C., Le-Mogne, T. and Epicier, T., Superlubricity of molybdenum disulphide, *Physical Review B*, **48** (14), 1993, 10583–10586.
7. Peterson, M. B. and Johnson, R. L., Friction and wear investigations of molybdenum disulfide I: effect of moisture, NACA Technical Note 3055, 1953.
8. Lancaster, J. K., A review of the influence of environmental humidity and water on friction, lubrication and wear, *Tribology International*, **23** (6), 1990, 371–389.
9. Wahl, K. J. and Singer, I. L., Quantification of a lubricant transfer process that enhances the sliding life of a MoS_2 coating, *Tribology Letters*, **1**, 1995, 59–66.
10. Cohen, S. R., Rapoport, L., Ponomarev, E. A., Cohen, H., Tsirlina, T., Tenne, R. and Lévy-Clément, C., The tribological properties of type II textured MX_2 (M = Mo, W; X = S, Se) films, *Thin Solid Films*, **324**, 1998, 190–197.
11. Tenne, R., Homyonfer, M. and Feldman, Y., Nanoparticles of layered compounds with hollow cage structures (inorganic fullerene-like structures), *Chemistry of Materials*, **10** (11), 1998, 3225–3238.
12. Tenne, R., Advances in the synthesis of inorganic nanotubes and fullerene-like nanoparticles, *Angewandt Chemistry International Edition*, **42**, 2003, 5124–5132.
13. Azulay, D., Kopnov, F., Tenne, R., Balberg, I. and Millo, O., Observation of current reversal in the scanning tunneling spectra of fullerene-like WS_2 nanoparticles, *Nanoletters*, **6** (4), 2006, 760–764.
14. Feldman, Y., Frey, G. L., Homyonfer, M., Lyakhovitskaya, V., Margulis, L., Cohen, H., Hodes, G., Hutchison, J. L. and Tenne, R., Bulk synthesis of inorganic fullerene-like MS_2 (M = Mo, W) from the respective trioxides and the reaction mechanism, *Journal of the American Chemical Society*, **118**, 1996, 5362–5367.
15. Feldman, Y., Lyakhovitskaya, V. and Tenne, R., Kinetics of nested inorganic fullerene-like nanoparticle formation, *Journal of the American Chemical Society*, **120**, 1998, 4176–4183.
16. Feldman, Y., Zak, A., Popovitz-Biro, R. and Tenne, R., New reactor for production of tungsten disulfide hollow onion-like (inorganic fullerene-like) nanoparticles, *Solid State Sciences*, **2**, 2000, 663–672.
17. Hu, J. J., Bultman, J. E. and Zabinski, J. S., Inorganic fullerene-like nanoparticules produced by arc discharge in water with potential lubricating ability, *Tribology Letters*, **17** (3), 2004, 543–546.
18. Brooks, D. J., Douthwaite, R. E., Brydson, R., Calvert, C., Measures, M. G. and Watson, A., Synthesis of inorganic fullerene (MS_2, M = Zr, Hf and W) phases using H_2S and N_2/H_2 microwave-induced plasmas, *Nanotechnology*, **17**, 2006, 1245–1250.
19. Cizaire, L., Lubrification par les nanoparticules (in French), PhD Thesis in Génie des Matériaux, Ecole Centrale de Lyon No. 2003-26, 2003.
20. Saito, Y., Yoshikawa, T., Bandow, S., Tomita, M. and Hayashi, T., Interlayer spacings in carbon nanotubes, *Physical Review B*, **48**, 1993, 1907–1909.
21. Luttrell, R. D., Brown, S., Cao, J., Musfeldt, J. L., Rosentsveig, R. and Tenne, R., Dynamics of bulk versus nanoscale WS_2: local strain and charging effects, *Physical Review B*, **73**, 2006, 035410.
22. McMullan, W. G. and Irwin, J. C., Raman scattering from 2H and 3R-NbS_2, *Solid State Communications*, **45** (7), 1983, 557–560.
23. Hirata, T. and Ohuchi, F. S., Temperature dependence of the Raman spectra of 1T-TaS_2, *Solid State Communications*, **117**, 2001, 361–364.
24. Wieting, T. J. and J. L. Verble, J. L., Infrared and Raman studies of long-wavelength optical phonons in hexagonal MoS_2, *Physical Review B*, **3** (12), 1971, 4286–4292.
25. Zhu, Y. Q., Hsu, W. K., Firth, S.,Terrones, M., Clark, R. J. H., Kroto, H. W. and Walton, D. R. M., (2001), Nb-doped WS_2 nanotubes, *Chemical Physics Letters*, **342**, 2001, 15–21.
26. Coehoorn, R., Haas, C. and de Groot, R. A., Electronic structure of $MoSe_2$, MoS_2 and WS_2. II. The nature of the optical band gaps, *Physical Review B*, **35** (12), 1987, 6203–6206.
27. Frey, G.L., Elani, S., Homyonfer, M., Feldman, Y., and Tenne, R., Optical-absorption spectra of inorganic fullerenelike MS_2 (M = Mo, W), *Physical Review B*, **57** (11), 1998, 6666–6671.
28. Frey, G. L., Tenne, R., Matthews, M. J., Dresselhaus, M. S. and Dresselhaus, G., Raman and resonance Raman investigation of MoS_2 nanoparticles, *Physical Review B*, **60** (4), 1999, 2883–2892.
29. Cizaire, L., Vacher, B., Le-Mogne, T., Martin, J. M., Rapoport, L., Margolin, A. and Tenne, R., Mechanisms of ultra-low friction by hollow inorganic fullerene-like MoS_2 nanoparticles, *Surface and Coatings Technology*, **160**, 2002, 282–287.
30. Srolovitz, D. J., Safran, S. A., Homyonfer, M. and Tenne, R., Morphology of nested fullerenes, *Physical Review Letters*, **74** (10), 1995, 1779–1782.
31. Srolovitz, D. J., Safran, S. A. and Tenne, R., Elastic equilibrium of curved thin films, *Physical Review E*, **49** (6), 1994, 5260–5270.

32. Isshiki, T., Nishio, K., Aoyagi, I., Yabuuchi, Y., Takahashi, N., Saijo, H. and Shiojiri, M., High resolution transmission electron microscopy of layer structure and stacking faults in tungsten disulfide lubricants, *Wear*, **170** (1), 1993, 55–61.
33. Panich, A. M., Kopnov, F. and Tenne, R., Nuclear magnetic resonance study of fullerene-like WS_2, *Journal of Nanoscience and Nanotechnology*, **6** (6), 2006, 1678–1683.
34. Sourisseau, C., Cruege, F., Fouassier, M. and Alba, M., Second-order Raman effects, inelastic neutron scattering and lattice dynamics in 2H-WS_2, *Chemical Physics*, **150**, 1991, 281–293.
35. Bhushan, B., Gupta, B. K., Cleef, G. W. V., Capp, C. and Coe, J. V., Fullerene (C_{60}) films for solid lubrication, *Tribology Transactions*, **36** (4), 1993, 573–580.
36. Schwarz, U. D., Zwörner, O., Köster, P. and Wiesendanger, R., Quantitative analysis of the frictional properties of solid materials at low loads. I – carbon compounds, *Physical Review B*, **56** (11), 1997, 6987–6995.
37. Liang, Q., Tsui, O. K. C., Xu, Y., Li, H. and Xiao, X,, Effect of C_{60} molecular rotation on nanotribology, *Physical Review Letters*, **90** (14), 2003, 1461021–1461024.
38. Miura, K. and Kamiya, S., C_{60} molecular bearings, *Physical Review Letters*, **90** (5), 2003, 055509.
39. Rapoport, L., Leshchinsky, V., Volovik, Y., Lvovsky, M., Nepomnyashchy, O., Feldman, Y., Popovitz-Biro, R. and Tenne, R., Modification of contact surfaces by fullerene-like solid lubricant nanoparticles, *Surface and Coatings Technology*, **163–164**, 2003, 405–412.
40. Rapoport, L., Feldman, Y., Homyonfer, M., Cohen, H., Sloan, J., Hutchison, J. L. and Tenne, R., Inorganic fullerene-like material as additives to lubricants: structure–function relationship, *Wear*, **225–229** (2), 1999, 975–982.
41. Rapoport, L., Bilik, Y., Feldman, Y., Homyonfer, M., Cohen, S. R. and Tenne, R., Hollow nanoparticles of WS_2 as potential solid-state lubricants, *Nature*, **387** (6635), 1997, 791–793.
42. Rapoport, L., Leschchinsky, V., Lapsker, I., Volovik, Y., Nepomnyashchy, O., Lvovsky, M., Popovitz-Biro, R. and Tenne, R., Tribological properties of WS_2 nanoparticles under mixed lubrication, *Wear*, **255** (7–12), 2003, 785–793.
43. Hu, J. J. and Zabinski, J. S., Nanotribology and lubrication mechanisms of inorganic fullerene-like MoS_2 nanoparticles investigated using lateral force microscopy (LFM), *Tribology Letters*, **18** (2), 2005, 173–180.
44. Greenberg, R., Halperin, G., Etsion, I. and Tenne, R., The effect of WS_2 nanoparticles on friction reduction in various lubrication regimes, *Tribology Letters*, **17** (2), 2004, 179–186.
45. Drummond, C., Alcantar, N., Israelachvili, J., Tenne, R. and Golan, Y., Microtribology and friction-induced material transfer in WS_2 nanoparticle additives, *Advanced Functional Materials*, **11** (5), 2001, 348–354.
46. Rapoport, L., Nepomnyashchy, O., Lapsker, I., Verdyan, A., Moshkovich, A., Feldman, Y. and Tenne, R., Behavior of fullerene-like WS_2 nanoparticles under severe contact conditions, *Wear*, **259**, 2005, 703–707.
47. Fleischer, N., Genut, M., Zak, A., Rapoport, L. and Tenne, R., Superior performance of inorganic fullerene-like nanospheres as EP/AW additives for oils and grease, in *Additives 2005*, Dublin, 5–7 April 2005.
48. Chhowalla, M. and Amaratunga, G. A. J., Thin films of fullerene-like MoS_2 nanoparticles with ultra-low friction and wear, *Nature*, **407**, 2000, 164–167.
49. Katz, A., Redlich, M., Rapoport, L., Wagner, H. D. and Tenne. R., Self-lubricating coatings containing fullerene-like WS_2 nanoparticles for orthodontic wires and other possible medical applications, *Tribology Letters*, **21** (2), 2006, 135–139.
50. Zou, T. Z., Tu, J. P., Zhang, S. C., Chen, L. M., Wang, Q., Zhang, L. L.and Ne, D. H., Friction and wear properties of electroless Ni-P-(IF-MoS_2) composite coatings in humid air and vacuum, *Materials Science and Engineering A*, **426**, 2006, 162–168.
51. Seguin, L., Figlarz, M., Cavagnat, R. and Lassègues, J. C., Infrared and Raman spectra of MoO_3 molybdenum trioxides and MoO_3 $x H_2O$ molybdenum trioxide hydrates, *Spectrochimica Acta Part A: Molecular and Biomolecular Spectroscopy*, **51**, 1995, 1323–1344.
52. Frey, G.L., Rothschild, A., Sloan, J., Rosenttsveig, P., Popovitz-Biro, R. and Tenne, R., Investigations of non-stoichiometric tungsten oxide nanoparticles, *Journal of Solid State Chemistry*, **162**, 2001, 300–314.
53. Joly-Pottuz, L., Dassenoy, F., Belin, M., Vacher, B., Martin, J. M. and Fleischer, N., Ultra-low friction and wear properties of IF-WS_2 under boundary lubrication, *Tribology Letters*, **18** (4), 2005, 477–485.
54. Cameron, A., *Basic Lubrication Theory*, Ellis Horwood Ltd., 1981.
55. Moshkovith, A., Perfiliev, V., Verdyan, A., Lapsker, I., Popovitz-Biro, R., Tenne, R. and Rapoport, L., Sedimentation of IF-WS_2 aggregates and a reproducibility of the tribological data, *Tribology International*, **40**, 2007, 117–124.
56. Perfiliev, V., Moshkovith, A., Verdyan, A., Tenne, R. and Rapoport, L., A new way to feed nanoparticles to friction interfaces, *Tribology Letters*, **21** (2), 2006, 89–93.

57. Bogner, A., Thollet, G., Basset, D., Jouneau, P. H. and Gauthier, C., Wet STEM: a new development in environmental SEM for imaging nano-objects included in a liquid phase, *Ultramicroscopy*, **104**, 2005, 290–301.
58. Grossiord, C., Mécanismes tribochimiques de réduction du frottement limite. Cas des additifs au molybdène (in French), PhD Thesis in Matière Condensée, Surfaces et Interfaces, Ecole Centrale de Lyon, No. 1999–11, 1999.
59. Moshkovich, A., Perfiliev, V., Yutujyan, K. and Rapoport, L., Friction and wear of solid lubricant films deposited by different types of burnishing, *Wear*, **263** (7–12), 2007, 1467–1469.
60. Iwaki, M., Le-Mogne, T., Fontaine, J., Martin, J. M., Watanabe, S. and Noshiro, J., Superlow friction of WS_2 coatings in UHV at low temperature, in *WTC III*, Washington, 2005.
61. Schuffenhauer, C., Popovitz-Biro, R. and Tenne, R., Synthesis of NbS_2 nanoparticles with (nested) fullerene-like structure (IF), *Journal of Materials Chemistry*, **12**, 2002, 1587–1591.
62. Schuffenhauer, C., Synthesis and properties of closed cage nanostructures (IF) from group 5 transition-metal chalcogenides, PhD Thesis, Weizmann Insitute of Science (Israel), 2005.
63. Jamison, W. E., Structure and bonding effects on the lubricating properties of crystalline solids, *ASLE Transactions*, **15**, 1972, 296–305.
64. Jamison, W. E. and Cosgrove, C. L., Friction characteristics of transition-metal disulfides and diselenides, *ASLE Transactions*, **14**, 1971, 62–72.
65. Waghray, H., Lee, T. S. and Tatarchuk, B. J., A study of the tribological and electrical properties of sputtered and burnished transition metal dichalcogenide films, *Surface and Coatings Technology*, **76–77** (2), 1995, 415–420.
66. Feldman, Y., and Tenne, R., High rate, gas phase growth of MoS_2 nested inorganic fullerenes and nanotubes, *Science*, **267**, 1995, 222–225.
67. Nath, M., Mukhopadhyay, K. and Rao, C. N. R., $Mo_{1-x}W_xS_2$ nanotubes and related structures, *Chemical Physics Letters*, **352**, 2002, 163–168.
68. Chen, J., Li, S. L. and Tao, Z. L., Novel hydrogen storage properties of MoS_2 nanotubes, *Journal of Alloys and Compounds*, **356–357**, 2003, 413–417.
69. Remskar, M., Mrzel, A., Skraba, Z., Jesih, A., Ceh, M., Demsar, J., Stadelmann, P., Levy, F. and Mihailovic, D., Self-assembly of subnanometer-diameter single-wall MoS_2 nanotubes, *Science*, **292**, 2001, 479–481.
70. Annal-Therese, H., Li, J., Kolb, U. and Tremel, W., Facile large scale synthesis of WS_2 nanotubes from WO_3 nanorods prepared by hydrothermal route, *Solid State Sciences*, **7** (1), 2005, 67–72.
71. Nath, M. and Rao, C. N. R., Nanotubes of the disulfides of groups 4 and 5 metals, *Pure and Applied Chemistry*, **74** (9), 2002, 1545–1522.
72. Wu, X., Tao, Y., Ke, X., Zhu, J. and Hong, J., Catalytic synthesis of long NbS_2 nanowire strands, *Materials Research Bulletin*, **39**, 2004, 901–908.
73. Kaplan-Ashiri, I., Cohen, S. R., Gartsman, K., Rosentsveig, R., Seifert, G. and Tenne, R., Mechanical properties of individual WS_2 nanotubes, *Journal of Materials Research*, **19** (2), 2004, 454–459.
74. Kaplan-Ashiri, I., Cohen, S. R., Gartsman, K., Ivanovskaya, V., Heine, T., Seifert, G., Wiesel, I., Wagner, H. D. and Tenne, R., On the mechanical behavior of WS_2 nanotubes under axial tension and compression, *PNAS*, **103** (3), 2006, 523–528.
75. Erdemir, A., Erylmaz, O. L. and Fenske, G. R., Synthesis and tribology of super-low friction coefficient carbon films, *Journal of Vacuum and Surface Technology*, **18**, 2000, 1987.
76. Martin, J. M., Pascal, H., Donnet, C., Le-Mogne, T., Loubet, J. L. and Epicier, T., Superlubricity of MoS_2: crystal orientation mechanisms, *Surface and Coatings Technology*, **68/69**, 1994, 427–432.
77. Dienwiebel, M., Verhoeven, G. S., Pradeep, N.,. Frenken, J. W. M., Heimberg, J. A. and Zandbergen, H. W., Superlubricity of graphite, *Physical Review Letters*, **92** (12), 2004, 126101.
78. Nemanic, V., Zumer, M., Zajec, B., Pahor, J., Remskar, M., Mrzel, A., Panjan, P. and Mihailovic, D., Field emission properties of molybdenum disulfide nanotubes, *Applied Physics Letters*, **82** (25), 2003, 4573–4575.
79. Verstraete, M. and Charlier, J. C., *Ab initio* study of MoS_2 nanotube bundles, *Physical Review B*, **68**, 2003, 045423.
80. Vrbanic, D., Remskar, M., Jesih, A., Mrzel, A., Umek, P., Ponikvar, M.. Jancar, B., Meden, A., Novosel, B., Pejovnik, S., Venturini, P., Coleman, J. C. and Mihailovic, D., Air-stable monodispersed $Mo_6S_3I_6$ nanowires, *Nanotechnology*, **15**, 2004, 635–638.
81. Holmgren, D. J., Demers, R. T., Klein, M. V. and Ginsberg, D. M., Raman study of phonons in Chevrel-phase crystals, *Physical Review B*, **36** (4), 1987, 1952–1955.
82. Vrbanic, D., Synthesis, characterization and various properties of low-dimensional materials of transitional metal chalcogenides and carbon, PhD Thesis, Faculty of Chemistry and Chemical Technology, Ljubljana, 2004.

83. Joly-Pottuz, L., Dassenoy, F., Martin, J. M., Vrbanic, D., Mrzel, A., Mihailovic, D., Vogel, W. and Montagnac, G., Tribological properties of Mo-S-I nanowires used as additive in oil, *Tribology Letters*, **18** (3), 2005, 385–393.
84. Belin, M., Triboscopy: a new quantitative tool for microtribology, *Wear*, **168**, 1993, 7–12.
85. Nicolosi, V., Vrbanic, D., Mrzel, A., McCauley, J., O'Flaherty, S., McGuiness, C., Compagnini, G., Mihailovic, D., Blau, W. J. and Coleman, J. N., Solubility of $Mo_6S_{4.5}I_{4.5}$ nanowires in common solvents: a sedimentation study, *Journal of Physical Chemistry B*, **109**, 2005, 7124–7133
86. Sliney, H. E., Dynamics of solid lubrication as observed by optical microscopy, *ASLE Transactions*, **21** (2), 1977, 109–117.
87. Scharf, T. W. and I. L. Singer, I. L., Role of third bodies in friction behavior of diamond-like nanocomposite coatings studied by *insSitu* tribometry, *Tribology Transactions*, **45**, 2002, 363–371.
88. Descartes, S. and Berthier, Y., Rheology and flow of solid third bodies: background and application to an $MoS_{1.6}$ coating, *Wear*, **252**, 2002, 546–556.
89. Martin, J. M., Belin, M., Le-Mogne, T., Donnet, C., Millet, J. M. and Millard-Pinard, N., (1995), Towards the *in-situ* analysis of tribological surfaces, in *ECASIA*, Montreux (Suisse), 1995.
90. Mansot, J. L., Etude des pressions dans un interface sphère/plan en présence d'une couche mince organique, Ecole Centrale de Lyon, 1986.
91. Jubault, I., Mansot, J. L., Vergne, P. and Mazuyer, D., *In-situ* pressure measurements using Raman microspectroscopy in a rolling elastohydrodynamic contact, *Journal ot Tribology*, **124**, 2002, 114–120.
92. Zhu, Y. Q., Sekine, T., Brigatti, K. S., Firth, S., Tenne, R., Rosentsveig, R., Kroto, H. W. and Walton, D. R. M., Shock-wave resistance of WS_2 nanotubes, *Journal of American Chemical Society*, **125**, 2003, 1329.
93. Zhu, Y. Q., Sekine, T., Li, Y. H., Wang, W. X., Fay, M., Edwards, H., Brown, P., Fleischer, N. and Tenne, R., WS_2 and MoS_2 inorganic fullerenes – super shock absorbers at very high pressures, *Advanced Materials*, **17**, 2005, 1500–1503.
94. Schwarz, U. S., Komura, S. and Safran, S. A., Deformation and tribology of multi-walled hollow nanoparticles, *Europhysics Letters*, **50** (6), 2000, 762–768.
95. Joly-Pottuz, L., Martin, J. M., Dassenoy, F., Belin, M., Montagnac, G., Reynard, B. and Fleischer, N., Pressure-induced exfoliation of inorganic fullerene-like WS_2 particles in a Hertzian contact, *Journal of Applied Physics*, **99** (2), 2006, 023524.
96. Sugai, S. and Ueda, T., High pressure Raman spectroscopy in the layered materials, *Physical Review B*, **26** (12), 1982, 6554–6558.
97. Johnson, K.L., *Contact Mechanics*, Cambridge University Press, Cambridge, 1985.
98. Grossiord, C., Varlot, K., Martin, J. M., Le-Mogne, T., Esnouf, C. and Inoue, K., MoS_2 single sheet lubrication by molybdenum dithiocarbamate, *Tribology International*, **31** (12), 1998, 737–743.
99. Schwarz, U. S., Safran, S. A. and Komura, S., Mechanical, adhesive and thermodynamic properties of hollow nanoparticles, *Materials Research Society Symposium Proceedings*, **651**, 2001, T5.3.1–T5.3.6.
100. Prasad, S. V. and Zaenski, J. S., Tribology of tungsten disulphide (WS_2): characterization of wear-induced transfer films, *Journal of Materials Science Letters*, **12**, 1993, 1413–1415.
101. Hirano, M. and Shinjo, K., Superlubricity and frictional anisotropy, *Wear*, **168** (1–2), 1993, 121–125.
102. Kataoka, K., Smart polymeric micelles as nanocarriers for gene and siRNA delivery, in 12th International Symposium on *Recent Advances in Drug Delivery Systems*, Salt Lake City, Utah, 2005.
103. Moshkovith, A., Perfiliev, V., Gindin, D., Parkansky, N., Boxman, R. and. Rapoport, L., Surface texturing using pulsed air arc treatment, *Wear*, **263** (7–12), 2007, 1467–1469..
104. Margolin, A., Popovitz-Biro, R., Albu-Yaron, A., Rapoport, L. and Tenne, R., Inorganic fullerene-like nanoparticles of TiS_2, *Chemical Physics Letters*, **411**, 2005, 162–166.

3

Carbon-Based Nanolubricants

Lucile Joly-Pottuz and Nobuo Ohmae

The most known natural carbon structures are graphite and diamond. Graphite has been studied for its tribological properties since 1950 [1]. In the macroscopic scale, the friction coefficient of 0.1 was obtained with graphite, without any other lubricant and even at high temperatures [2]. During the 1960s, Spreadborough studied the lubrication properties of bulk graphite samples and used transmission electron microscopy (TEM) to understand the lubrication mechanism [3]. Graphite has been used as a solid lubricant for a long time. Diamond is known as the hardest material in nature. On the macroscopic scale, the friction coefficient for an unlubricated diamond/diamond contact was found to be 0.05, with an excellent wear resistance [2] and a much lower friction coefficient (0.001) was found for the same contact lubricated by water [4]. The same results were obtained for a contact diamond/metal surface; thus diamond is currently used for cutting tools. Furthermore, diamond films can be deposited by chemical vapour deposition (CVD) and diamond films are used to coat all kinds of tools and machinery parts. Diamond-like carbon (DLC) coatings, introduced in 1973, are considered as amorphous or nanocrystalline diamond, and are now widely studied for their excellent tribological properties [5–7]. A survey by Miyoshi and Street attracted the interest of tribologists for carbon compound materials [8]. The discovery of fullerenes by Kroto, Smalley and Curl in 1985 [9] opened a new field of research for carbon compounds in tribology. Indeed, carbon can also form nanometre sized materials: carbon nanotubes, carbon onions and carbon nanohorns, which present the advantage of easily entering the confined contact area. These nanolubricants are still in the early stage of tribological investigations but some potential applications are already envisaged, for example the lubrication of micromechanical systems (MEMSs). This chapter focuses on promising and encouraging results obtained with carbon onions and carbon nanotubes used as lubricant additives in a base oil.

Nanolubricants Edited by J. M. Martin and N. Ohmae

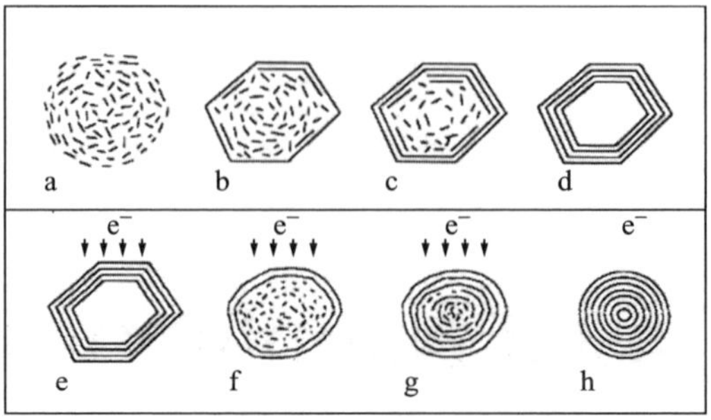

Figure 3.1 Schematic representation of the formation of a graphitic particle (a to d) and the transformation of such graphitic nanoparticles into a carbon onion under an electron beam. Reproduced by permission of Elsevier from Ugarte [11], Figure 1, © (1995) Elsevier

3.1 Graphite Onion Synthesis and Characterization

Carbon onions were synthesized for the first time by Ugarte in 1992 by degradation of carbon soot under the electron beam irradiation in a transmission electron microscope [10]. The growth mechanism of carbon onions was not obvious and several theories were considered [11]. Graphite sheets could be formed, starting on the surface of the nanoparticle before proceeding into the core (see Figure 3.1). Kuznetsov *et al.* also suggested the formation of an outer graphite shell first by the transformation of the (111) diamond planes into the (001) graphite planes, as shown in Figure 3.2 [12]. Thus the transformation occurs from the surface to the centre in the diamond nanoparticle. This growth mechanism resembles the one described for MS_2 fullerene nanoparticles (see Chapter 2).

Ozawa *et al.* proposed another growth mechanisms of carbon onions based on TEM observations of carbon black irradiated by an electron beam [13]. 'Spiroids' are first formed and then carbon onions (Figure 3.3). Qin and Iijima also studied the formation of spheroid graphitic particles by irradiating diamond crystals inside a TEM [14]. Since the mechanism of carbon onion formation was not known, several studies were carried out to follow the transformation of diamond nanoparticles into carbon onions directly inside a transmission electron

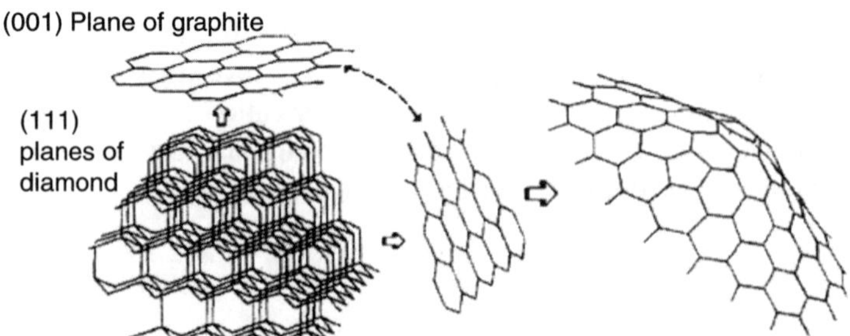

Figure 3.2 Scheme of the transformation of nanodiamond into a carbon onion. The (111) planes of diamond are transformed into (001) graphite planes. Reproduced by permission of Elsevier from Kuznetsov *et al.* [12], © (1994) Elsevier

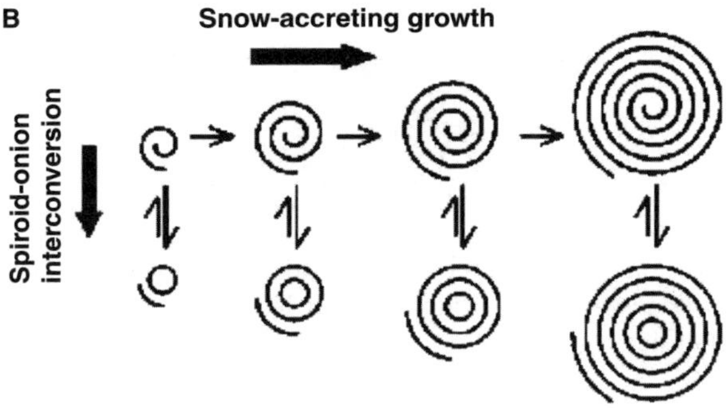

Figure 3.3 Growth mechanism of carbon onion from 'spiroid' proposed by Ozawa *et al.* Reproduced by permission of American Chemical Society from Ozawa *et al.* [13], Figure 3, © (2002) American Chemical Society

microscope. Roddatis *et al.* followed the transformation by irradiating the diamond nanoparticles and by performing high-resolution transmission electron microscopy (HRTEM) coupled with electron energy-loss spectroscopy (EELS) [15]. Recently, Mykhaylyk *et al.* studied the same transformation by coupling results of several analytical techniques: HRTEM and EELS, X-ray diffraction (XRD), small-angle X-ray scattering and ultraviolet Raman spectroscopy [16]. These two studies confirmed that the transformation of nanodiamond into carbon onions proceeds from the outside shell and progressively continues into the particle diamond core, thus confirming the theory earlier reported by Ugarte. This suggests that the size of the carbon onions is directly related to the size of the initial diamond nanoparticle. Hiraki *et al.* used an ultrahigh vacuum transmission electron microscope to avoid carbon contamination, which can disturb the observation at high resolution [17]. They highlighted the fact that large-diameter nanodiamond particles are transformed into graphite because the reaction rate is not sufficient and the first layers are not finished when others begin to form, eventually leading to a graphite structure.

Tomita *et al.* studied the transformation of diamond nanoparticles by X-ray diffraction, according to the temperature to which they were subjected [18]. They showed that after annealing at 1400 °C, graphitization first occurred on the surface of the diamond nanoparticles. After annealing at 1700 °C, nanoparticles are converted into carbon onions, but a residual core of the nanoparticles is made of diamond. After annealing at 2000 °C, the diamond core is reduced in size but carbon onions become faceted. They consist of well-ordered graphite stacks and have a more important size. The increase in volume of these onions is attributed to the difference in density between pure diamond (3.515 g/cm^3) and highly oriented pyrolytic graphite (HOPG) (2.267 g/cm^3) [19].

A 'zipper-like' transformation mechanism was tentatively proposed to explain the transformation of diamond into graphite [20]. It is based on the opening of three cubic (111) diamond planes to form graphite sheets (Figure 3.4). The inside layer should take part equally in the formation of both graphite sheets, so that the lengths for the two individual graphite sheets are the same. The transformations from either the (111) or the (100) planes were compared in order to understand the mechanism fully. Transformation from the (111) plane was the thermodynamically preferred one.

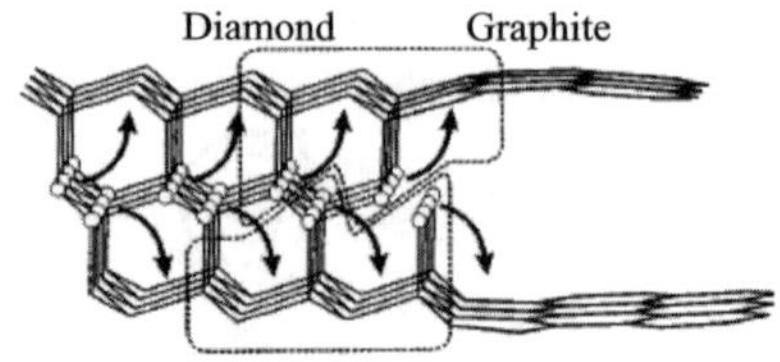

Figure 3.4 'Zipper-like' mechanism of the transformation of diamond into graphite. Reproduced by permission of the American Institute of Physics from Kuznetsov *et al.* [20], Figure 6, © (1999) American Institute of Physics

Several synthesis methods for carbon onions were reported in the literature. Xu and Tanaka described the synthesis of carbon onions by the conversion of an amorphous carbon film under Al nanoparticles and onions were induced by electron beam irradiation in a transmission electron microscope [21]. The first graphite layers are created at the interface between Al nanoparticles and amorphous carbon, a carbon onion structure containing the Al nanoparticle is then obtained and eventually Al nanoparticles move outside the carbon onion. Here Al acts as a catalyst, but it is important to notice that some Al atoms remain between the carbon sheets. This was clearly shown by chemical analysis (EDS), which involves an expansion of the interlayer distance (0.36 nm instead of 0.34 nm in HOPG or a well-shaped carbon onion). Same observations were made by using Pd clusters [22]. Based on the same synthesis principle, Troiani *et al.* developed a synthesis of carbon onions on gold nanoparticles with a production 10 times faster than in the synthesis reported by Xu [23]. Sano *et al.* described another synthesis technique of carbon onions by arc discharge in water [24]. The onions float on the surface of the water whereas a deposit is formed at the bottom of the goblet used for the synthesis. This segregation leads to a pure sample being obtained composed of carbon onions 25 to 30 nm in diameter. Hu *et al.* also studied this synthesis method and adapted it to the synthesis of IF-MoS_2 [25]. Another method based on thermal reduction of glycerin in the presence of magnesium allows the synthesis of carbon onions with a larger diameter (60 to 90 nm) but they contain defects [26].

Recently, a new synthesis method based on the decomposition of CH_4 on NiO/Al composite powder led to a massive production of carbon onions of about 1g per hour [27, 28]. These carbon onions have a diameter from 5 to 50 nm but some carbon onions contain an Ni core. Wang *et al.* developed a synthesis method of carbon onions containing Fe core by CVD [29]. C_2H_2 is decomposed on ferrocene and leads to the formation of carbon onions 15 to 40 nm in diameter. The presence of a metallic core (Fe or Ni) can affect the tribological properties of carbon onions, as previously demonstrated in the case of carbon nanotubes. Carbon onions containing an Ni core would present interesting tribological properties similar to those of carbon nanotubes (see Section 3.5.1.1).

So far, very few studies on the tribological properties of carbon onions are reported in the literature. Most of the studies were carried out on carbon onions concern their synthesis and the determination of their structures. Cabioc'h *et al.* studied the tribological properties of silver films containing carbon onions [30]. The addition of onions does not have an effect on the friction coefficient, but an increase in lifetime of the coating by a factor of 15 was observed. Cabioc'h *et al.* explained friction-reducing properties of carbon onions by the fact that they are embedded inside the metal, so they cannot roll. Hirata *et al.* studied tribological properties of carbon onions used as solid lubricant between a steel ball and a silicon wafer [31]. Wear observed with carbon onions is lower than that observed with graphite. Low friction coefficients were measured. Hirata also studied the influence of their size compared to the roughness of surfaces. The carbon onions are effective when their diameter is larger than the surface roughness. In the presence of moisture, carbon onions usually agglomerate, and the effect of their size is then less important. Carbon onions form a layer between two surfaces involving low friction and low wear. Addition of carbon onions in Krytox 143B, a perfluorinated polyether (PFPE) oil, leads to a strong reduction of the friction coefficient (0.05 instead of 0.13) with a long lifetime of the oil, for tests run in air [32]. In vacuum, no significant differences were observed between pure oil and oil containing carbon onions. Street *et al.* used carbon onions because they are relatively stable against oxidation [32]. Graphite is known to be oxidized rapidly due to the edges of crystals. The temperature of oxidation is influenced by the surface specific area. The thermal stability is lost during the reduction in size of the particles, which occurs during grinding for application or during lubrication of the contact parts. However, carbon onions are more stable against oxidation because of their nested structure with less amount of dangling bond. Figure 3.5 presents a comparison of thermogravimetric analysis (TGA) of carbon onions and carbon black. The results indicate that carbon onions are more resistant to oxidation than carbon black. The temperature of oxidation is 792 °C for carbon onions while it is 588 °C for carbon black. These results strongly suggest the use of carbon onions as lubricant additives.

Carbon onions studied here were synthesized by Ohmae at Kobe University, by annealing of diamond nanoparticles using the experimental device shown in Figure 3.6. Two kinds of sample were synthesized. The first one was obtained by annealing the diamond nanoparticles inside a graphite crucible. After annealing at 1600 °C for 13 minutes, carbon onions containing a residual diamond core were preferentially formed (Figure 3.7(a)). They had an average diameter of 5 to 10 nm. The presence of a residual diamond core was attributed to an incomplete graphitization. Thus, a second set of carbon onions was synthesized by annealing

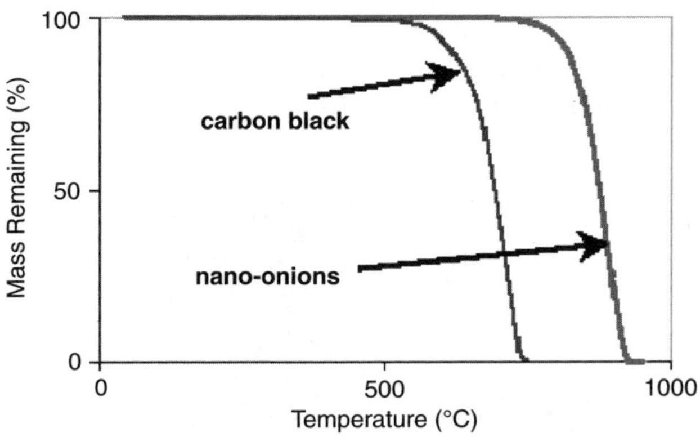

Figure 3.5 TGA analyses of carbon onion and carbon black. The temperature at which 5 % of the materials are oxidized indicates the temperature where rapid oxidation occurs. This temperature is 588 °C for carbon black and 792 °C for carbon onions. With kind permission from Springer Science and Business Media from Street *et al.* [32], Figure 3, © (2004) Springer Science and Business Media

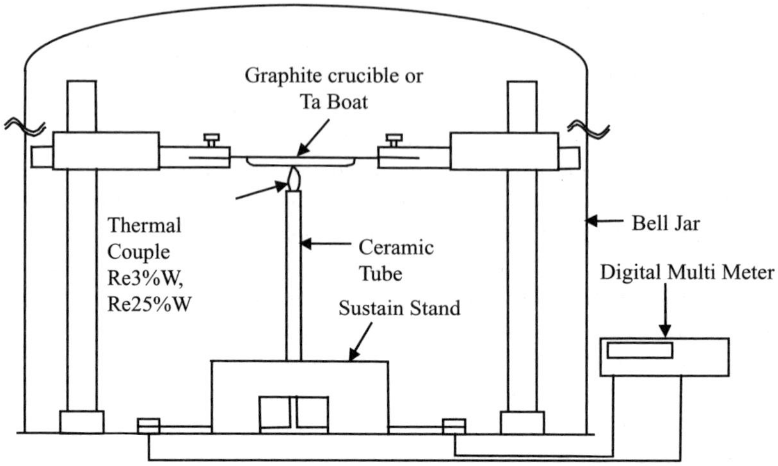

Figure 3.6 Schematic of the synthesis apparatus used for the carbon onion synthesis. Two different crucibles were used to obtain carbon onions: a graphite crucible (carbon onions with diamond core) and a tantalum boat

nanodiamond particles inside a tantalum boat. Annealing was then performed at 1700 °C to obtain a carbon onion without a diamond core, as shown in the HRTEM image (Figure 3.7(b)). The interlayer distance can be estimated as 0.34 nm for both samples, which corresponds to the distance between two graphene planes in pure graphite.

Electron energy-loss spectroscopy (EELS) characterization was also performed on these two samples to complete the HRTEM observations. EELS in transmission electron microscopy is a useful technique to distinguish the different forms of carbon. EELS consists in measuring the energy loss of accelerated electrons when they pass through the thin sample. An EELS spectrum is typically composed of three parts: the zero-loss peak at 0 eV, the low-loss region (<100 eV) and the high-loss region (>100 eV).

The low-loss peak corresponds to the excitation of valence electrons ($\pi + \sigma$). Recent studies showed that the plasmon peak is of particular interest in the study of mechanical and physical properties of materials [33]. A correlation between the plasmon maximum energy

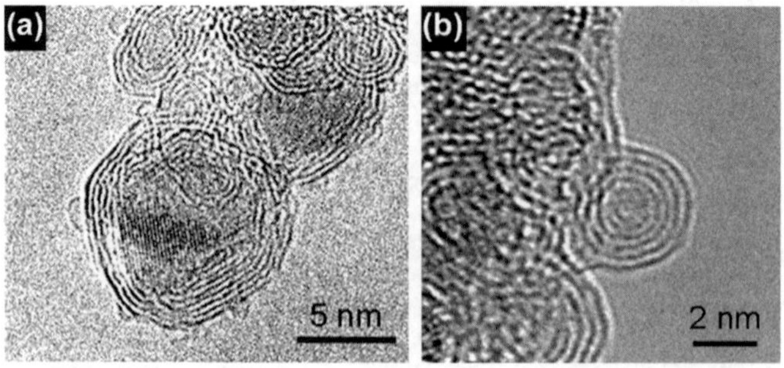

Figure 3.7 TEM image of carbon onions (a) with a diamond core and (b) without a diamond core

and mechanical properties such as the Young modulus and even the hardness was established. Monthioux *et al.* also studied the correlation between plasmon energy and mechanical properties for carbon fibres [34].

Another way of possible interaction of electrons with the sample is atomic ionization, leading to the high-loss regions EELS spectrum. Characteristic ionization edges of the elements in the sample are observed. They correspond to electrons of inner shells (i.e. K, L, M, etc.) which are excited into unoccupied states. This process thus allows the chemical semi-quantitative analysis of the sample. At any ionization edge onset, the energy loss near the edge structure (ELNES) gives valuable information on chemical bonding. Each transition can be decomposed into two components: the matrix element between the initial state and the final state, and the density of states (DOS). The matrix element gives the basic shape of the ionization edge, for example a 'saw-tooth' shape for the K edge. The density of states informs the atomic environment (number of neighbouring atoms, bond angles, local electronic states).

EELS has been widely used to study carbon nanoparticles: single-walled [35] or multi-walled carbon nanotubes [36], and C_{60} and C_{70} fullerenes [37]. For carbon onions, it was used to study the transformation of diamond in carbon onions [16, 38]. Redlich *et al.* used EELS to study the transformation of carbon onions into diamond under irradiation-induced compression [39]. In all these studies, both the plasmon peak and carbon ionization edge were used because they are useful to distinguish different carbon forms. In the case of the plasmon peak, its energy differs for each carbon form (Figure 3.8). It is located at 27 eV for graphite and 33 eV for diamond [40]. For graphite, a peak at 6 eV corresponding to the excitation of π electrons can also be observed. The coupling ionization edge and fine structure is also very useful. In the case of diamond and graphite, since graphite is sp^2 hybridized and diamond is sp^3 hybridized, an easy feature to distinguish them is the π^* feature (at 285 eV), which is present only for graphite (see Figure 3.9). The σ^* features are more complicated to

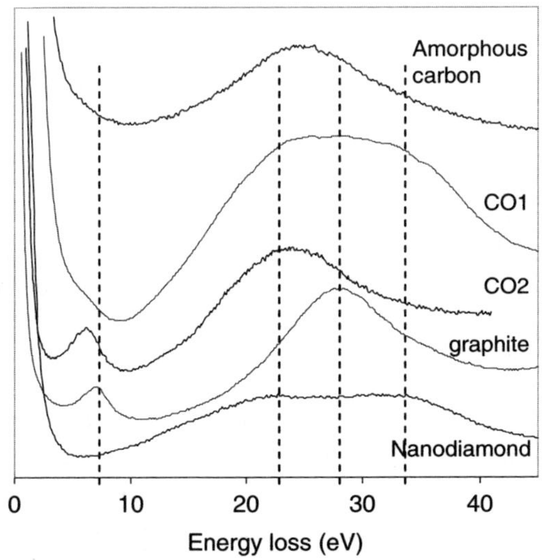

Figure 3.8 Plasmon peaks of nanodiamond, graphite and a carbon onion with a diamond core (CO1) or without a diamond core (CO2), and amorphous carbon

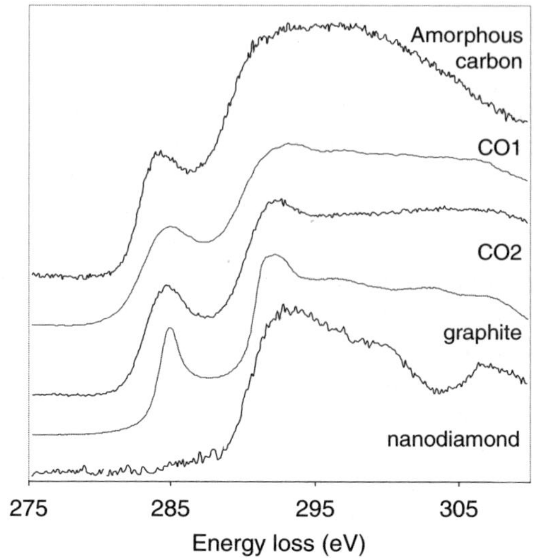

Figure 3.9 C-K edge EELS of diamond, graphite and carbon onions with a diamond core (CO1), carbon onions without a diamond core (CO2), and amorphous carbon

interpret and a more precise study of the DOS of the two structures is necessary. These DOS have already been studied for both structures [41–43], but usually it is not necessary to understand all the features of spectra. Spectra can be used as fingerprints. For example, the π^* feature is very useful in the study of carbon onions because it is characteristic of sp^2 bondings. It also allows diamond-like carbon (DLC) coatings to be characterized because the sp^2/sp^3 ratio can be calculated in this case [44, 45].

In the present case, the study will essentially be focused on the π^* feature and the energy of the plasmon peak to probe the presence of the diamond phase in carbon onion samples. EELS experiments were performed using an LEO 912 microscope with an omega filter in the column. Nanodiamond particles used as precursors for the carbon onion synthesis were also analysed for comparison. The plasmon peak can be decomposed into two peaks: the first one for amorphous carbon (22 eV) and the second one for diamond (33 eV). This is due to the presence of adventitious carbon around nanodiamond particles, which has already been observed by Tomita *et al.* [38]. The plasmon peak of a carbon onion with a diamond core also presents several contributions: one for diamond nanoparticles and one for graphite (Figure 3.8). This confirms the presence of a diamond core inside the carbon onion, as observed by HRTEM. A small peak can be seen at 6 eV on the spectrum of the carbon onion, which is characteristic of graphitic carbon. The low-loss spectrum of carbon onions without the diamond core (CO2) differs from that of the CO1 sample. The π plasmon peak is more intense. It is centred at 6.1 eV compared to 7 eV for graphite. This difference has already been observed by Cabioc'h *et al.* [46]. This might be explained by the theoretical work by Yannouleas *et al.* [47]. Due to the curvature of shells, the coupling of electrons on the spherical shells is different from the coupling in the planar case. No contribution of diamond is observed on the $(\sigma + \pi)$ plasmons peak of CO2, showing that these carbon onions are perfectly graphitized. Analyses of the carbon K-edge EELS spectra are summarized in Figure 3.9. The K-edge

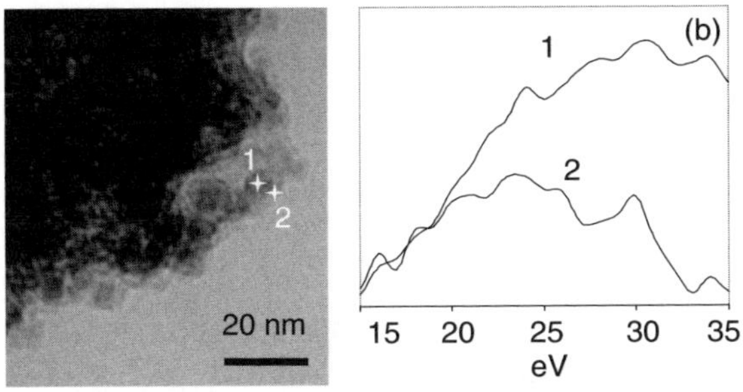

Figure 3.10 TEM picture of carbon onions and reconstructed spectra of two selected areas: centre of a diamond core with carbon onions and its periphery (ESI method)

observed for nanodiamond is very similar to that of diamond. A small peak at 285 eV can be observed. This peak is not present in the case of pure diamond but is due to the presence of amorphous carbon around nanodiamond particles [38]. The π^* peak is observed in the spectrum of the carbon onion. The broadening of this peak compared to graphite can be explained by the absence of a long-range graphite-like order in the carbon structure [16].

The electron spectroscopic imaging (ESI) method was used to study the core–shell structure of a carbon onion with a diamond core. This image spectrum method is based on the acquisition of a set of energy-loss images; then spectra can be reconstructed from any selected area. This method was described by Reimer in 1995 [48]. Figure 3.10 presents a TEM picture with two spectra reconstructed at two positions (images between 15 and 35 eV). Position 1 corresponds to the centre of a carbon onion and position 2 to the periphery. Spectra obtained have a different maximum energy in the low-loss region. This confirms that the core–shell structure of the first carbon onion sample is composed of a diamond core (energy of the low-loss peak at 30 eV) surrounded by graphitic shells (maximum energy of the plasmon peak is at 25 eV).

A comparison of images obtained at 25 and 30 eV clearly shows the two features of these carbon onions: graphitic shells are observed in Figure 3.11(a), while cores are observed in Figure 3.11(b). On the image at 25 eV, diamond cores can be easily observed (black round)

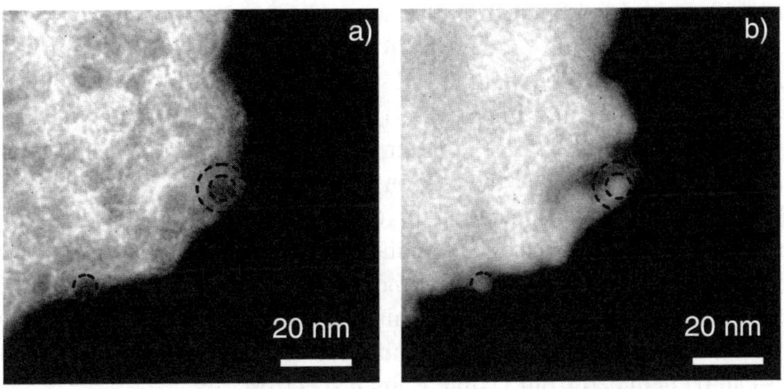

Figure 3.11 Energy-loss images at 25 and 30 eV. The diamond core is visible on the two images (black round in (a) and white round in (b))

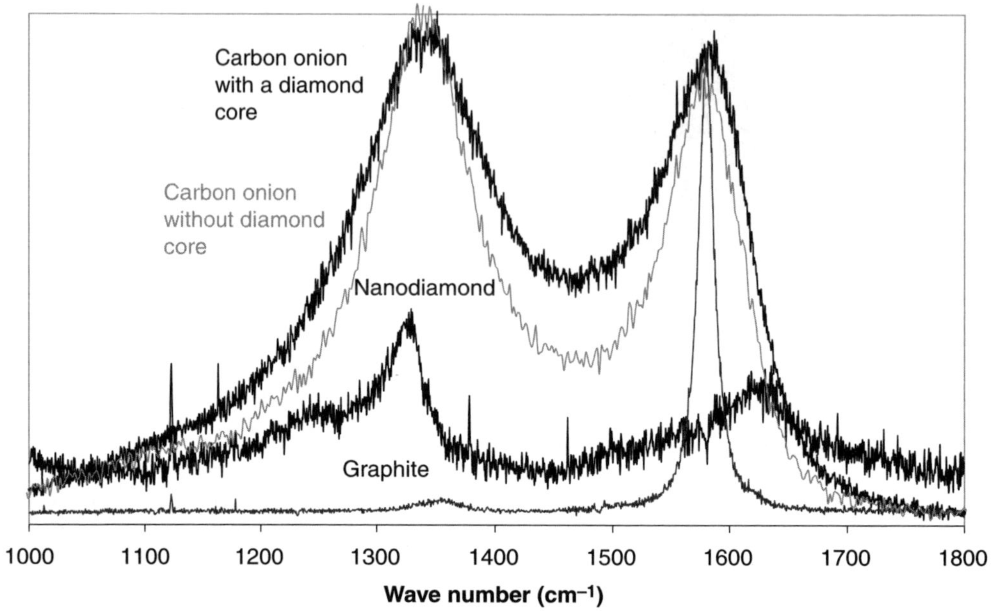

Figure 3.12 Raman spectra of carbon onions compared with those of nanodiamonds and graphite ($\lambda = 514.5$ nm). The nanodiamond spectrum was plotted after suppression of the luminescence background approximated by a cubic polynomial

and appear in white on the image at 30 eV. Since the maximum plasmon peak of nanodiamond particles is at 33 eV, nanodiamond particles will not appear on a 25 eV-loss image but will be visible on a 30 eV-loss image.

Raman spectra of carbon onions and diamond nanoparticles are compared in Figure 3.12. Peaks of carbon onions are very broad. Roy *et al.* realized a Raman study of carbon onions at different laser wavelengths [49]. They observed a shift of the peak due to the G band on the spectrum of carbon onions (1575 cm^{-1}) compared to the peak of graphite (1582 cm^{-1}) and attributed this shift to strains imposed on the layers due to their curvature. In Figure 3.12, a slight shift of the peak of carbon onions (1574 cm^{-1}) is observed compared to the graphite peak (1579 cm^{-1}). Tomita showed that the full width at half maximum of the peaks decreases with the temperature of annealing (Figure 3.13), which can be explained by a reduction in the disorder [50]. Indeed, at a higher annealing temperature, graphite planes are well ordered in the stack structure. The spectrum of the carbon onion sample with a diamond core is similar to the one of carbon onions after annealing at 1600 °C, but the peak at 1350 cm^{-1} is broader. This can be explained by the presence of nanodiamond in the core of the carbon onions (Figure 3.7(a)). The spectrum of the carbon onion without a diamond core presents two peaks with a full width narrower than the other sample. This can be explained by the fact that they were annealed at 1700 °C, as demonstrated by Tomita. However, no clear difference between the two structures can be deduced from visible Raman spectroscopy.

Multiwavelength Raman spectroscopy is currently used to study carbon compounds. There is a large diversity of carbon compounds depending on the ratio of sp^2 to sp^3 that can be

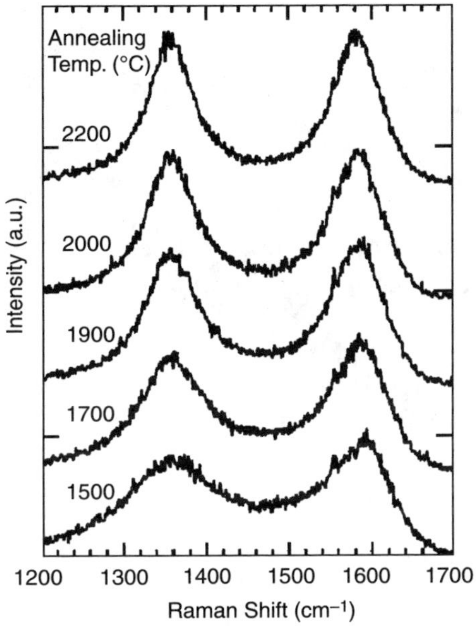

Figure 3.13 Raman spectra of carbon onions synthesized at different temperatures of annealing ($\lambda =$ 457.9 nm) [50]

studied using this technique. Raman scattering with carbon compounds is always a resonant process. This means that the configurations whose energy corresponds to the gap energy are preferentially excited. Any mixture of carbon containing sp^3, sp^2 or sp^1 bonding whose band gap is between 0 and 5.5 eV has energy corresponding to IR(infrared)–visible–UV(ultraviolet) Raman systems. However, visible Raman spectroscopy is known to be more sensitive to π bonding, which exists in carbon structures formed by atomic bonds from sp^2 hybridized orbitals, whereas UV Raman spectroscopy is more sensitive to σ bonding present in all carbon structures. Indeed, the energy of photons with UV excitation is equal to 5.1 eV, allowing the excitation of both σ and π states. There it probes both sp^2 and sp^3 bonding, allowing a direct probe of sp^3 bonding [51]. Multiwavelength Raman spectroscopy has recently been applied to study the disorder in carbon structures [52].

Here, UV Raman spectroscopy is chosen in order to give a better Raman signal of diamond and to confirm its presence in the centre of the nanoparticles. As first shown by Sun *et al.*, the spectrum of nanodiamond excited by a visible radiation contains a luminescent background which should be removed [53]. Mykhaylyk *et al.* recently compared spectra obtained at 633, 488 and 244 nm (Figure 3.14) [16]. This comparison clearly demonstrated the advantage of using UV excitation, which also allows the intense diamond optic mode to generate in the spectrum.

Spectra obtained at 244 nm are presented in Figure 3.15. This comparison clearly shows the presence of nanodiamond only in the carbon onions containing a diamond core. These results thus confirm observations made by TEM and EELS at the nano scale. It also proves the usefulness of UV Raman spectroscopy to characterize carbon compounds at the macroscopic scale.

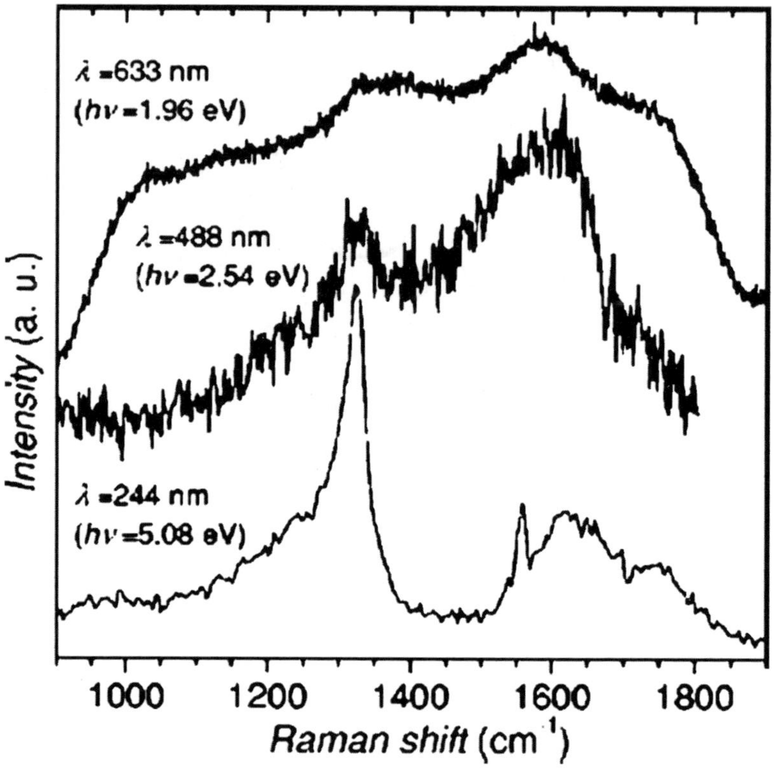

Figure 3.14 Multiwavelength spectra of nanodiamond. The Raman spectrum at 488 nm was plotted after subtraction of a luminescent background approximated by a cubic polynomial. This comparison clearly demonstrates the advantage of using UV radiation to analyse nanodiamond. Reproduced by permission of the American Institute of Physics from Mykhaylyk *et al.* [16], Figure 11, © (2005) American Institute of Physics

From these results, the average size of the diamond core can also be determined. In fact, as observed first by Nemanich *et al.* in the case of h-BN crystallites [54], a shift and a broadening of the peak are observed as the size of the crystallites decreases. This was explained by the phonon-confinement effect [55]. It is independent of the wavelength of the exciting radiation used and was already observed with visible radiation [56]. The Lorentzian line width Γ of the diamond peak is dependent on the particle size and can be expressed as

$$\Gamma = \alpha + \beta/L \tag{1}$$

where α and β are two coefficients and L is the diamond particle size. In the case of a visible radiation (514.5 nm), α and β were calculated as 2.990 and 0.052 respectively (with L expressed in μm) [56]. Recently, Sun *et al.* calculated these coefficients in the case of UV radiation of 244 nm and determined them to be 9.45 and 275.93 (with L expressed in nm) [53]. Here these coefficients were used to calculate the sizes of pristine nanodiamond and

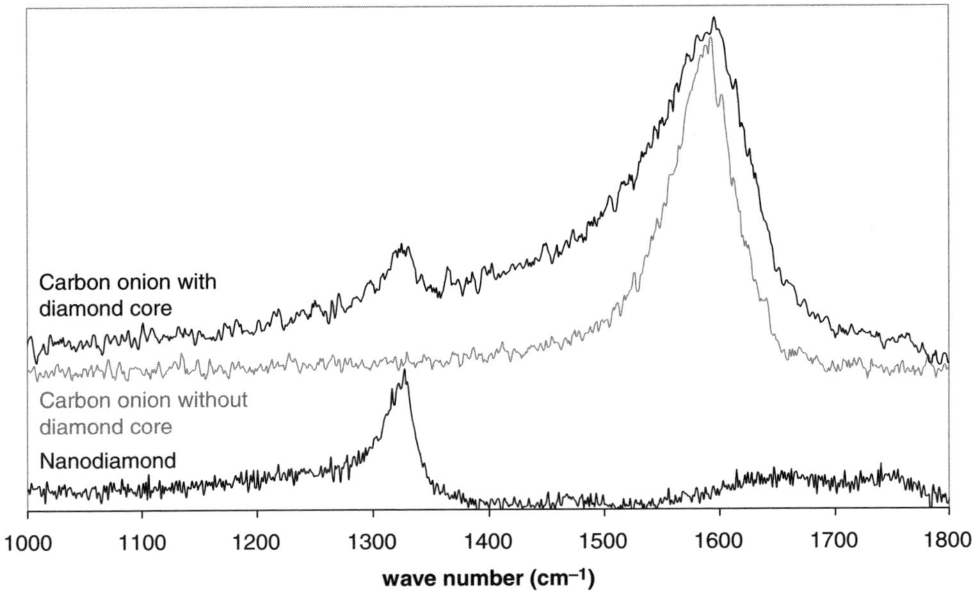

Figure 3.15 Raman spectrum of carbon onions compared with those of nanodiamonds ($\lambda = 244$ nm)

nanodiamond inside the carbon onions and sizes of 12 and 9 nm were found, respectively. These values seem quite large if it is considered that the nanodiamond particles used had a diameter of 5 to 10 nm. Mykhaylyk *et al.* compared the results obtained using several couples of α and β coefficients and observed differences [16]. Furthermore, the presence of amorphous carbon around the diamond nanoparticles or graphite sheets around the carbon onions can have an influence on the results, since the phonons can propagate into the amorphous carbon or graphite sheet layers as well, and thus the strong confinement phonon model used here could not be applicable. XRD analyses on the two samples would be interesting to enable a more precise determination of the size of the diamond phase. Nevertheless, from these results the slight diminution in size of the diamond nanoparticles can be estimated after annealing and the formation of graphite sheets to form carbon onions.

Analyses performed by TEM, EELS and Raman spectroscopy confirm that one of the carbon onion samples is composed of a diamond core surrounded by graphite sheets while the other is composed only of fullerene-like sheets.

3.2 Tribological Properties of Different Carbon Onions

Contrary to the other nanoparticles studied before, no effect of the concentration of onions in PAO on the friction coefficient is observed. At a low concentration (0.1 wt %), the same friction coefficient is observed as the one obtained with a concentration of 1 wt % (Figure 3.16). This is very interesting because it means that even a low concentration is sufficient to obtain good tribological properties for this type of additive.

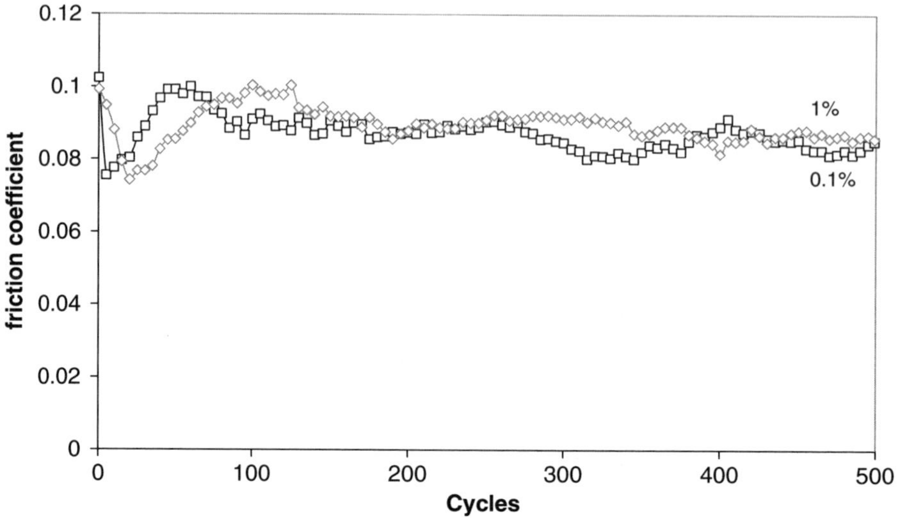

Figure 3.16 Effect of the onion concentration in PAO on the dynamic friction coefficient (P =1.12 GPa, v = 2.5 mm/s, T = 20 °C)

A study of the effect of the contact pressure on the friction-reducing properties was performed with a concentration of 0.1 wt % (Figure 3.17). In a way similar to the other nanoparticles studied, the friction coefficient decreases with the contact pressure. A critical pressure of 1.42 GPa is highlighted, a pressure at which the friction coefficient becomes lower (0.06). A comparison of the tribological performances of carbon onions containing a diamond core and without a diamond core indicates a similar behaviour for the two samples (Figure 3.17). Quite similar friction coefficients are obtained with the two samples for each contact pressure studied.

To understand tribological performances of carbon onions better, these results were compared with those obtained with pure PAO, PAO containing 0.1 wt % graphite, PAO containing 0.1 wt % of nanodiamond used as precursors for carbon onions synthesis and PAO containing 0.1 wt % of C_{60}, which can be considered as a single-walled carbon onion.

PAO is found to be efficient at high contact pressures (above 1.12 GPa), with the lowest friction coefficient lying between 0.08 and 0.1 (Figure 3.18). However, such experiments are not always reproducible. In some experiments PAO is active and friction becomes low, while in some other cases friction is high or a long induction period is necessary to obtain the low friction regime (as in the case of 1.12 GPa). Friction coefficients obtained with graphite are very similar to those obtained with carbon onions (Figure 3.19). Nevertheless, it is important to mention that dispersion of graphite in oil is not stable. It can be explained by the micrometric size of the graphite particles, which leads to a rapid sedimentation of the particles. Friction coefficients obtained with nanodiamond are similar to those obtained with PAO alone after 100 cycles at the two contact pressures tested (Figure 3.20). There is no effect of the addition of nanodiamond on the friction coefficient. Again, it was found that friction with nanodiamond decreases when the contact pressure increases. At 0.83 GPa, addition of C_{60}

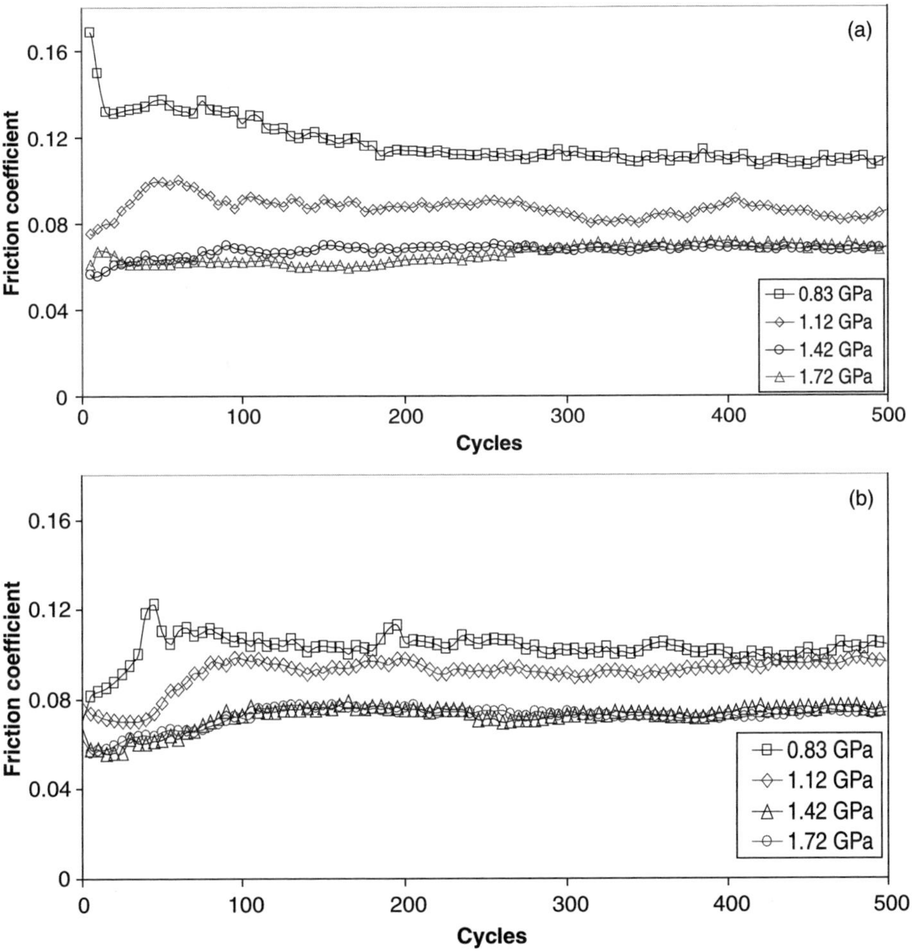

Figure 3.17 Effect of the contact pressure on the friction coefficient for a dispersion PAO + 0.1 wt % carbon onions (a) containing a diamond core and (b) without a diamond core

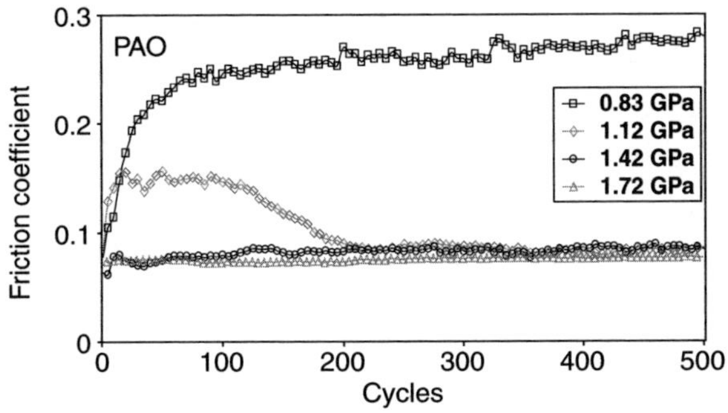

Figure 3.18 Friction coefficient obtained at four contact pressures (0.83 to 1.72 GPa) with pure PAO as a lubricant

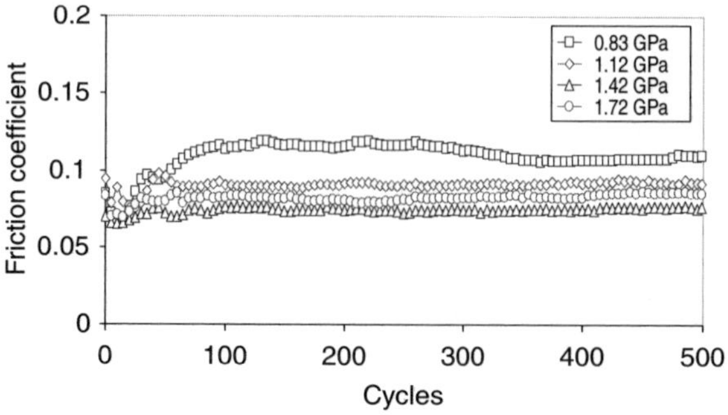

Figure 3.19 Friction coefficient obtained at four contact pressures (0.83 to 1.72 GPa) with PAO containing 0.1 wt % graphite

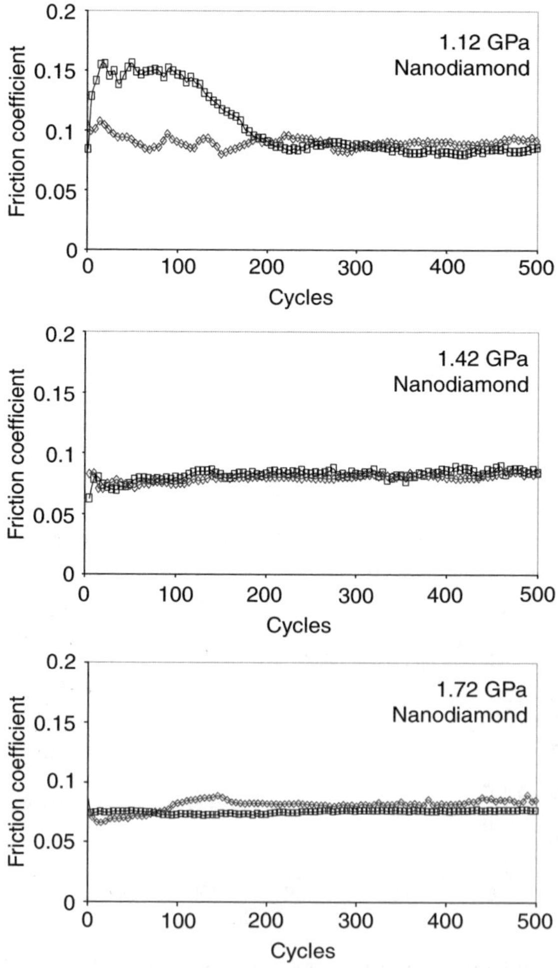

Figure 3.20 Friction coefficients obtained with PAO or PAO with 0.1 wt % of nanodiamond at 1.12 , 1.42 and 1.72 GPa (friction coefficients with pure PAO are represented by a black square)

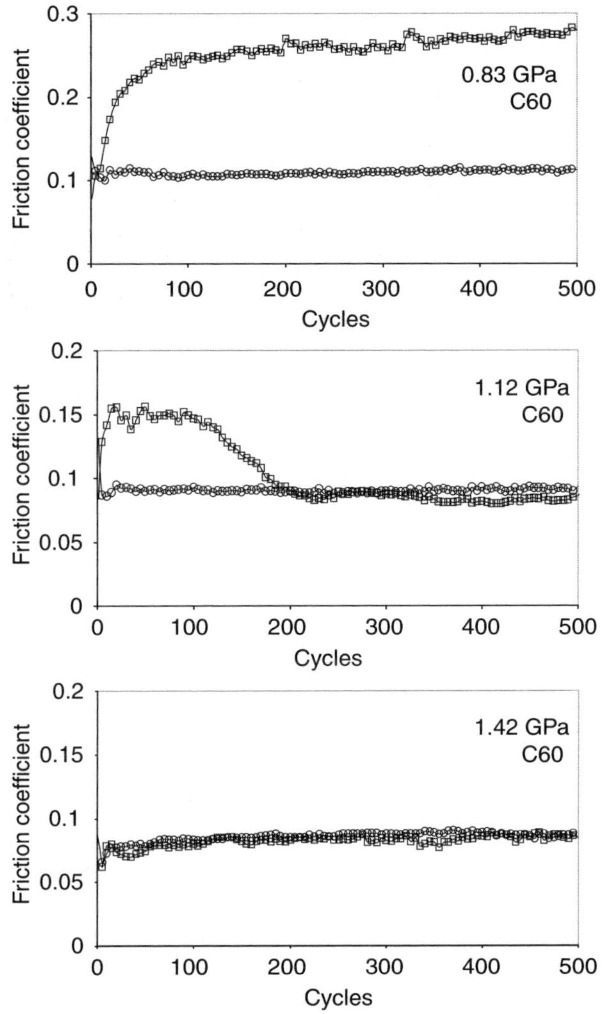

Figure 3.21 Friction coefficients obtained with PAO or PAO with 0.1 wt % of C_{60} at 0.83, 1.12 and 1.42 GPa (friction coefficients with pure PAO are represented by a black square)

leads to a strong reduction in the friction of PAO (Figure 3.21), as observed with carbon onions.

The general trend is that all carbon-based additives tested here reinforce the friction- reducing performance of the PAO base. A friction coefficient of 0.1 is observed for both structures.

Wear observed on pins after friction tests with a dispersion of 0.1 wt % of carbon onions in PAO is lower than that with pure PAO (Figure 3.22). Wear is slightly lower with a concentration of 0.1 wt %. For a contact pressure of 0.83 GPa, the wear scar diameter is 130 μm with a concentration of 1 wt % and 120 μm with a concentration of 0.1 wt %. This confirms that a low concentration of onions is sufficient to obtain interesting antiwear properties. Wear is

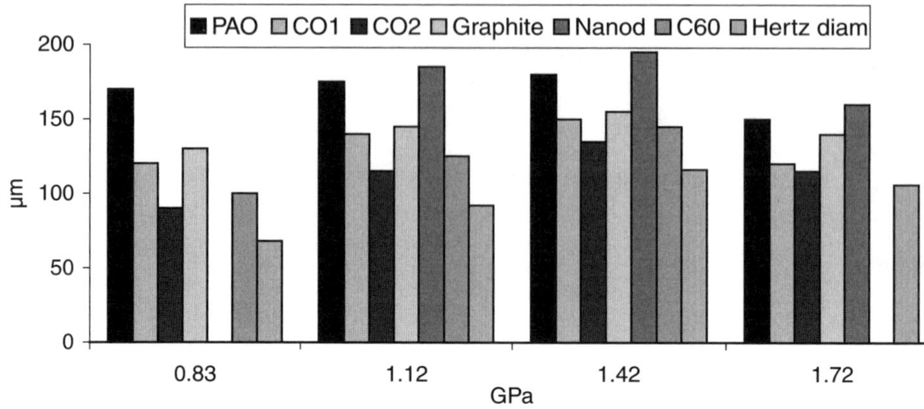

Figure 3.22 Summary of data of the wear scar diameters (μm) observed after the friction test with a concentration of 0.1 wt % of the two carbon onion samples at different contact pressures. These diameters are compared with those observed with pure PAO, graphite, C_{60} and nanodiamonds at a concentration of 0.1 wt %

higher than that observed with Ni/Y-SWNTs. A comparison of the two carbon onion samples indicates a lower wear for the sample without the diamond core. This can be explained by the presence of nanodiamonds in the sample, these nanodiamonds being able to scratch the steel surfaces. Wear observed after friction experiments performed with a dispersion of 0.1 wt % of nanodiamond is more important than that observed with pure PAO. Interestingly, wear observed with graphite is higher than carbon onions without a diamond core and slightly higher than that of carbon onions with a diamond core. Addition of graphite at 0.1 wt % leads to a reduction in wear compared to pure PAO. Nevertheless, wear increases with the concentration of graphite particles in oil. For example, at 0.83 GPa the wear scar diameter is equal to 130 μm with a concentration of 0.1 wt % and increases to 155 μm with a concentration of 1 wt %. Wear observed with C_{60} is slightly higher than that observed with CO2. This suggests very good properties of C_{60} when they are added to PAO. When C_{60} was added to a mineral oil, Gupta and Bhushan showed that there were only slight improvements [57]. As already mentioned in Section 2.3, studies on the tribological properties of C_{60} led to contradictory conclusions. No study has yet been reported on the tribological properties of C_{60}. These results show that the dispersion of C_{60} in PAO, which leads to a purple colour of PAO, presents interesting properties.

A characterization of wear particles collected at the end of the friction test with a carbon onion containing a diamond core confirms that wear occurs during friction. EELS chemical mapping reveals that wear particles are composed of carbon and iron (Figure 3.23).

To understand the behaviour of carbon onions inside the contact area better, diffraction patterns (selected area electron diffraction, or SAED) were compared of pristine material (carbon onions with a diamond core, see Figure 3.24) and of wear particles (Figure 3.25). First a comparison of the diffraction obtained with pure nanodiamond (Figure 3.24(a)) and carbon onions (Figure 3.24(b)) proves that these carbon onions contain a residual diamond core. On the diffraction pattern of nanodiamond, three rings are observed, which correspond to the distances of 2.02 Å (4.94 nm^{-1}on the diffraction pattern), 1.24 Å (8.02 nm^{-1}) and

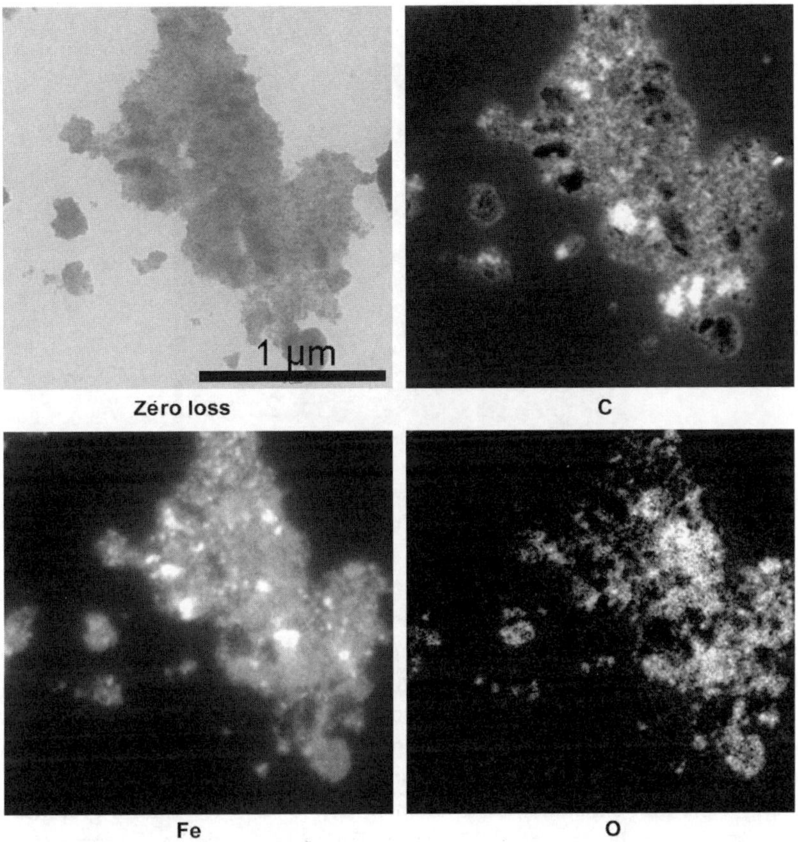

Figure 3.23 TEM image on wear particles collected after friction with carbon onions and the corresponding chemical mapping of C, Fe and O

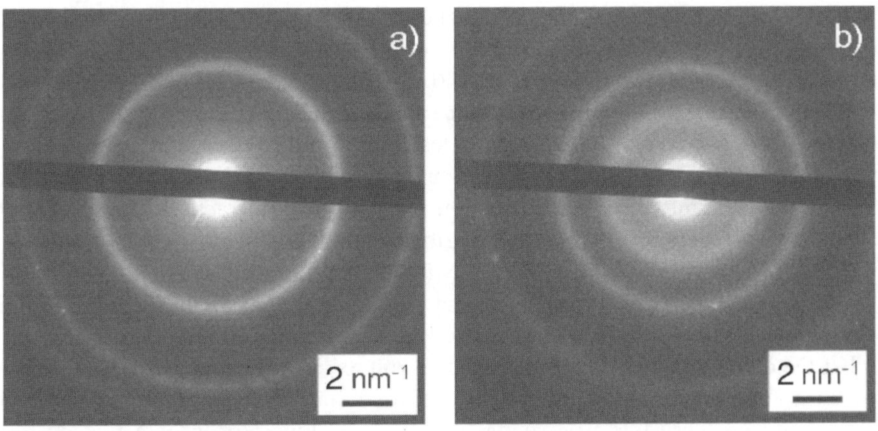

Figure 3.24 SAED patterns of (a) nanodiamond and (b) carbon onions with a diamond core

Figure 3.25 SAED patterns of wear particles observed

1.07 Å (9.33 nm^{-1}). These distances are in good agreement with results obtained by XRD analyses (JCPDS data 02-1248). For carbon onions, these three rings are also observed but a new ring corresponding to the graphite (002) is observed at 2.6 nm^{-1} (about 3.8 Å). The results obtained with graphite (not shown here) confirm that the distance between two graphite layers is 0.34 nm (0.3 $Å^{-1}$ on the diffraction pattern). These results suggest an expansion of the interlayer distance in carbon onions. A thermal expansion anisotropy of graphite-like structures has already been reported by Abe *et al.* [58] and Mykhaylyk *et al.* [16]. By using a thermal expansion coefficient of 2.7×10^{-5} K^{-1} and a graphite interlayer distance of 0.344 nm, a distance of 0.355 nm is found for a synthesis temperature of 1700 K. A diffraction pattern was also obtained from wear particles (Figure 3.25). Six rings are observed: two rings corresponding to nanodiamond (2.05 Å and 1.25 Å) and four other rings (2.99, 2.52, 1.60 and 1.48 Å). The distances obtained are in good agreement with magnetite iron oxide, Fe_3O_4 (JCPDS data 19-0629), and/or maghemite iron oxide, γ-Fe_2O_3 (JCPDS data 39-1346), which is an iron-deficient magnetite. These results clearly show the presence of nanodiamond and iron oxide in the wear particles.

The same experiments were performed for carbon onions without a diamond core (Figure 3.26). For carbon onions, the distance is found to be 0.38 nm (0.26 $Å^{-1}$) and the same distance is observed for the wear particles. This could suggest that no structural modifications of carbon onions occur during friction tests. Figure 3.27 presents the histogram of intensity calculated from these two diffraction patterns. The diffraction pattern obtained on the wear particle presents several continuous rings centred at 3.41, 3.95, 4.76 and 6.16 nm^{-1} (Figure 3.26(b)). One ring corresponds to pristine carbon onions (very weak contribution of the (002) basal plane distances of graphite) and several new rings are observed. Interreticular distances calculated for these new rings fit well with distances of both magnetite iron oxide, Fe_3O_4 (JCPDS data 19-0629), and/or maghemite iron oxide, γ-Fe_2O_3 (JCPDS data 39-1346), which is an iron-deficient magnetite. Because of their similar distances, it is not possible to distinguish between these two iron oxide species. It is important to note that no hematite structure, which is the stable structure of iron oxide, is observed.

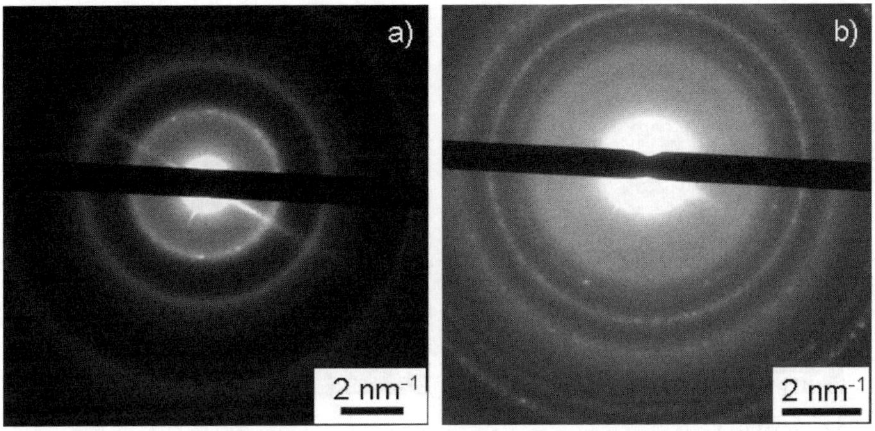

Figure 3.26 SAED patterns of (a) pristine carbon onions and (b) wear particles

To understand the nanostructure of the wear particle better, zero-loss dark-field images on the same particle (Figure 3.28(a)) were performed on the different rings observed. The use of the energy filter (present in the column of the microscope to perform energy filtered transmission electron microscopy (EFTEM) imaging) allowed the elimination of inelastic electrons, which greatly improves the spatial resolution in the images and the sharpness of rings in diffraction patterns. The position of the objective aperture used is indicated on the corresponding diffraction pattern in Figure 3.28(b). The first dark field image (Figure 3.29(a)) was performed on a part of the diffuse ring, which corresponds to the (002) reflection of graphite basal planes. Very small white dots are observed in the particle material, which may correspond to carbon onions. The same experiments (not shown here) were performed on a sample of pure carbon onions and similar images were obtained. An EELS spectrum at the carbon K-edge was performed on the same area and the comparison of this spectrum with that of pristine onions confirms the presence of intact onions in the wear debris material. Dark-field images performed on the other rings corresponding to iron oxide species (Figure 3.29(b)) clearly indicate the presence of iron oxide nanoparticles (10 nm) well distributed in the carbon onion network.

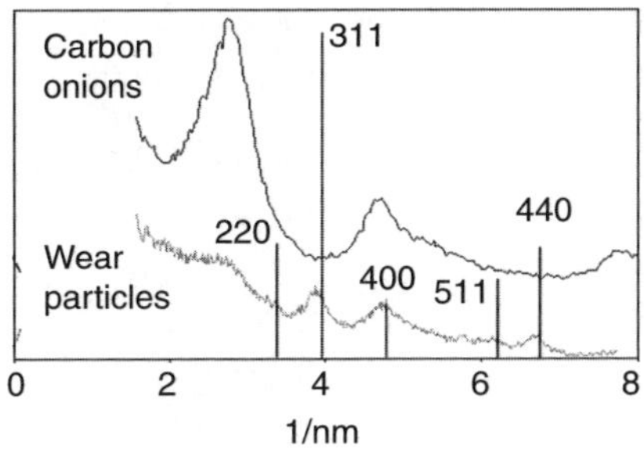

Figure 3.27 Intensity profile of the SAED patterns. Diffraction patterns of maghemite iron oxide is indicated

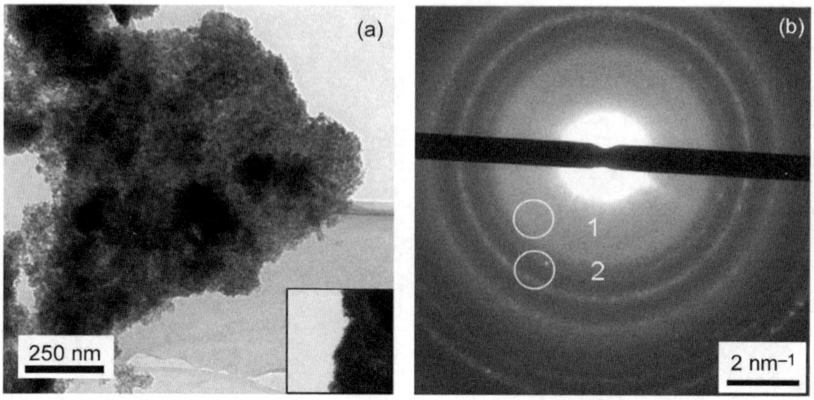

Figure 3.28 (a) TEM image of a wear particle (in the frame, the part of the wear particle analysed in detail is presented). (b) SAED pattern of the whole particle. Positions of the objective aperture used to obtain the dark-field images are represented

HRTEM images of wear particles confirm the presence of iron nanoparticles and intact carbon onions. Figure 3.30(a) presents a magnetite nanoparticle observed under a $[1\ -1\ 0]$ axis zone. A diffractogram was calculated from the image of this nanoparticle. Distances measured on this diffractogram ($d = 4.85\ nm^{-1}$) are in good agreement with the theoretical magnetite structure observed under a $\{1\ -1\ 0\}$ azimuth. This confirms that observed nanoparticles have a magnetite (or maghemite) structure. Intact carbon onions can also be observed in the wear particles (Figure 3.30(b)).

The tribofilm formed on the flat during the friction test was studied using atomic force microscopy (AFM) and scanning tunneling microscopy (STM). AFM images were performed using a Nanoscope® III and STM using an Omicron® AFM-UHV(ultrahigh vacuum) and STM. AFM pictures (Figure 3.31) indicate the presence of small amounts of carbon onions in the wear scar [59]. It is important to notice that the friction is low on these agglomerated

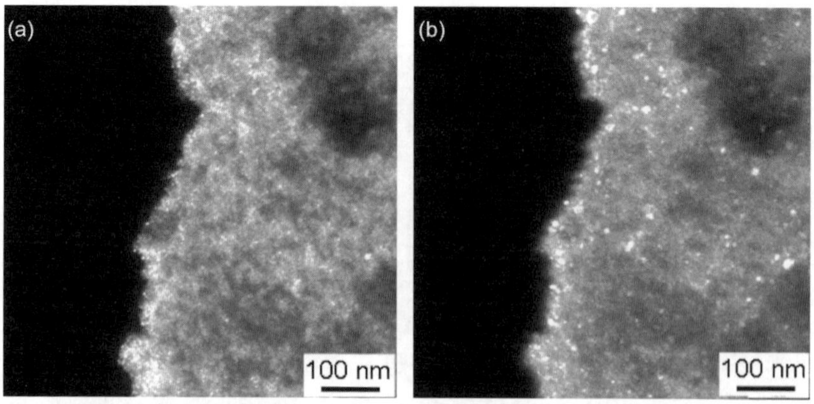

Figure 3.29 Dark field images obtained for several positions of the objective aperture in Figure 3.27

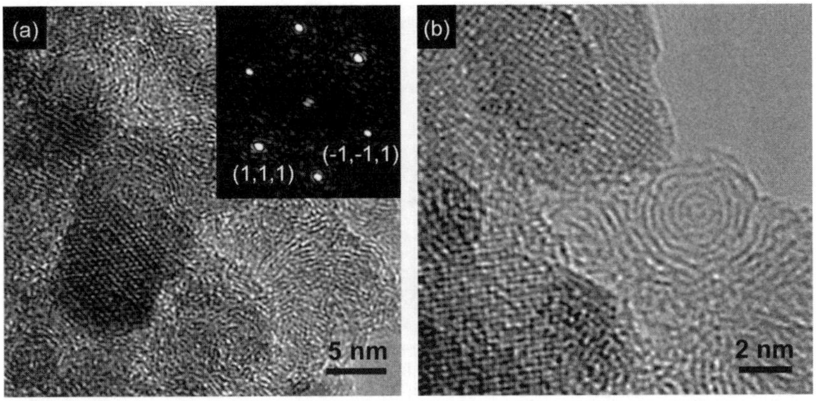

Figure 3.30 HRTEM of a wear particle. (a) Maghemite nanoparticle observed under a [1, −1, 0] azimuth. Distances measured on the calculated diffractogram ($d = 4.85\ \text{nm}^{-1}$) are in good agreement with the theoretical magnetite structure observed under a [1, −1, 0] azimuth. (b) Intact carbon onions are observed

small amounts of carbon onions. These amounts are aligned along the direction of friction. The maximum height of the surface is only 20 nm. Thus the surface is covered by a smooth tribofilm.

UHV STM pictures also show that the agglomerates are aligned along the direction of friction (Figure 3.32). Aligned narrower bands of 0.1 μm are observed on the surface. The surface was found to be extremely smooth with a height of 12 nm. The maximum height was found to be 10 nm in Figure 3.32(b). An analysis of one of these agglomerates reveals

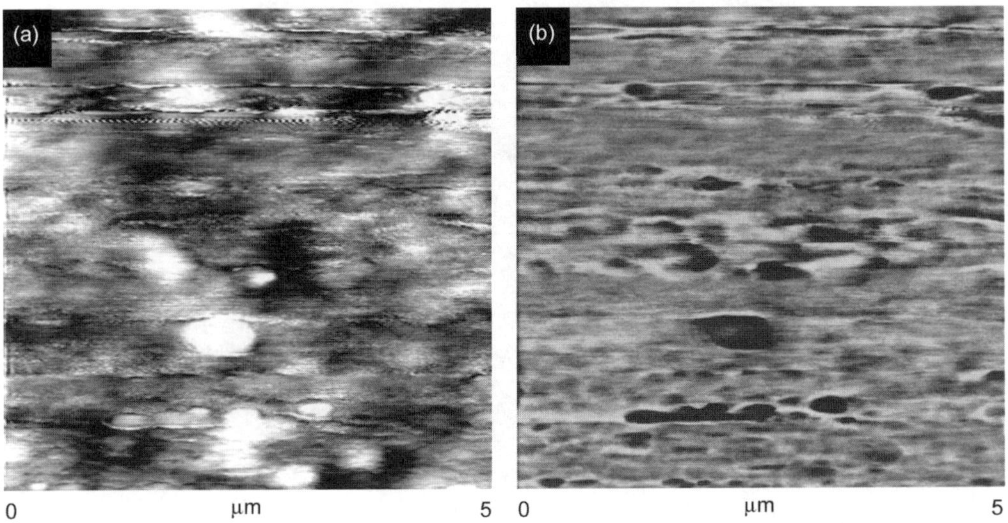

Figure 3.31 AFM images of the wear scar observed on the flat after the friction test: (a) height and (b) friction. The direction of friction was horizontally

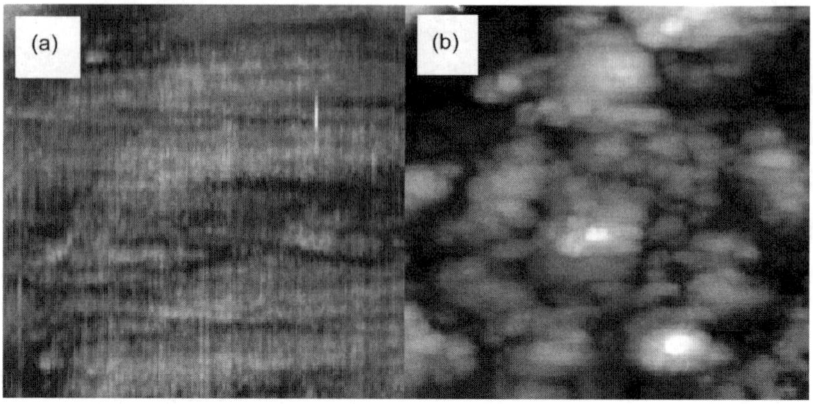

Figure 3.32 UHV STM pictures of the wear scar (scale 700 × 700 nm) and at higher magnification (scale 200 × 200 nm). The direction of friction was vertically

the presence of an individual carbon onion (Figure 3.33). It is difficult from this picture to know whether the carbon onion is intact or crushed. An attempt was made to estimate the distance between two atoms by studying the difference in height observed along a line (see Figure 3.34). The distance between two spots is equal to 0.48 nm, which is double the atomic distance of the basal plane.

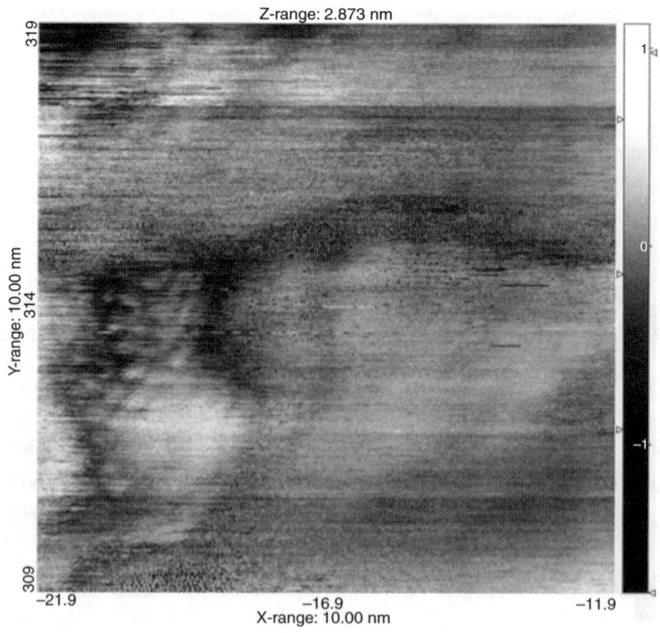

Figure 3.33 STM images of one of the small amounts (scale 10 × 10 nm). The atoms can clearly be distinguished. It may be a carbon onion

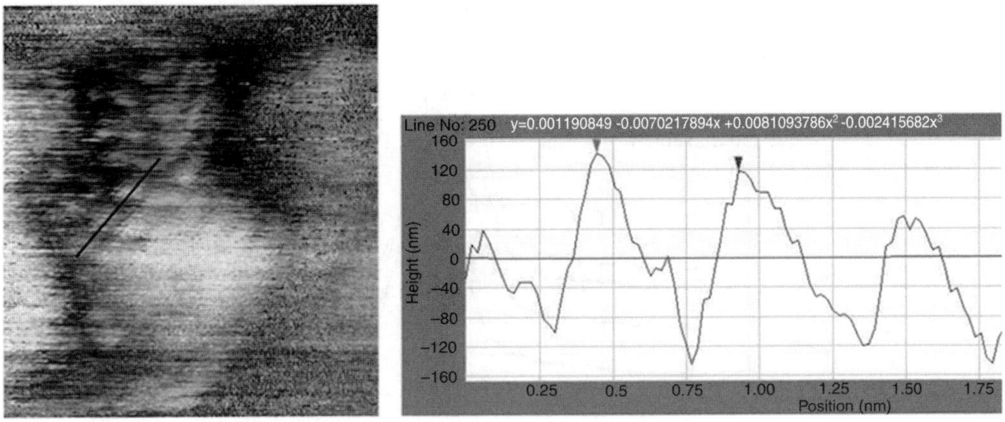

Figure 3.34 Zoom on the carbon onion. On the right, contrast on the line traced on the picture to determine the distance between the spots observed

FFM experiments on carbon onions deposited on graphite reveal a higher friction coefficient on carbon onions than on graphite (Figure 3.35). The AFM image (Figure 3.35(a)) shows the presence of carbon onions on the graphite (marked with black circles). A comparison of FFM and AFM images clearly shows a higher friction on carbon onions. The friction coefficient was measured for several normal loads (25 to 130 nN). A fit of the values obtained shows an average friction coefficient value of 0.017 for graphite and 0.050 for carbon onions. These experiments suggest that intact carbon onions need to be structurally modified in order to be active in the contact area. This may explain why the carbon onions are only active at high contact pressures. Due to their multiwalled structure, they are more resistant

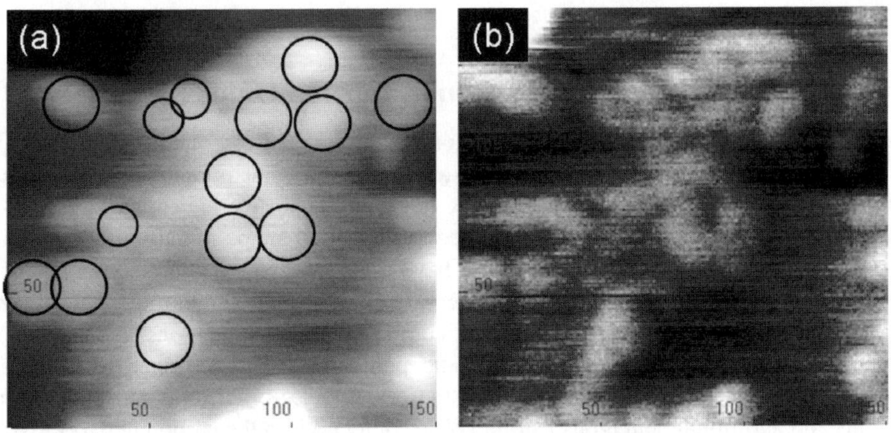

Figure 3.35 (a) AFM image of carbon onions deposited on graphite (150 nm × 150 nm, the white level indicated a height of 10 nm). Black circles show the presence of carbon onions. (b) Corresponding FFM image. High friction (white areas) corresponds to the location of carbon onions

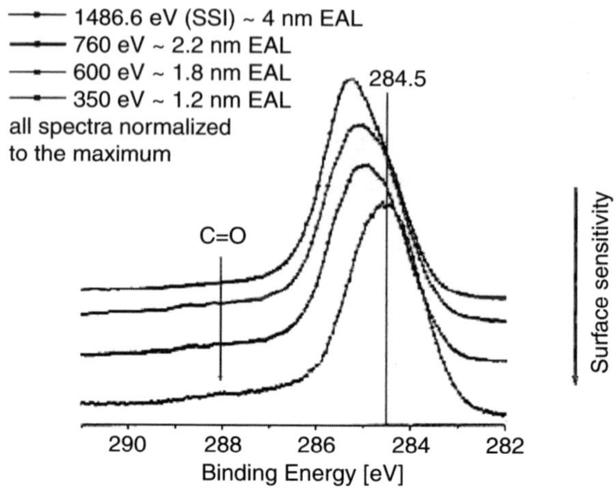

Figure 3.36 XPS spectra obtained at different energies on a DLC coating. It shows the accuracy in detecting a difference in the structure of carbon compounds depending on the depth analysed [60]

to the pressure, so high pressures are necessary to obtain graphite. However, as discussed below, it is quite difficult to observe the formation of graphite on the friction surface. Furthermore, TEM results show the presence of intact carbon onions in the wear particles. To study whether or not carbon onions are exfoliated like IF-MS_2 to produce graphene sheets inside the contact area, variable XPS energy would be a very powerful technique. Indeed, the presence of graphite on the surface could be monitored. Figure 3.36 presents results obtained on a diamond-like carbon coating analysed at different beam energies. With an energy of 320 eV, the extreme surface can be probed (1.2 nm) and the 1s peak of carbon is characteristic of the graphite structure [60]. These experiments are not easily performed since they require a synchrotron beam, but it would be very interesting to characterize graphite formed on the samples.

3.3 Possible Lubrication Mechanism of Carbon Onions

Even if it is difficult to know whether the onions are exfoliated or not and whether graphene sheets are produced inside the contact area, an important point is that carbon onions play more important tribological roles than graphite (particularly as an antiwear additive). The first reason to explain this difference can be related to the round shape of carbon onions without edges. Furthermore, in carbon onions, sheets have a nested shape and are composed of both pentagons and hexagons. Pentagons are necessary to obtain the curvature of the graphene plane (Figure 3.37(a)). For example, C_{60} is composed of 12 pentagons and 20 hexagons (Figure 3.37(b)). The number of hexagons and pentagons are determined by Euler's law. Furthermore, Terrones and Terrones showed that the sphericity is improved by introducing heptagons and additional pentagons (Stone–Wales defects) into the carbon network [61].

Another explanation can be formulated by using recent results obtained by Oleshko and Howe. They postulated that it is possible to determine the physical properties of materials from the low-loss peak obtained by EELS measurements [33, 62]. For isotropic or amorphous

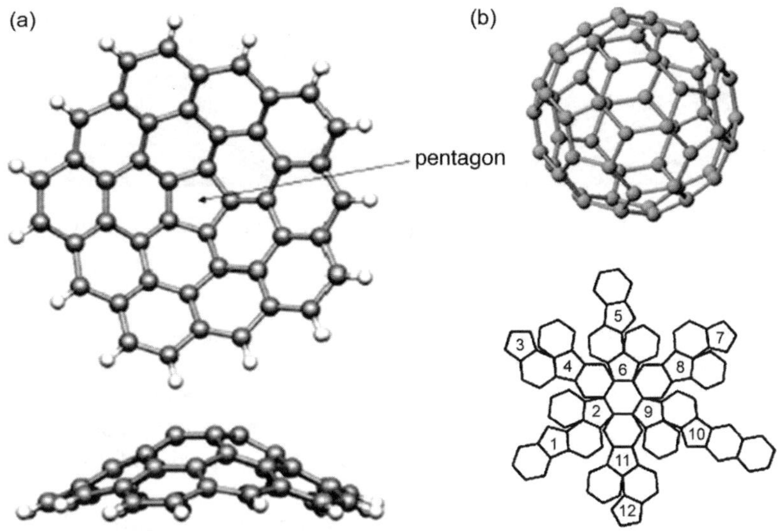

Figure 3.37 (a) A pentagon is necessary to obtain the curvature of the graphene plane. (b) Schematic representation of C_{60} fullerene composed of 12 pentagons and 20 hexagons

carbon compounds, an empirical relation between the maximum of the plasmons loss peak (E_p) and the so-called 'electronic' hardness (H_m) was obtained as follows:

$$\log H_m = -7.44 + 6.1 \log E_p \tag{2}$$

Using this relation, a hardness of about 25 GPa is found for graphite and 10 GPa for carbon onions. It should be noticed that, in these conditions, the hardness given by Oleshko's relation for graphite is maximum and corresponds to the contribution of 'in-plane' covalent carbon bonds. There is a huge discrepancy between this value and those determined from classical nanohardness measurements in the literature. Graphite is one of the softest solid materials and its measured hardness is found to lie between 0.1 GPa [63] and 2.4 GPa [64]. This kind of disagreement was also observed on carbon black and soot material. However, it is well known that soot can abrade steel like a cutting tool. Therefore some graphite particles might be able to scratch steel as well.

Density can also be determined from EELS measurements. Xu *et al.* used electron energy-loss spectroscopy (EELS) to characterize hydrogen-free tetrahedral amorphous carbon films [65]. They studied several carbon film materials deposited with different bias voltages. From their results, it is possible to fit a good correlation between the density (D) of the carbon films tested and the maximum of the low loss plasmon peak (E_p). Thus, Xu obtained an empirical relation as follows:

$$\log d = 1.0519 \log E_p + 1.9563 \tag{3}$$

Using this relation, the densities of the graphite and carbon onions were tentatively determined at 2897 and 2593 kg/m^3, respectively. These values are higher than data from the

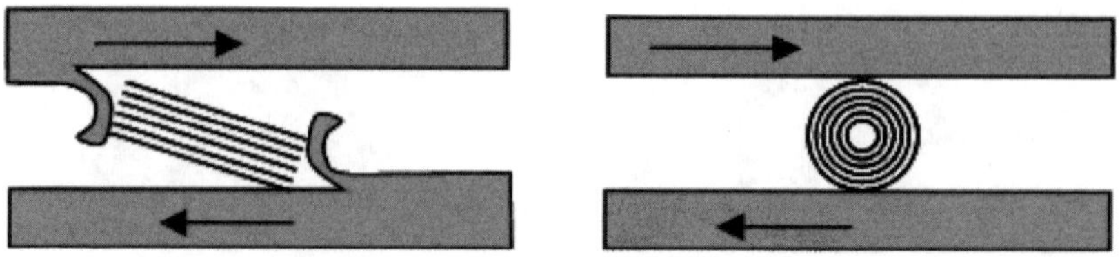

Figure 3.38 Schematic representations of the behaviour of carbon onions or graphite in the contact area

literature. However, the density calculated by EELS for carbon onions indicates a 10 % decrease compared to graphite. On the other hand, SAED revealed a 10 % expansion of the basal plane distance between carbon sheets in the carbon onions. These results are in good agreement with the decrease in the density of carbon onions compared to graphite and prove that the EELS measurement can be used to determine physical properties of preferably isotropic carbon compounds.

Based on the results of EELS, graphite can be described as 'electronically' harder than carbon onions. This may explain their abrasive behaviour inside the contact area. Figure 3.38 schematically summarizes the behaviour of graphite and carbon onions inside the contact area.

From tribological results obtained for several carbon forms in PAO, a general trend is observed. The addition of carbon compounds seems to reinforce the friction-reducing performance of the PAO base. Raman analyses were performed on the film formed during the friction test with pure PAO and PAO containing carbon onions (Figure 3.39). After friction tests with pure PAO, the Raman spectrum is composed of a peak at 670 cm^{-1}, which can be attributed to iron oxide (magnetite) [66], and two small peaks at 1330 and 1600 cm^{-1} indicating the presence of a carbon film. The Raman spectrum after the friction test with carbon onions shows the two peaks of the carbon film with a higher intensity and no peak of magnetite. Sometimes, peaks due to maghemite (350, 500, 655 and 714 cm^{-1}) were observed [67].

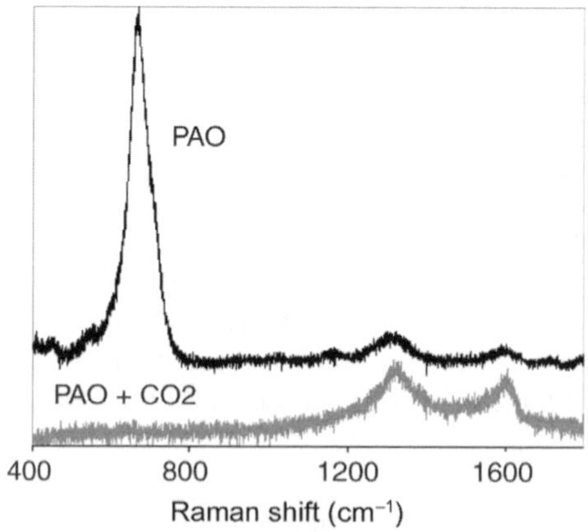

Figure 3.39 Raman analyses performed inside the wear scar on the flat after the friction test with pure PAO and PAO containing carbon onions ($\lambda = 514.5$ nm). A carbon film is formed on the surfaces in both cases

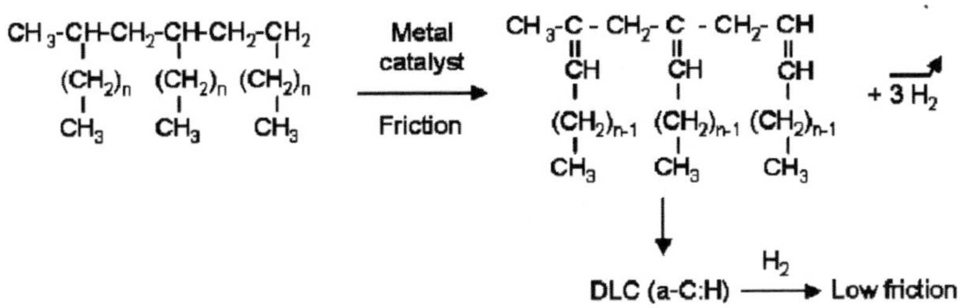

Figure 3.40 Possible mechanism of lubrication by PAO and PAO containing carbon additives. The presence of carbon additives may reinforce the DLC film produced

Iron oxide in the case of pure PAO is due to wear occurring during the friction test. This did not indicate that carbon onions are added to PAO. The origin of the tribological efficiency of PAO will possibly be attributed to the branched backbone of the molecule in which dehydrogenation is likely to occur, liberating some sp^2 carbon-like material progressively forming the tribofilm (Figure 3.40) [68].

The excellent antiwear properties of carbon onions can also be explained by the structures of the wear particles formed during the friction test. A few iron oxide particles produced during the friction test are actually trapped in the tribofilm. The maghemite or magnetite structure of these particles proves that they are trapped and do not react, otherwise the structure of hematite (the most stable structure of iron oxide) would have been found. It is well known that hematite particles (actually rust) are highly abrasive. Thus, the absence of hematite may explain the low wear observed. On the other hand, a surface covered with magnetite coating was found to have a good friction response under boundary lubrication [69]. The iron oxide particles have a nanometre size and are practically embedded in the carbon structure. An epitaxial stabilization of iron oxide by graphite can be envisaged to explain the presence of maghemite or magnetite. Indeed, their interreticular distances are very close. A model for the onion-induced tribofilm can then be proposed (Figure 3.41). It is composed of a mosaic of graphitic sheets and intact onions bound by lubricious iron oxides and iron hydroxide nanoparticles. Low friction was observed with Fe_3O_4 and FeOOH nanoparticles dispersed on an Fe-C matrix [70]. In the model proposed by Yuansheng and Shenghua, Fe_3O_4 acts as lubricious oxides and FeOOH supplies hydrogen on the counter surface. A similar mechanism can be involved in the case of carbon onions with the formation of a carbon film containing lubricious iron oxide and with the presence of OH groups on the surface. Hydroxylation of diamond-like carbon coatings was also found to be the origin of a superlow friction coefficient during friction between surfaces coated with DLC coatings and lubricated by glycerol mono-oleate (GMO) [7]. More work is necessary to confirm this mechanism.

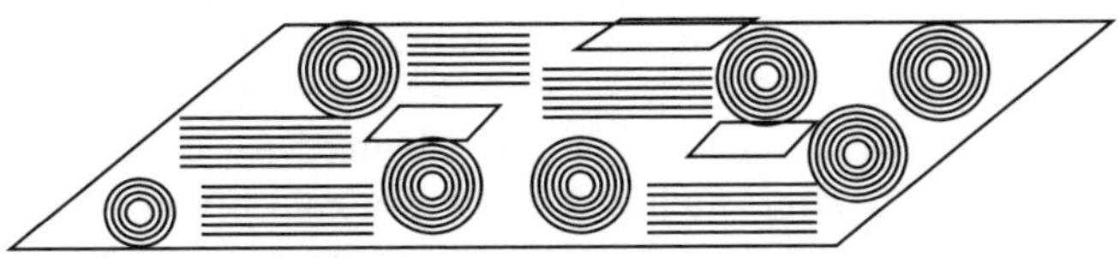

Figure 3.41 Schematic representation of the composition of the tribofilm formed with carbon onions

Carbon onions present very interesting tribological properties when used as lubricant additives. An optimization of their synthesis conditions would lead to an improvement of their tribological properties. Another interesting use of these nanoparticles will be as coatings. Several studies report low friction obtained on C_{60} films. Carbon onion films may also present interesting properties. As developed by Nakagawa *et al.* [71], molecular beam epitaxy could be envisaged to make such coatings. Very thin C_{60} films were obtained by this technique and friction force measured by FFM was very low.

3.4 Nanotube Synthesis and Characterization

Observed in 1991 by Iijima [72] by transmission electron microscopy as a by-product of the reaction of the synthesis of fullerenes, carbon nanotubes can be described as a rolling-up of graphite planes. Because of its relation to graphite as well as its nanometer size and one dimensionality, this new material is expected to present unique properties. Several applications have been envisaged: for electronics [73–75], for energy storage [76, 77] or in medicine [78]. They also present unique mechanical properties [79] and many studies are presently being performed with the aim to use them as reinforcements in polymers [80–82].

Nanotubes can be distinguished as single-walled nanotubes and multiwalled nanotubes. The first nanotubes observed in 1991, multiwalled nanotubes can be described as nested nanotubes that can slip one inside the other. Single-walled nanotubes were observed in 1993 and can be described as a single sheet of graphene rolled up [83]. They have a mean diameter equal to 1 nm and a length of several micrometres, and often gather in bundles. Another important parameter to define a nanotube is its helicity: 'armchair', 'zig-zag' or 'chiral', according to the way that the graphene sheet is rolled up. In this study, the chirality of the nanotubes was not taken into account.

Two main categories of synthesis methods of carbon nanotubes can be distinguished: high-temperature methods and average-temperature methods.

The high-temperature methods consist of vaporizing the carbon, whose temperature of sublimation is 3200 °C, and condensing it in a chamber housing a strong gradient of temperature. Three devices are available: arc discharge [84], laser ablation [85] and the solar reactor [86]. For these methods, the use of a catalyst is essential to obtain single-walled nanotubes. These catalysts are present in the samples after the synthesis.

In the average-temperature methods, a carbonaceous gas is decomposed on a metal catalyst. These techniques are named chemical vapour deposition (CVD). The temperature of the furnace is between 800 and 1100 °C and gases used are carbon monoxide, methane or acetylene. The metal catalyst used is a transition metal, such as iron, nickel or cobalt. It is difficult to remove these metal catalysts after the synthesis without damaging the structure of the nanotubes because they are imprisoned in the nanotubes or are surrounded by amorphous carbon. They are therefore present in the nanotube samples after synthesis.

Until now, the macroscopic tribological properties of the nanotubes had not been studied. The only studies carried out used the nanotubes as reinforcement in various matrices: diamond thin film [87], polyimide [88], Ni [89], Ni-P [90], carbon/carbon composites [91] and alumina [92, 93]. These composite coatings present a better wear resistance and a lower friction. In nanotribology, Ohmae *et al.* studied friction of a gold tip on a nanotube forest. The friction coefficient obtained was high (1.2 to 1.5) and independent of humidity [94]. A very weak adhesion of the tip on the nanotubes was observed. During friction, no distortion of the nanotubes intervenes and there is no transfer of nanotubes on the tip. Recently, Miyoshi *et al.*

studied carbon nanotubes as solid lubricant. Nanotubes were aligned on a surface or dispersed in a solvent and deposited by solvent evaporation [95]. The friction coefficients obtained were close to 0.05 in air and 0.009 in vacuum and a good resistance to wear was observed. These results are explained by the presence of cut or broken multiwalled nanotubes (MWNTs) inside the contact area, and friction resistance is minimized by the rolling of these nanotubes. The presence of the nanotubes leads to a decrease in the surface energy of the interface and the real contact area is minimized by the high elastic modulus of the MWNTs. Several simulations on the tribological properties of the nanotubes were also carried out. Buldum and Lu studied the movement of nanotubes on a graphite surface and the influence of the orientation of the nanotube compared to graphite [96]. Ni and Sinnott studied the behaviour of nanotubes submitted to hydrostatic pressures or to shear [97].

Several kinds of nanotubes were studied to highlight the influence of different parameters on tribological properties of nanotubes: the number of walls, the presence and the nature of a catalyst. Thereafter, acronyms will be used to describe the various samples of nanotubes (single-walled or multiwalled) studied:

- Single-walled nanotubes (SWNTs) containing a mixture of nickel/yttrium catalyst. These nanotubes were synthesized by electric arc discharge at the University of Shizuoka (Ni/Y-SWNTs). A $J \times B$ arc method has been developed [98] where a magnetic field is applied perpendicular to the arc current. The magnetic field enhances the efficiency of the electric discharge method synthesis by a factor of 2.5 [99]. The use of this technique was first used for the synthesis of fullerenes [98]. Furthermore, a new reactor has been developed. The anode is progressively introduced inside the reactor by a motor and presently Mieno is working on a large reactor to produce a high amount of nanotubes.
- Single-walled nanotubes purified from Meijo University at Nagoya (SWNTs). These nanotubes were synthesized by electric arc discharge (iron was used as catalytic particles) and then purified by annealing at 693 K and a mild hydrochloric acid treatment [100].
- Double-walled nanotubes (DWNTs) synthesized by CVD at CIRIMAT at Toulouse University. These nanotubes were synthesized by the CVD process. Catalytic nanoparticles are formed from solid solutions of $Mg_{1-x}Co_xMo_yO$. The synthesis method used is the same as that described by Bacsa *et al.* [101] but with small differences in the catalytic nanoparticles synthesis.
- Multiwalled nanotubes containing iron particles, used for the synthesis, from Nanocyl Company (Fe-MWNTs).
- Multiwalled nanotubes (MWNTs) without catalytic particles from the Laboratory of Ohmae, Kobe University. Nanotubes were synthesized by CVD assisted by plasma [102]. They grow vertically on a surface to form a 'forest' of nanotubes (Figure 3.42). At the end of the synthesis phase, iron nanoparticles necessary to the synthesis are encapsulated at the top of the nanotubes. These nanotubes are then purified by a new specific method developed by Ohmae (Figure 3.43). After an ion bombardment which allows the nanotubes to be opened, a chemical attack using strong acid finally eliminates all the metal particles. With a thermal CVD synthesis using acetylene, the metal nanoparticles are located at the 'foot' of the nanotubes. By separating the nanotubes from the substrate, the end containing the catalyst is opened. A chemical attack using a mixture of acid eliminates the catalytic particles. Ohmae's laboratory improved the growth rate of vertically aligned carbon nanotubes to a value of 280 nm/s, and with previous plasma-enhanced CVD techniques [103,104] or thermal CVD [105] a growth of 100 nm/s was obtained.

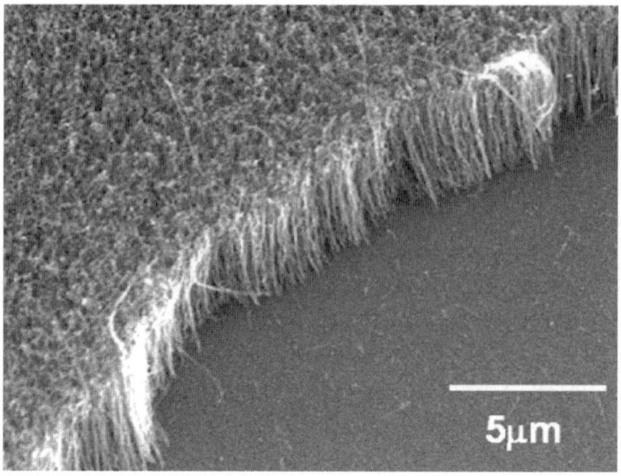

Figure 3.42 SEM image of vertically aligned nanotubes on silicone substrate. These nanotubes are like a 'forest'

Transmission electron microscopy was used to characterize the different kinds of nanotubes used: nanotubes arrangement, length, purity (the presence of catalyst, the amorphous phase). Figure 3.44 shows the difference between the studied samples. Ni/Y-SWNTs are gathered in bundles with an average diameter of 10 nm (Figure 3.44(a)). Catalytic particles from 5 to 15 nm in diameter are surrounded of pseudo-graphitic carbon in this sample. SWNTs are gathered in bundles from 30 to 60 nm (Figure 3.44(b)). No catalytic particles are observed. DWNTs are also gathered in bundles. These nanotubes have a diameter of 2.5 nm for a length of several hundreds of μm (Figure 3.44(c)). MWNTs have a diameter from 20 to 120 nm (Figure 3.44(d)). The characteristic of this sample is the absence of catalytic particles. XPS and TOF-SIMS (time-of-flight secondary ion mass spectrometry) analyses showed the very great purity of this sample made only of carbon.

Raman spectroscopy is a method of characterization very often used for carbonaceous compounds. In the case of nanotubes, two principal modes are very useful to their characterization (Figure 3.45). The first is the radial 'breathing' mode (RBM) at 150 cm^{-1}. It is characteristic of sp^2 carbons of carbon nanotubes because it corresponds to the vibration of all the carbon

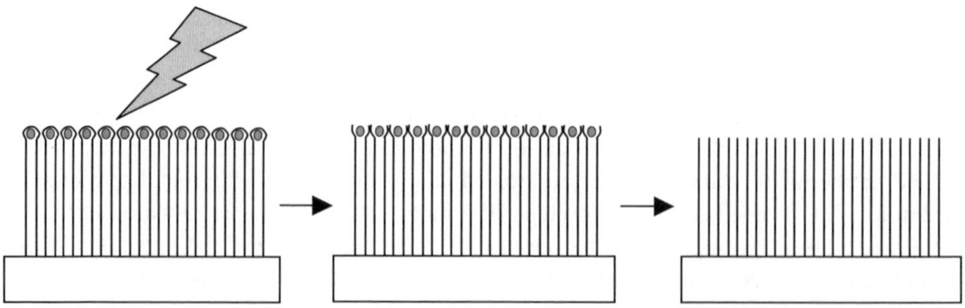

Figure 3.43 Purification method of multiwalled carbon nanotubes without catalytic particles

Figure 3.44 TEM images of the different samples tested: (a) Ni/Y-SWNTs, (b) SWNTs, (c) DWNTs, (d) MWNTs

atoms in the radial direction of the nanotube in a kind of breathing. The determination of the diameter of nanotubes can be done at the RBM frequency. If the single- walled nanotube is isolated, ω $(\mathrm{cm}^{-1}) = 222/d$ (nm) [106]. If the nanotube is in a bundle, van der Waals interaction between tubes must be considered and thus ω $(\mathrm{cm}^{-1}) = 238/d^{0.93}$ (nm). The tangential mode (TM) at 1600 cm^{-1} is useful to determine the electronic properties of the nanotubes.

Raman spectroscopy of single-walled nanotubes is known to be resonant. This means that the wavelength of the laser used influences the intensity and the shape of the characteristic modes. With a wavelength of 632.817 nm the intensity of the RBM is higher (Figure 3.46) and the shape of the TM differs from the one obtained at 514.5 nm. Resonant Raman spectroscopy of single-walled carbon nanotubes can be explained by the electronic structure of carbon

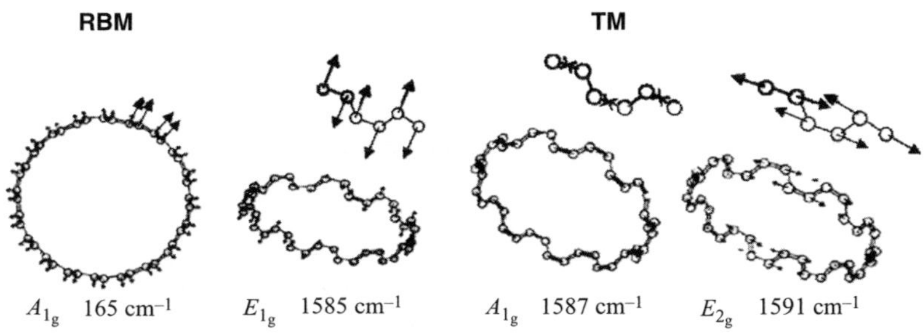

Figure 3.45 Different vibrational modes of nanotubes; RBM and TM are more important [107]

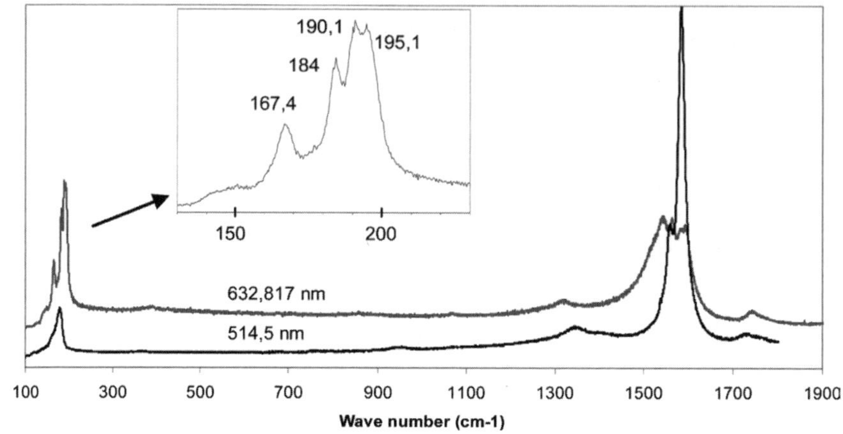

Figure 3.46 Raman spectrum of Ni/Y-SWNTs at two different laser wavelengths: 632.817 and 514.5 nm. A 632.817 nm wavelength allows a better RBM signal to be obtained. The frame presents the RBM allowing calculations of the different nanotube diameters (between 1.26 and 1.47 nm)

nanotubes and particularly the Van Hove singularities [108]. Even if the RBM signal is more intense at 632.817 nm, for practical reasons of the *in situ* Raman experiments a wavelength of 514.5 nm was used. By using the formula indicated above, the average diameter of the nanotubes was determined at 1.3 nm.

3.5 Friction-Reducing and Antiwear Properties of Different Nanotubes

3.5.1 SWNTs

A preliminary study with single-walled nanotubes determined the optimal concentration of about 1 % for this additive in a synthetic PAO base oil (Figure 3.47). The friction coefficient

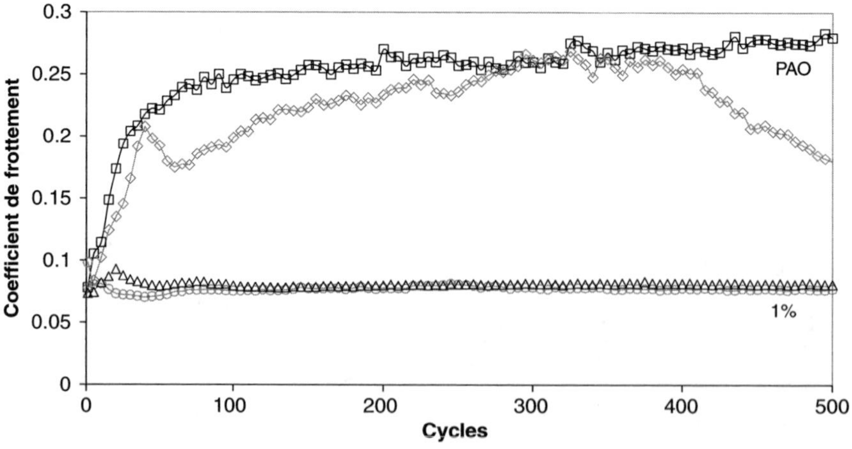

Figure 3.47 Effect of the nanotube concentration in PAO on the friction coefficient at 0.83 GPa. These results show an optimal concentration of about 1 wt %

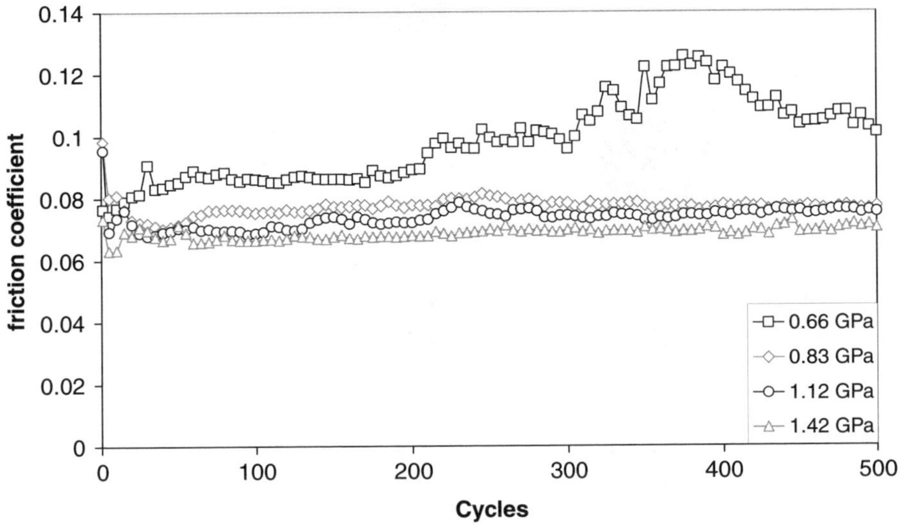

Figure 3.48 Contact pressure effect on the friction coefficient for a 1 wt % Ni/Y-SWNT dispersion

is very low (0.07) from the first cycles and remains very stable throughout the test. The effect of pressure is very similar to that observed in the case of fullerenes (Figure 3.48). Friction decreases when the contact pressure increases [109].

A comparison of the results obtained with other carbon relatives (graphite, C_{60}) studied under the same conditions highlights the very interesting tribological properties of Ni/Y-SWNTs (Figure 3.49). The friction coefficient obtained with Ni/Y-SWNTs is lower than those obtained with C_{60} and graphite. The friction coefficient obtained with C_{60} is also low (0.1). These

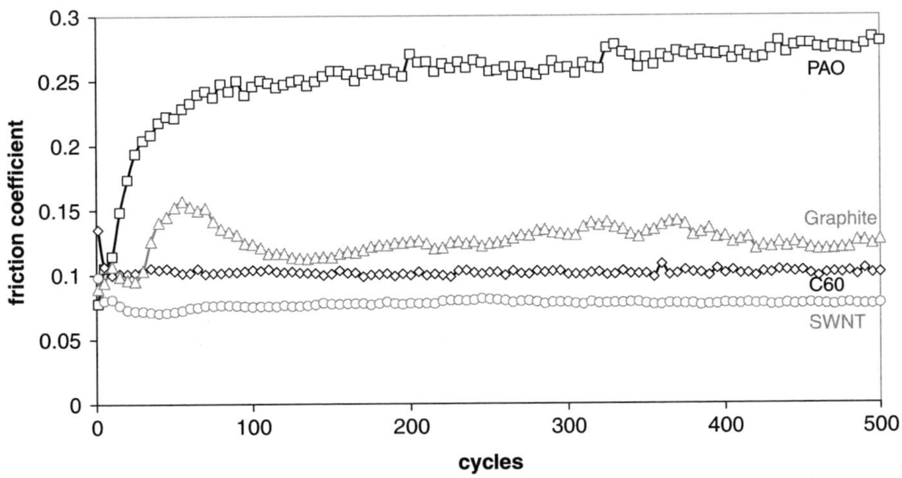

Figure 3.49 Comparison of the friction reducing properties of Ni/Y-SWNTs and other carbon forms tested in the same conditions (1 wt % concentration, $P = 0.83$ GPa, $v = 2.5$ mm/s, $T = 20\,°C$)

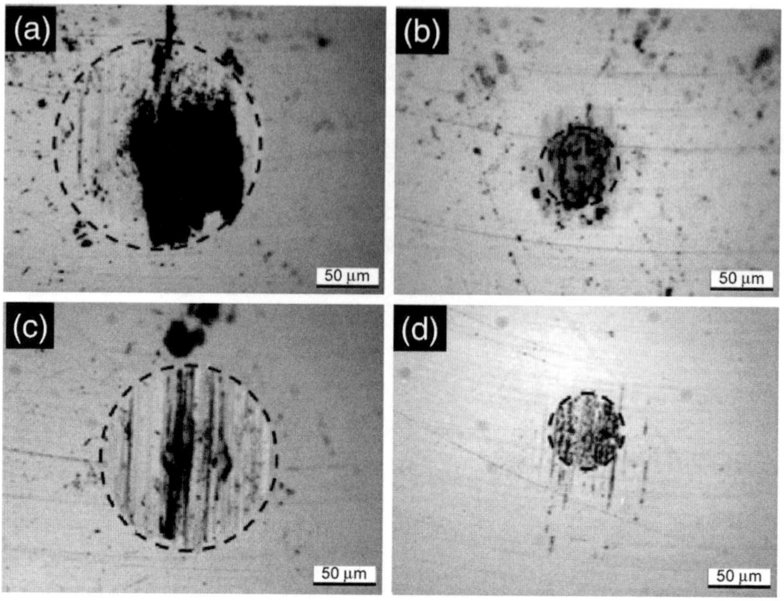

Figure 3.50 Wear scars observed on pins after the friction test with (a) pure PAO, (b) 1wt % Ni/Y-SWNTs, (c) 1wt % graphite and 1wt % C_{60}

results show that C_{60} can be considered as an additive in synthetic PAO base oil. Ni/Y-SWNTs also present interesting antiwear properties (Figure 3.50) with a diameter of the wear scar observed on the pin lower than those observed with PAO only and graphite. This diameter is equal to the Hertzian diameter (68 μm), which indicates that the pin did not wear and truncated during the friction test.

A comparison of nanotubes without a catalyst (SWNTs) showed a key role for catalysts in the tribological properties of the single-walled nanotubes. Addition of purified nanotubes (SWNTs) in PAO does not lead to a reduction in the friction coefficient (Figure 3.51). A slight diminution of the wear on the pin is observed with SWNTs, the wear scar diameter being 115 μm, against 170 μm in the case of PAO. However, this reduction is nevertheless lower than the one observed with Ni/Y-SWNTs (Figure 3.52).

3.5.1.1 Role of the Catalyst

The presence of catalytic particles is necessary for the synthesis of single-walled nanotubes. The nature of the catalyst and in particular the content of rare earth (Y, Ce) determines the morphology of nanotubes: long ropes of nanotubes or short nanotubes forming a sea urchin structure [110]. In the case of a catalyst made of Ni and a small percentage of rare earth, long nanotubes are obtained, while a strong percentage of rare earth produces nanotubes of the sea urchin type (Figure 3.53).

The growth mechanism of carbon nanotubes from a metal catalyst is based on the solvation of carbon vapour into metal clusters. This is due to the ability of metals, such as Ni and Co, to dissolve carbon when liquid [111]. Quantum molecular dynamics (QMD) simulations

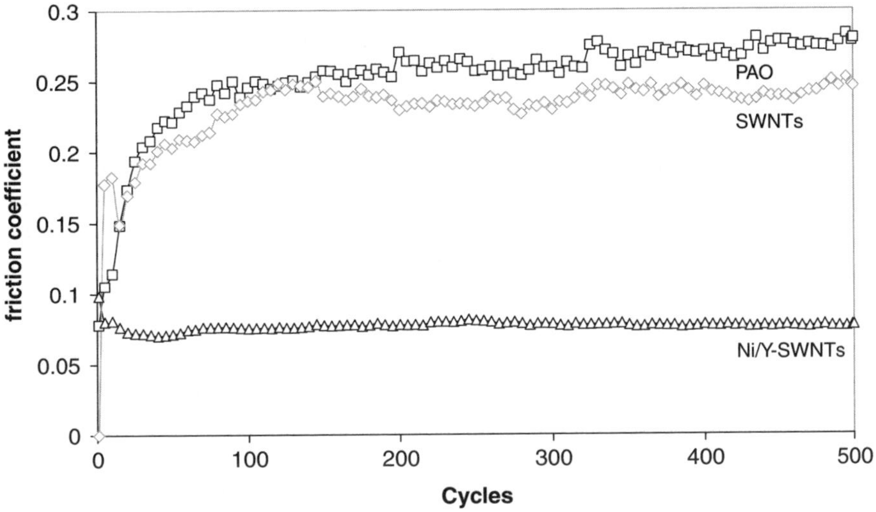

Figure 3.51 Comparison of friction coefficients obtained with Ni/Y-SWNTs and SWNTs dispersed at 1 wt % in PAO

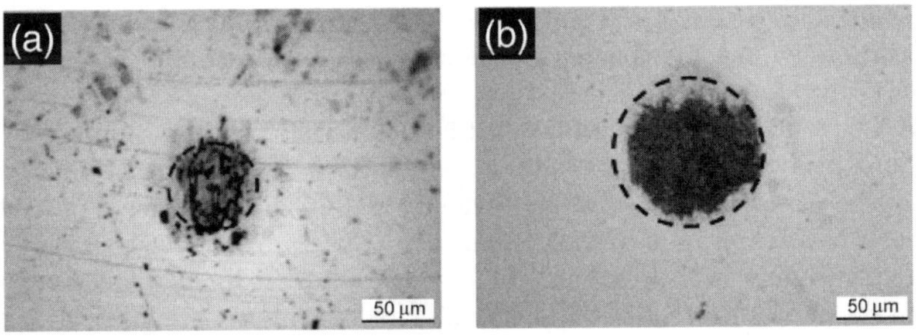

Figure 3.52 Wear scars observed on pins after the friction test with (a) Ni/Y-SWNTs and (b) SWNTs. The addition of the two types leads to a reduction of wear, more significant in the case of Ni/Y-SWNTs

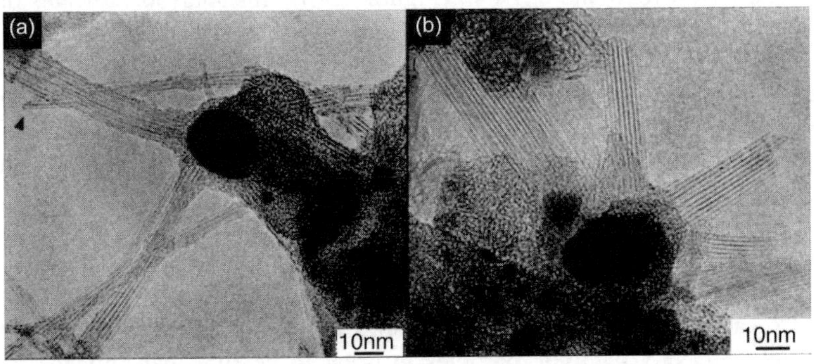

Figure 3.53 TEM images showing two types of single-walled carbon nanotubes obtained with different catalysts: (a) nanotubes in long bundles, (b) sea urchin nanotubes. Reproduced by permission of Elsevier from Gavillt *et al.* [110], Figure 1, © (2002) Elsevier

demonstrate that carbon tends to segregate on the surface of the catalytic particle when the temperature decreases. Then the atoms of carbon move quickly on the surface to form aromatic rings [112]. The content of rare earth in the catalytic particles has an influence on two parameters of the nanotube synthesis: nucleation sites of nanotubes and time for formation. The rare earth is at the periphery of the catalytic particles. Nucleation sites are rare earth carbides present at the surface of catalytic particles [113]. With a high content of rare earth, nucleation sites cover all the particles involving the formation of a sea urchin structure. The content of rare earth also influences the solidification temperature of the catalytic particles and thus the time for formation of nanotubes. Gavillet *et al.* studied precisely the nucleation and growth of single-walled nanotubes from metallic particles and particularly nickel-rare earth metal catalytic particles [114]. This study was based on experimental results (TEM observations) and Monte Carlo simulation of atomic interactions C-C and C-M. To understand the growth mechanism of nanotubes better, new techniques have been developed. One of them consists of *in situ* laser analyses performed during the synthesis inside a laser ablation reactor. Spatial evolution of the catalyst species and C2 and C3 clusters inside the reactor are followed [115]. The role of catalytic particles for the formation of single-walled carbon nanotubes is not yet completely understood, but they are necessary and remain in the nanotube sample after the synthesizing process.

It is difficult to remove these catalytic particles from the sample because they are at the end of the nanotubes or are surrounded by amorphous carbon. Energy filtered transmission electron microscopy (EFTEM) realized on Ni/Y-SWNTs shows the presence of catalytic particles not only at the end of nanotubes but also inside the amorphous carbon present in the sample (Figure 3.54).

Ni/Y-SWNTs contain a mixture of catalyst Ni/Y (38 wt %) with an atomic ratio of 85/15. With this ratio, nanotubes in long bundles are obtained [110]. With an XPS analysis carried out on nanotubes deposited on a substrate covered by a thin gold film, it was possible to check the hypothesis of a segregation of yttrium on the periphery of the catalytic particle (Figure 3.55). Indeed, yttrium was observed only in the oxide form (Y 3d3/2 at 159 eV) while nickel was only in metallic form (Ni 2p3/2 at 853 eV). This confirms the growth mechanism of carbon nanotubes previously described.

An analysis of the wear scar on the flat after the friction test indicates the presence of yttrium oxide, Y_2O_3, and especially the presence of nickel in the metal form (853 eV) and oxide forms, NiO and Ni_2O_3 (856 and 861 eV) (Figure 3.56). This analysis indicates that the catalytic particles, which were initially protected inside carbon structures, react in the contact during the test of friction, because the nickel, located in the core of nanoparticles, is oxidized during friction. Nickel nanoparticles prepared by a chemical route and dispersed in a 500 NS base oil were also found to have interesting tribological properties [116]. According to these results, it might be assumed that the presence of nickel has no beneficial effect on the tribological properties of carbon nanotubes. Thus, a purification of carbon nanotubes may not be necessary for future tribological applications.

In order to understand whether the nature of catalytic particles has an influence on the tribological results of carbon nanotubes, experiments were recently performed with nanotubes containing Co catalytic particles. These nanotubes were synthesized in the same conditions as those containing Ni-Y catalytic particles in order to be sure that only the composition of the catalytic particles changed. Figure 3.57 shows the results obtained with these nanotubes at three contact pressures (0.83, 1.12 and 1.42 GPa). Friction coefficients obtained lie

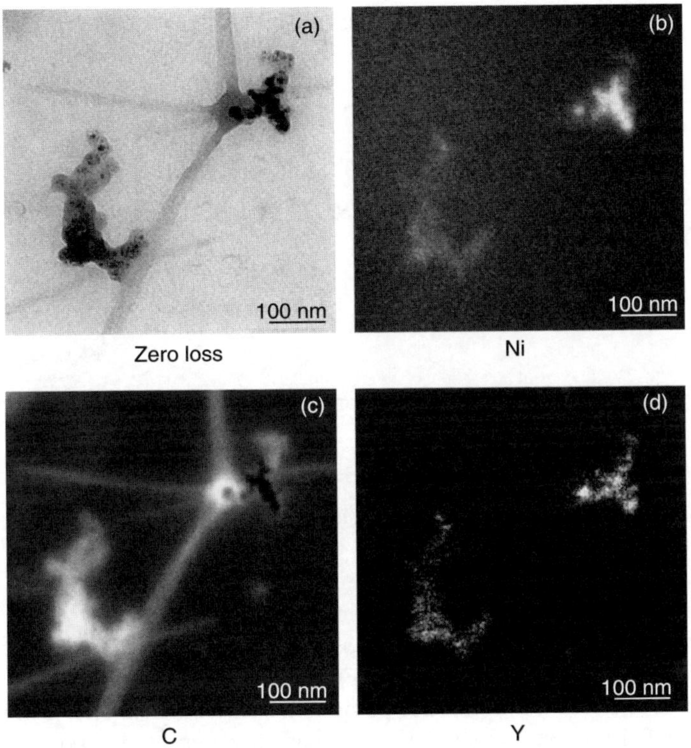

Figure 3.54 (a) TEM image of Ni/Y-SWNTs and (b) the corresponding chemical mappings of Ni, (c) C and (d) Y. These images show the presence of catalytic particles made of Ni and Y at the extremity of a nanotube and in the amorphous carbon phase

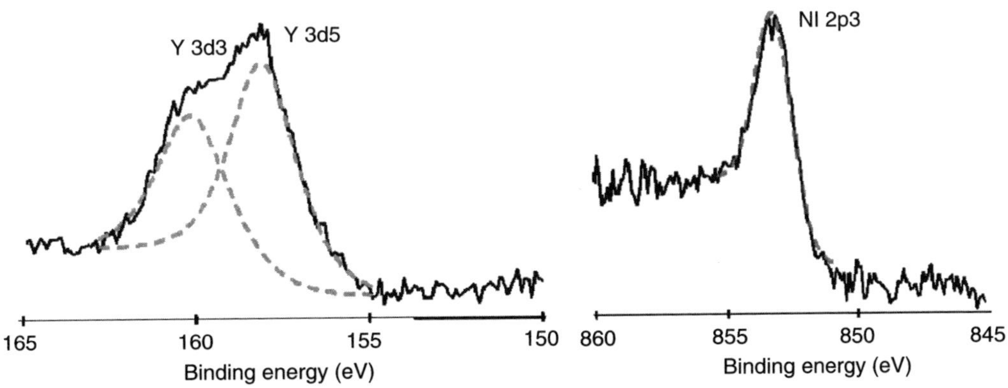

Figure 3.55 XPS analysis of Ni-Y/SWNTs indicating that yttrium is in the oxide form and nickel is only metallic

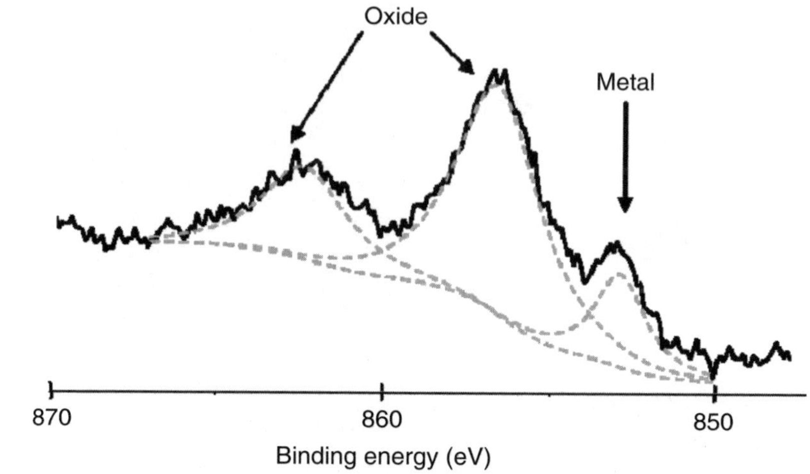

Figure 3.56 Fit of the Ni 2p3/2 of the XPS analysis performed on the wear track after the friction test with Ni/Y-SWNTs

between 0.08 and 0.1, while friction coefficients obtained with Ni/Y-SWNTs, tested in the same conditions, lie between 0.06 and 0.08 (Figure 3.48). From these recent results, it may be assumed that the nature of the catalytic particles has an influence on the tribological properties of an SWNT. More experiments are necessary to examine this point and analyses of the worn surfaces obtained with Co-SWNTs are envisaged to understand their lubrication mechanism.

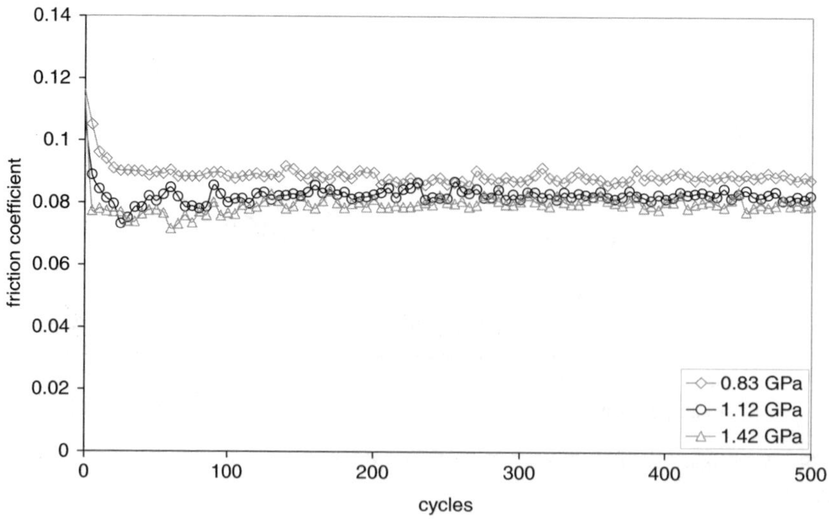

Figure 3.57 Contact pressure effect on the friction coefficient for a 1wt % Co-SWNT dispersion

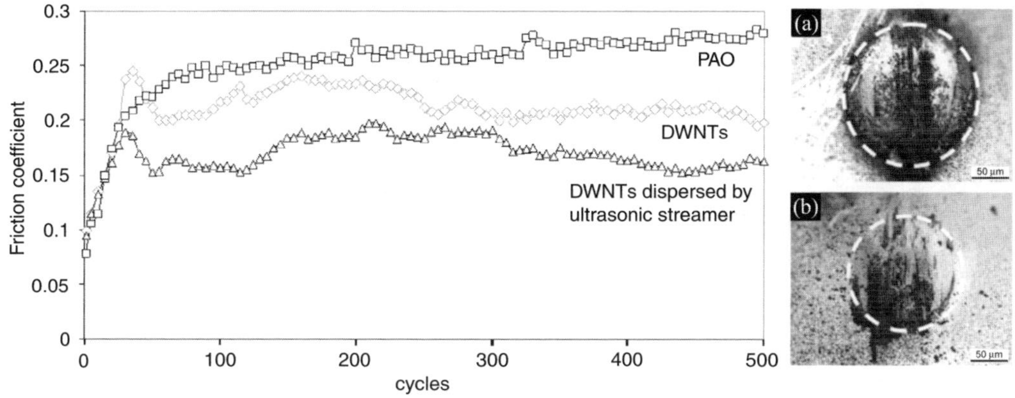

Figure 3.58 Friction coefficient and wear obtained with two dispersions of DWNTs at 1 wt % in PAO realized by using (a) an ultrasonic bath or (b) an ultrasonic streamer

3.5.2 DWNTs

Addition of DWNTs in PAO base oil does not cause a reduction of wear, and leads to a very weak reduction of the friction coefficient (Figure 3.58). These results are explained by a nonuniform dispersion of the nanotubes, which remain very intermingled. Because of their length of about several hundred micrometres, their dispersion in oil is very difficult. To obtain a better dispersion of nanotubes in oil, an ultrasonic streamer was used instead of an ultrasonic bath because a good agitation is obtained with an ultrasonic streamer. This technique improves the results obtained, but results are not as good as those obtained with single-walled nanotubes.

3.5.3 MWNTs

In the case of multiwalled nanotubes (MWNTs), a very weak concentration (0.1 %) is sufficient to obtain interesting tribological properties (Figure 3.59).

The contact pressure has an important influence on the friction coefficient (Figure 3.60). At 0.83 GPa, friction is relatively high (0.15). At a contact pressure of 1.12 GPa or higher, friction becomes low, reaching a value lower than 0.06 for contact pressures of 1.42 and 1.72 GPa. Friction coefficients at these pressures are lower than those observed in the case of Ni/Y-SWNTs tested under the same conditions.

Figure 3.61 presents wear scars observed on the pins after friction tests at various contact pressures compared with those obtained with PAO only. Addition of MWNTs to PAO leads to a reduction of wear at every contact pressure tested. This shows the good antiwear properties of MWNTs. Even when the friction coefficient is high, wear is low.

As in the case of single-walled nanotubes, a comparison of the tribological performances of multiwalled nanotubes with catalytic particles or without catalytic particles was performed. Figure 3.62 presents results obtained with MWNTs and Fe-MWNTs. The reduction of friction with Fe-MWNTs is very low compared with MWNTs. In the same way, a very low reduction of wear occurs with Fe-MWNTs (Figure 3.63).

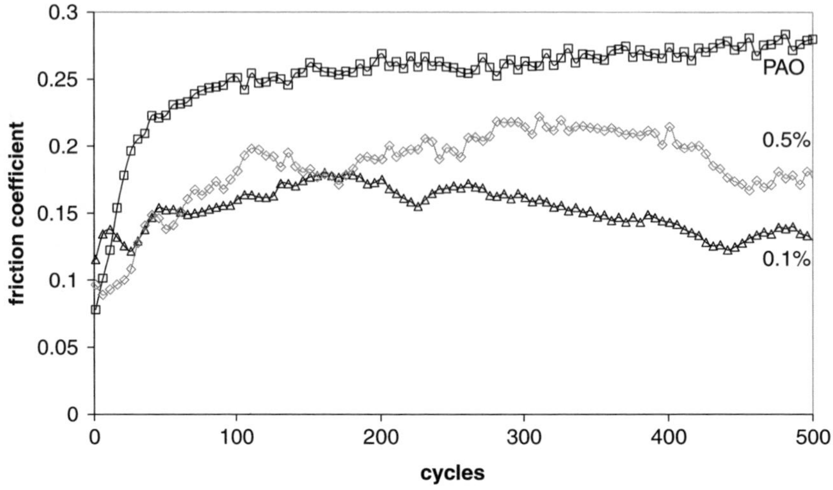

Figure 3.59 MWNT concentration effect on the friction coefficient at a contact pressure of 0.83 GPa. An optimal concentration of 0.1 wt % is determined

The comparison of the tribological performances of these two samples proves the necessity to eliminate iron catalytic particles of nanotubes samples. Indeed, contrary to nickel catalytic particles, which have a benefit effect on the tribological properties of single-walled nanotubes, iron catalytic particles seem to be less effective. To complete this study, it would be interesting to test multiwalled nanotubes containing Ni nanoparticles.

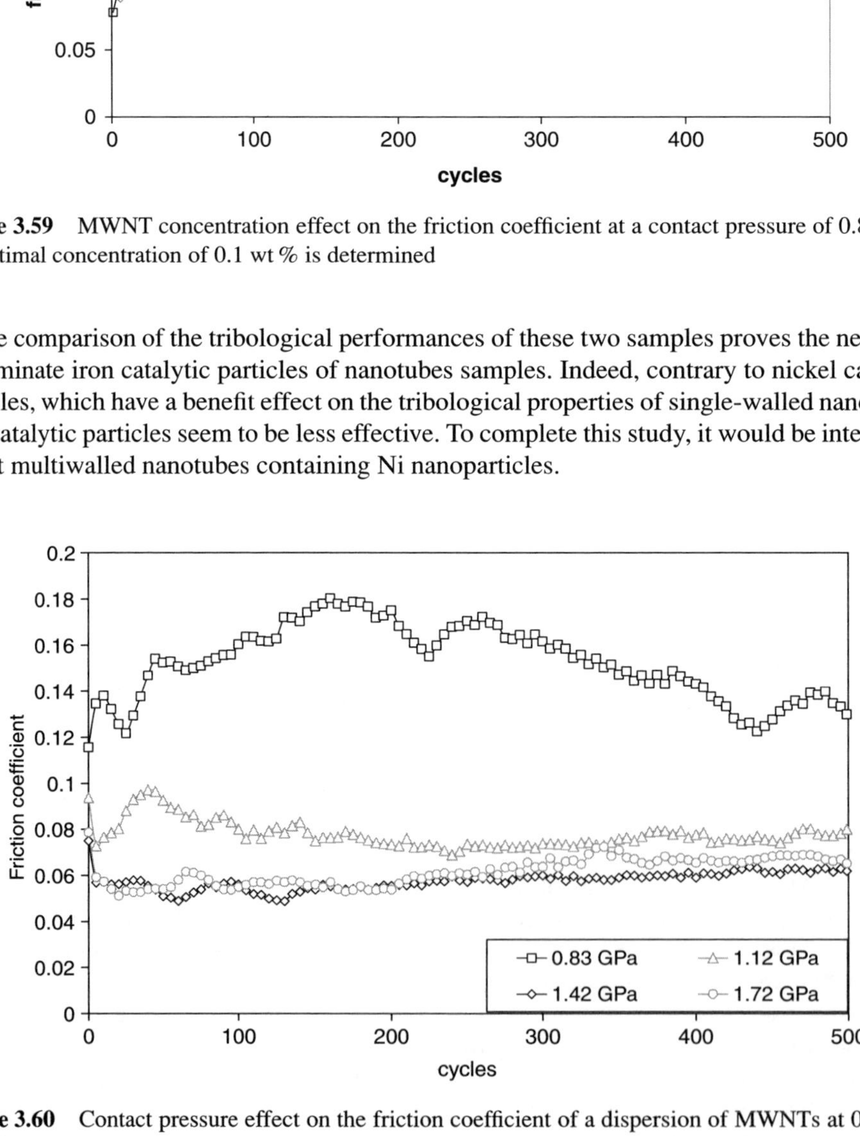

Figure 3.60 Contact pressure effect on the friction coefficient of a dispersion of MWNTs at 0.1 wt % in PAO

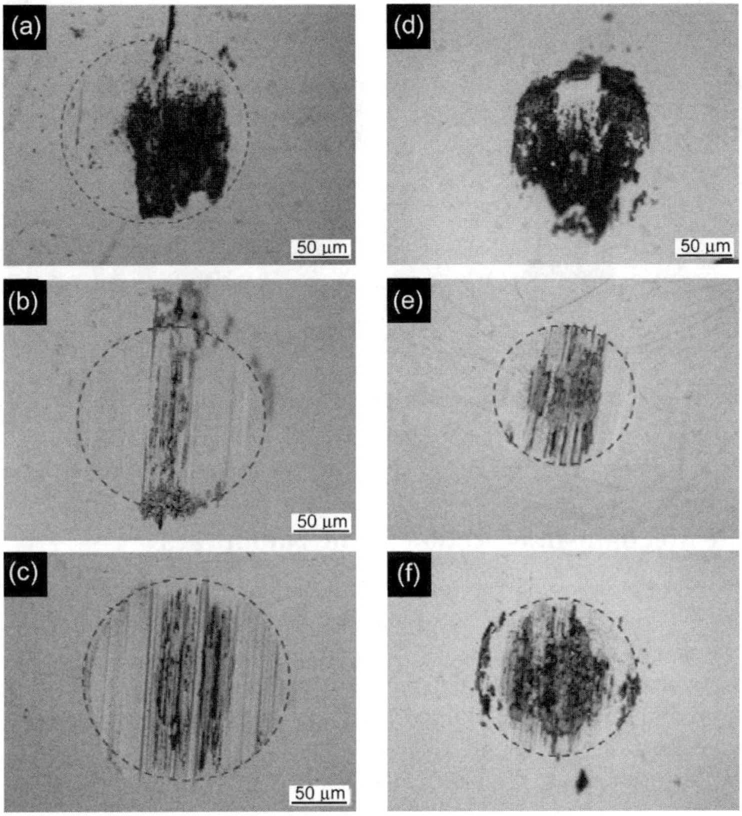

Figure 3.61 Wear scars observed with pure PAO (on the left) and PAO + 0.1wt % (on the right) at different contact pressures: 0.83 GPa (a and d), 1.12 GPa (b and e) and 1.42 GPa (c and f)

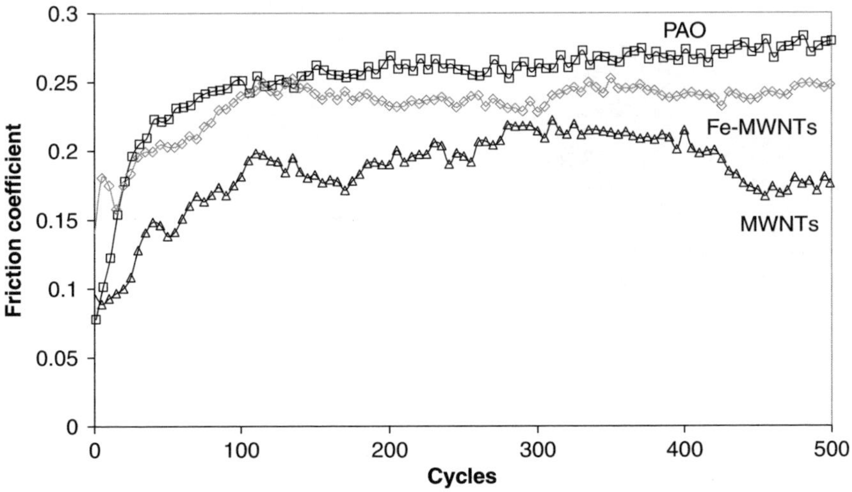

Figure 3.62 Comparison of friction coefficients obtained with Fe-MWNTs and MWNTs dispersed at 0.5 wt % in PAO ($v = 2.5$ mm/s, $P = 0.83$ GPa, $T = 20\,^{\circ}$C)

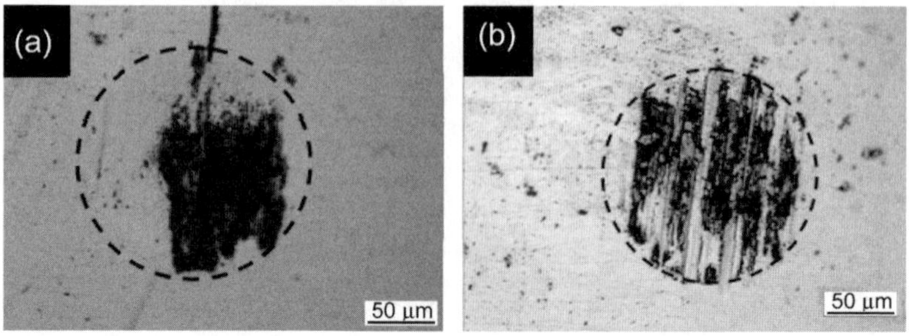

Figure 3.63 Wear scar observed on pins after friction with (a) pure PAO and (b) PAO + 0.5 wt % Fe-MWNTs. The addition of Fe-MWNTs leads to a very low wear reduction

3.6 Possible Mechanism of Action of the Nanotubes

Transmission electron microscopy performed on wear particles collected on the flat after the friction test reveals the presence of flake-like wear debris (Figure 3.64). EDS analyses of these flake-like debris show the presence of catalytic particles. An EELS analysis clarifies nature of the carbon constituting this debris. The peak characteristic of an sp^2 hybridization, visible on the spectra of graphite and nanotubes, does not exist on the spectrum of the wear particles (Figure 3.65). Therefore, this flake-like wear debris is made of amorphous carbon. An EELS analysis of the peak of inelastic losses provided the evaluation of the thickness to 60 nm [40].

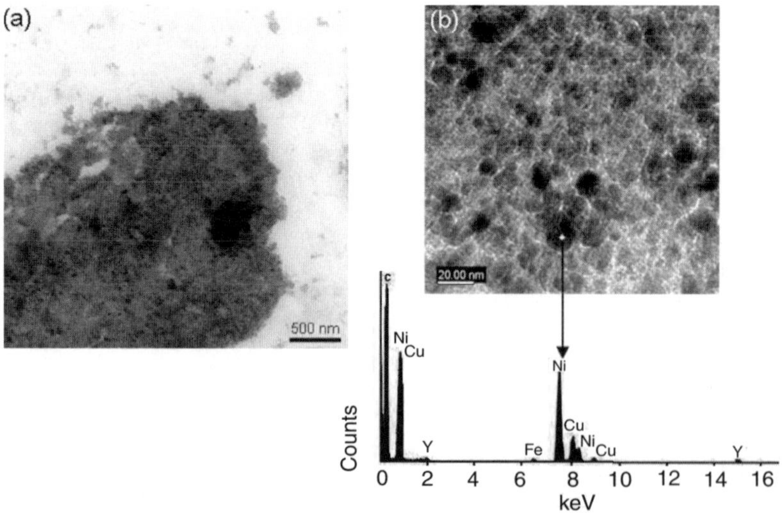

Figure 3.64 (a) TEM image of flake-like wear debris observed in wear particles after the friction test with Ni/Y-SWNTs. (b) TEM image on the wear-like carpet debris showing the presence of catalytic particles. An EDS analysis confirms the presence of catalytic particles

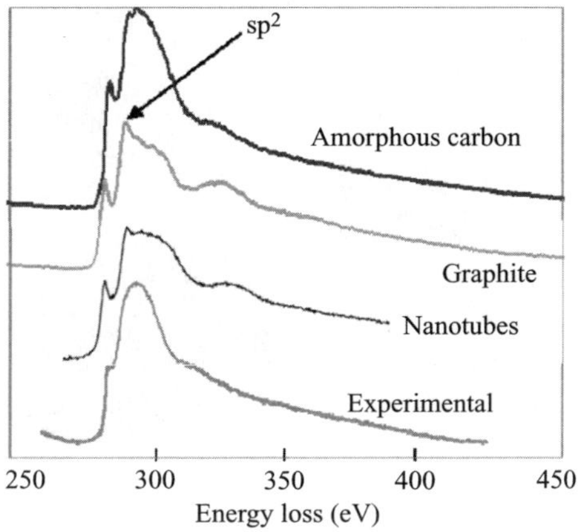

Figure 3.65 EELS analysis on the C K-edge of the wear particles. The disappearance of the sp^2 hybridization peak and the comparison with the spectrum of amorphous carbon show that they are composed of amorphous carbon

In situ Raman spectroscopy was performed during the friction test. An argon ion laser ($\lambda = 514.5$ nm) was used since with an He/Ne laser ($\lambda = 632.817$ nm) a large peak due to the sapphire flat was observed at 1400 cm^{-1}. This peak is due to the presence of chromium in the sapphire flat and prohibits analysis of the evolution of the carbon D band. Although the time of acquisition of Raman spectra was short (30 s), in order to follow the first minutes of friction, friction was stopped during acquisition (without opening the contact). These analyses show the disappearance of the RBMs and an amorphization of the carbon (Figure 3.66). This amorphization is determined by growth of the D band at 1400 cm^{-1}. These structural

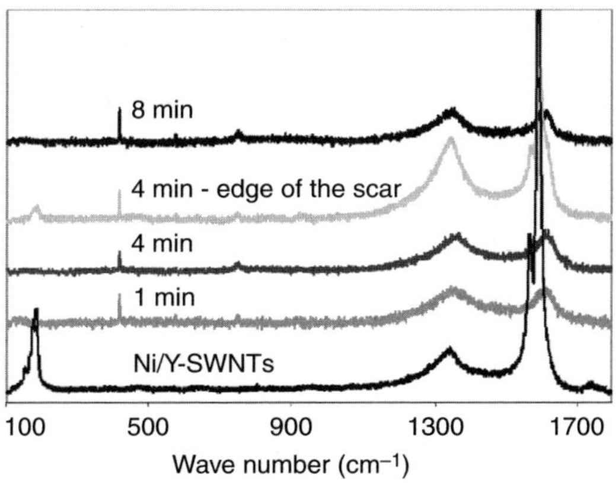

Figure 3.66 *In situ* Raman spectra obtained at different times during the friction test ($\lambda = 514.5$ nm). Spectra show the disappearance of the RBMs after 1 min of friction and an amorphization of the carbon in the contact area. At the edge of the contact area, the RBMs are present, indicating a structural modification only in the contact area

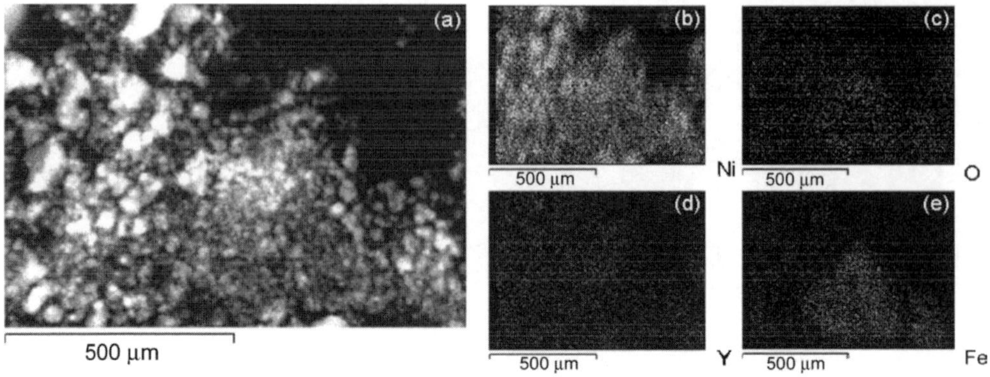

Figure 3.67 (a) TEM image obtained with an HAADF detector with corresponding chemical mapping of (b) Ni, (c) O, (d) Y and (e) Fe. These analyses show the presence of catalyst and iron oxide in these carpets

modifications are only observed inside the contact area, so it is clear that they only occur there.

A chemical mapping of flake-like wear debris confirms the presence of catalytic particles and also shows the presence of iron oxide (Figure 3.67). In short, this flake-like wear debris is made of amorphous carbon and contain catalytic particles and wear particles. The presence of wear particles in flake-like wear debris can explain the antiwear properties of Ni/Y-SWNTs. The few wear particles created during the friction test are actually trapped in flake-like debris. They are therefore no longer active in the contact, thus preventing important wear of the surfaces. This action mechanism can be compared to the action mechanism of phosphate glass created on surfaces by ZnDTP whose mode of action is 'to digest' the particles of wear [117]. The flake-like wear debris can also be compared to DLC doped with nickel. To confirm this hypothesis, it would be interesting to measure the mechanical properties (hardness) of this flake-like debris. DLC doped is known as hard material and is used to protect surfaces against abrasion [118, 119]. The formation of this flake-like wear debris could thus explain the low wear observed in the case of these nanotubes. EELS measurements and Oleshko's relation (2) can be used to estimate the hardness of the wear particles. Figure 3.68 presents

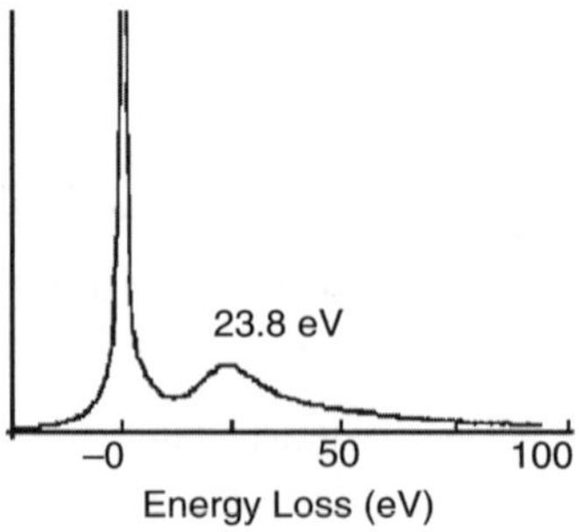

Figure 3.68 Plasmon peak obtained on the wear particle (Figure 3.64). A hardness of 9 GPa can be estimated from the value of $Ep = 23.8$ eV (relation (2))

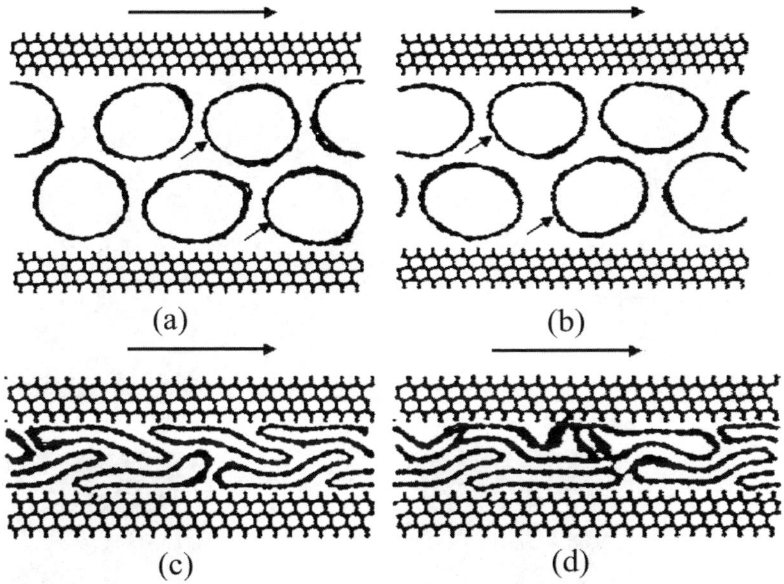

Figure 3.69 Simulation of carbon nanotube behaviour between two surfaces and under shear. At low pressure (a and b), moving nanotubes near the surface slide on the side opposite to the movement (see arrows). At high pressure (c and d), nanotubes are flattened and move like a tank belt [97]

the low-loss spectra obtained from the wear particle (Figure 3.64). The maximum of the low-loss energy peak is at 23.8 eV, which corresponds to a hardness of 9 GPa. Thus these particles are not very hard.

To explain the formation of flake-like debris, it is interesting to refer to the simulation works performed by Ni [97]. He simulated the behaviour of nanotubes horizontally aligned between two diamond surfaces and submitted to shear. At low pressure, the nanotubes slip in the direction opposite to the motion (Figure 3.69(a) and (b)). At high pressure, nanotubes are flattened and move like tank belt until meeting a nearby nanotube (Figure 3.69(c) and (d)). Thus nanotubes form flake-like debris made of flattened and intermingled nanotubes. Nevertheless, this flake-like debris is made of amorphous carbon, so it can be supposed that the nanotubes end up breaking under the effect of shearing.

Another hypothesis to explain the good friction-reducing properties of nanotubes is the formation of graphene single sheets inside the contact area, by unwinding of the nanotubes. Novoselov *et al.* have recently exfoliated graphite by a new cleavage technique and obtained a single crystal of graphene [120]. They showed that two-dimensional crystals remain stable. The hypothesis of the formation of single sheets of graphene in the contact is unfortunately difficult to confirm because they are very difficult to observe by TEM [121]. The only useful technique to study this type of structure is AFM.

Transmission electron microscopy of wear particles collected after the friction test with MWNTs at low contact pressure showed the presence of intact nanotubes (Figure 3.70). The wear particles observed are mainly composed of iron oxide.

The difference between multiwalled and single-walled nanotubes can be explained by the fact that multiwalled nanotubes are not flattened or that they are flattened at a higher load than

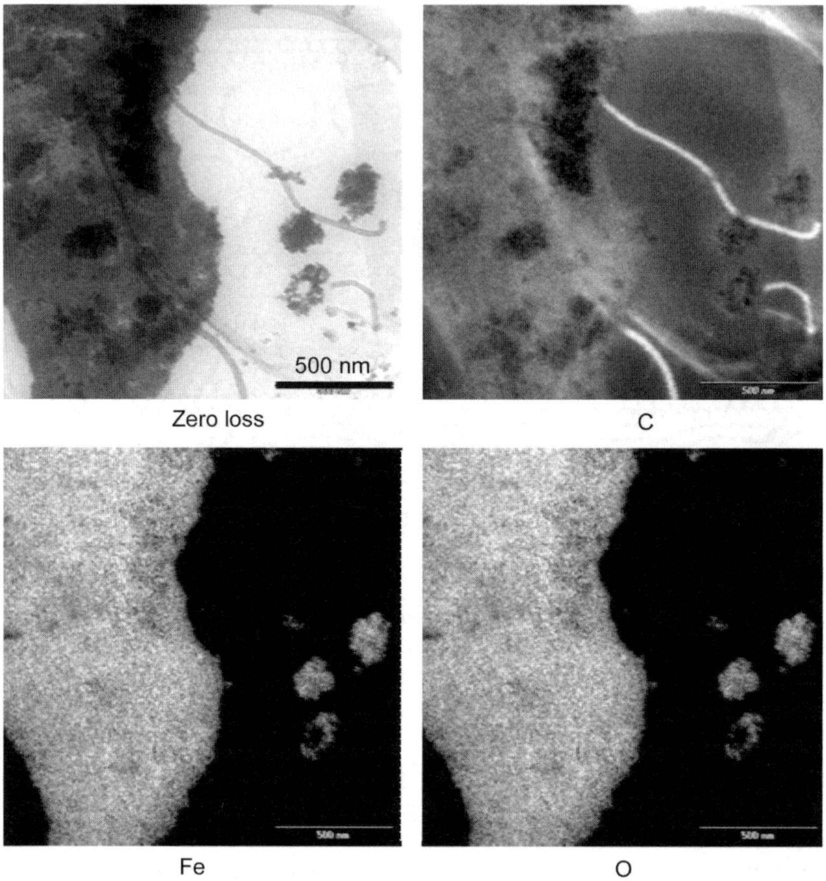

Figure 3.70 TEM image on wear particles collected after friction with MWNTs and the corresponding chemical mapping of C, Fe and O. Wear particles are mainly composed of iron oxide. Intact nanotubes are also observed

single-walled ones. A comparison of the effect of the pressure for Ni/Y-SWNTs and MWNTs indicates that the pressure is equal to 0.83 GPa with Ni/Y-SWNTs when a low friction coefficient is obtained, and 1.12 GPa for MWNTs. These contact pressures could be the pressures by which structural modifications of nanotubes occur.

The majority of the studied nanotubes presented interesting antiwear properties. Ni/Y-SWNTs also show friction-reducing properties. The presence of nickel in these nanotubes can explain these friction-reducing properties, even if the lubrication mechanism is not yet known. Dispersion of nanotubes in oil also seems to have an influence on their tribological properties. Indeed, an improvement of the dispersion of DWNTs in oil leads to better tribological results. To study the dispersion of nanotubes in oil, 'wet STEM' (see Section 2.3.2.1) would be a very interesting technique. The first results obtained on dispersion of MWNTs (Figure 3.71) show the possibility of analysing carbon nanotubes in oil by STEM.

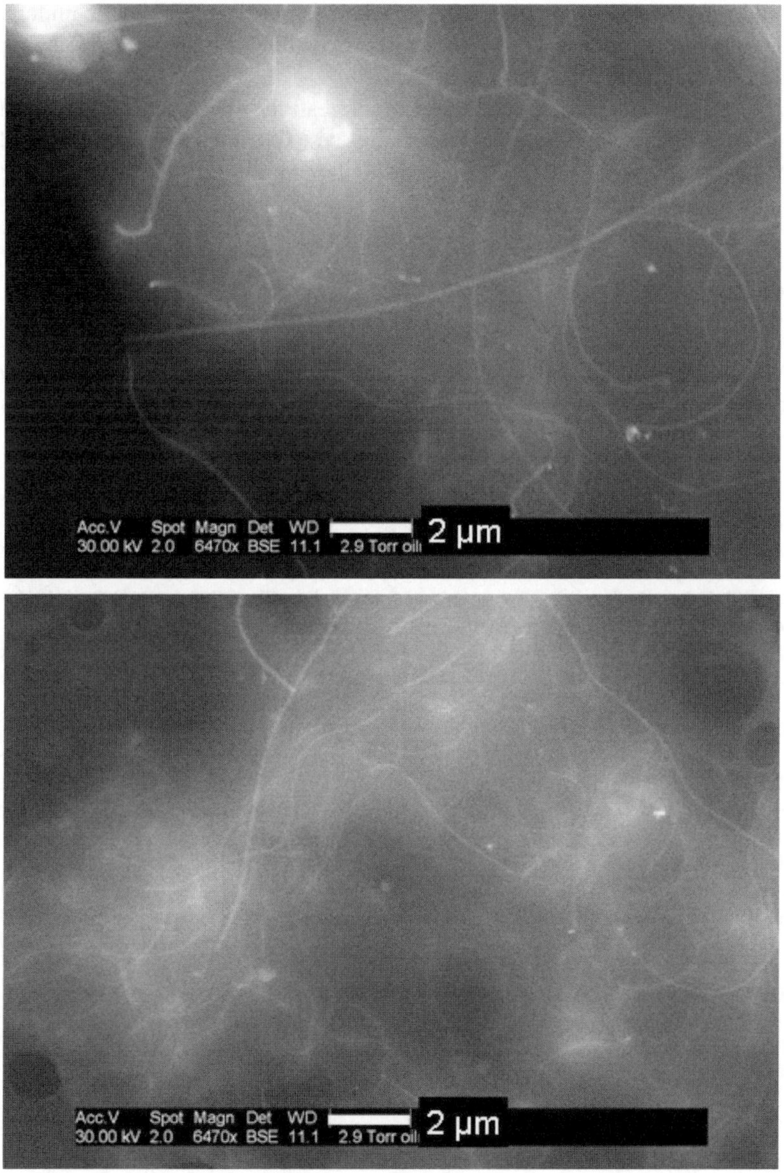

Figure 3.71 Wet STEM characterization of MWNTs in PAO base oil

3.7 Conclusion

Carbon nanoparticles are certainly of great interest as nanolubricants. They present excellent tribological properties. Their lubrication mechanism has not yet been completely elucidated, but they are mainly based on a structural modification of the nanoparticles. Thus, carbon nanoparticles present the same advantages that nanoparticles made on metal dichalcogenides: they are efficient even at ambient temperature. A better understanding of their lubrication mechanism will lead to an improvement in their use and their properties. Further studies are currently under consideration.

Other carbon nanoparticles may also present interesting tribological properties: nanohorns [122], peapods (nanotubes filled with C_{60}) [123] and C_{70}. Furthermore, the study of their tribological properties may also lead to a better understanding of the lubrication mechanism of carbon nanotubes and carbon onions.

Acknowledgements

The authors would like to thank Béatrice Vacher of the Ecole Centrale de Lyon for her support in the TEM experiments and Gilles Montagnac of the Ecole Normale Supérieure de Lyon for his help in the Raman spectroscopy measurements. Lucile Joly-Pottuz would like to thank Gilbert Thollet of INSA de Lyon for his help in the 'wet STEM' experiments.

References

1. Savage, R. H., Graphite lubrication, *Journal of Applied Physics*, **19** (1), 1948, 1–10.
2. Bowden, F. P., Tabor, D., *The Friction and Lubrication of Solids*, Clarendon Press, Oxford, 1950.
3. Spreadborough, J., The frictional behaviour of graphite, *Wear*, **5**(1), 1962, 18–30.
4. Tzeng, Y., Very low friction for diamond sliding on diamond in water, *Applied Physics Letters*, **63**, 1993, 3586.
5. Grill, A., Tribology of diamondlike carbon and related materials: an updated review, *Surface and Coatings Technology*, **94–95**, 1997, 507–513.
6. Erdemir, A. and Donnet, C., Tribology of diamond and diamond-like carbon films: an overview, in *Wear–Materials, Mechanisms and Practice* (ed. G. W. Stachowiak), John Wiley & Sons, Ltd, Chichester, 2005.
7. Kano, M., Yasuda, Y., Okamoto, Y., Mabuchi, Y., Hamada, T., Ueno, T., Ye, J., Konishi, S., Takeshima, S., Martin, J. M., De Barros Bouchet, M. I. and Le Mogne, T., Ultralow friction of DLC in presence of glycerol mono-oleate (GMO), *Tribology Letters*, **18** (2), 2005, 245–251.
8. Miyoshi, K., Novel carbons in tribology, *Tribology International*, **37** (11–12), 2004, 865–868.
9. Kroto, H. W., Heath, J. R., O'Brien, S. C., Curl, R. F. and Smalley, R. E., C_{60}: Buckminsterfullerene, *Nature*, **318**, 1985, 162–163.
10. Ugarte, D., Curling and closure of graphitic networks under electron-beam irradiation, *Nature*, **359**, 1992, 707–709.
11. Ugarte, D., Onion-like graphitic particles, *Carbon*, **33** (7), 1995, 989–993.
12. Kuznetsov, V. L., Chuvilin, A. L., Butenko, Y. V., Mal'kov, I. Y. and Titov, V. M., Onion-like carbon from ultra-disperse diamond, *Chemical Physics Letters*, **222**, 1994, 343–348.
13. Ozawa, M., Goto, H., Kusunoki, M. and Osawa, E., Continuously growing spiral carbon nanoparticles as the intermediates in the formation of fullerenes and nanoonions, *The Journal of Physical Chemistry B*, **106** (29), 2002, 7135–7138.
14. Qin, L. C. and Iijima, S., Onion-like graphitic particles produced from diamond, *Chemical Physics Letters*, **262**, 1996, 252–258.
15. Roddatis, V. V., Kuznetsov, V. L., Butenko, Y. V., Su, D. S. and Schlögl, R., Trasnformation of diamond nanoparticles into carbon onions under electron irradiation, *Physical Chemistry Chemical Physics*, **4**, 2002, 1964–1967.
16. Mykhaylyk, O. O., Solonin, Y. M., Batchelder, D. N. and Brydson, R., Transformation of nanodiamond into carbon onions: a comparative study by high-resolution transmission electron microscopy, electron energy-loss spectroscopy, X-ray diffraction, small-angle X-ray scattering, and ultraviolet Raman spectroscopy, *Journal of Applied Physics*, **97**, 2005, 074302.
17. Hiraki, J., Mori, H., Taguchi, E., Yasuda, H., Kinoshita, H. and Ohmae, N., Transformation of diamond nanoparticles into onion-like carbon by electron irradiation studied directly inside an ultra-high vacuum transmision electron microscope, *Applied Physics Letters*, **86**, 2005, 223101.

18. Tomita, S., Burian, A., Dore, J. C., LeBolloch, D., Fujii, M. and Hayashi, S., Diamond nanoparticles to carbon onions transformation: X-ray diffraction studies, *Carbon*, **40** (9), 2002, 1469–1474.
19. Andersson, O. E., Prasad, B. L. V., Sato, H. and Enoki, T., Structure and electronic properties of graphite nanoparticles, *Physical Review B*, **58** (24), 1998, 16387–16395.
20. Kuznetsov, V. L., Zilberberg, I.L., Butenko, Y. V., Chuvilin, A. L. and Segall, B., Theoretical study of the formation of closed curved graphite-like stuctures during annealing of diamond surface, *Journal of Applied Physics*, **86** (2), 1999, 863–870.
21. Xu, B. S. and Tanaka, S. I., Formation of giant onion-like fullerenes under Al nanoparticles by electron irradiation, *Acta Metallurgica*, **46** (15), 1998, 5249–5257.
22. Oku, T., Schmid, G. and Suganuma, K., Formation of carbon onions with Pd clusters in a high-resolution electron microscope, *Journal of Material Chemistry*, **8** (9), 1998, 2113–2117.
23. Troiani, H. E., Camacho-Bragado, A., Armendariz, V., Gardea-Torresday, J. L. and Jose-Yacaman, M., Synthesis of carbon onions by gold nanoparticles and electron irradiation, *Chemistry of Materials*, **15** (5), 2003, 1029–1031.
24. Sano, N., Wang, H., Chhowalla, M., Alexandrou, I. and Amaratunga, G. A. J., Nanotechnology: synthesis of carbon 'onions' in water, *Nature*, **414**, 2001, 506–507.
25. Hu, J. J., Bultman, J. E. and Zabinski, J. S., Inorganic fullerene-like nanoparticules produced by arc discharge in water with potential lubricating ability, *Tribology Letters*, **17** (3), 2004, 543–546.
26. Du, J., Liu, Z., Li, Z., Han, B., Sun, Z. and Huang, Y., Carbon onions synthesized via thermal reduction of glycerin with magnesium, Materials Chemistry and *Physics*, **93** (1), 2005, 178–180.
27. He, C., Zhao, N., Du, X., Shi, C., Ding, J., Li, J. and Li, Y., Low-temperature synthesis of carbon onions by chemical vapor deposition using a nickel catalyst supported on aluminium, *Scripta Materialia*, **54**, 2006, 689–693.
28. He, C., Zhao, N., Shi, C., Du, X., Li, J., Cui, L. and He, F., Carbon onion growth enhanced by nitrogen incorporation, *Scripta Materialia*, **54**, 2006, 1739–1743.
29. Wang, X., Xu, B., Liu, X., Guo, J. and Ichinose, H., Synthesis of Fe-included onion-like fullerenes by chemical vapor deposition, *Diamond and Related Materials*, **15**, 2006, 147–150.
30. Cabioc'h, T., Thune, E., Riviere, J. P., Camelio, S., Girard, J. C., Guérin, P., Jaouen, M., Henrard, L. and Lambin, P., Structure and properties of carbon onion layers onto various substrates, *Journal of Applied Physics*, **91** (3), 2002, 1560–1567.
31. Hirata, A., Igarashi, M. and Kaito, T., Study of solid lubricant properties of carbon onions produced by heat treatment of diamond clusters or particles, *Tribology International*, **37** (11–12), 2004, 899–905.
32. Street, K. W., Marchetti, M., Wal, R. L. V. and Tomasek, A. J., Evaluation of the tribological behavior of nano-onions in Krytox 143B, *Tribology Letters*, **16** (1–2), 2004, 143–149.
33. Oleshko, V. P., Murayama, M. and Howe, J. M., Use of plasmon spectroscopy to evaluate the mechanical properties of materials at the nanoscale, *Microscopy and Microanalysis*, **8**, 2002, 350–364.
34. Monthioux, M., Soutric, F. and Serin, V., Recurrent correlation between the electron energy loss spectra and mechanical properties for carbon fibers, *Carbon*, **35**, 1997, 1660–1664.
35. Kuzuo, R., Terauchi, M., Tanaka, M. and Saito, Y., Electron energy loss spectra of single-shell carbon nanotubes, *Japan Journal of Applied Physics*, **33**, 1994, L1316–1319.
36. Ajayan, P. M., Iijima, S. and Ichihashi, T., Electron-energy-loss spectroscopy of carbon nanometer-size tubes, *Physical Review B*, **47** (11), 1993, 6859–6862.
37. Sohmen, E., Fink, J. and Krätschmer, W., Electron energy-loss spectroscopy studies on C_{60} and C_{70} fullerite, *Zeitschrift für Physik B: Condensed Matter*, **86**, 1992, 87–92.
38. Tomita, S., Fujii, M., Hayashi, S. and Yamamoto, K., Electron energy-loss spectroscopy of carbon onions, *Chemical Physics Letters*, **305**, 1999, 225–229.
39. Redlich, P., Banhart, F., Lyutovich, Y. and Ajayan, P. M., EELS study of the irradiation-induced compression of carbon onions and their transformation to diamond, *Carbon*, **5–6**, 1998, 561–563.
40. Egerton, R.F., *Electron Energy-Loss Spectroscopy*, Plenum Press, 1986, p.410.
41. Morar, J. F., Himpsel, F. J., Hollinger, G., Hugues, G. and Jordan, J. L., Observation of a C-1s core exciton in diamond, *Physical Review B*, **54** (17), 1985, 1960–1963.
42. Jackson, K. A., and Pederson, M. R., New theoretical model for the diamond 1s core exciton, *Physical Review B*, **67** (18), 1991, 2521–2524.
43. Batson, P. E., Carbon 1s near-edge-absorption fine structure in graphite, P*hysical Review B*, **48** (4), 1993, 2608–2610.

44. Berger, S. D., McKenzie, D. R. and Martin, P. J., EELS analysis of vacuum arc-deposited diamond-like films, *Philosophical Magazine Letters*, **57**,1988, 285–290.
45. Ponsonnet, L., Donnet, C., Varlot, K., Martin, J. M., Grill, A. and Patel, V., EELS analysis of hydrogenated diamond-like carbon films, *Thin Solid Films*, **319**, 1998, 97–100.
46. Cabioc'h, T., Girard, J. C., Jaouen, M., Denanot M. F. and Hug, G., Carbon onions thin film formation and characterization, *Europhysics Letters*, **38**(6), 1997, 471–475.
47. Yannouleas, C., Bogachek, E. N. and Landman, U., Collective excitations of multishell carbon microstructures: multishell fullerenes and coaxial nanotubes, *Physical Review B*, **53** (15), 1996, 10225–10236.
48. Reiner, L., *Energy-Filtering Transmission Electron Microscopy*, Springer-Verlag, New York, 1995.
49. Roy, D., Chhowalla, M., Wang, H., Sano, N., Alexandrou, I., Clyne, T. W. and Amarantunga, G. A. J., Characterisation of carbon nano-onions using Raman spectroscopy, *Chemical Physics Letters*, **373** (1–2), 2003, 52–56.
50. Tomita, T., Formation and physical properties of novel carbonaceous nano-materials, PhD Thesis, Kobe University, 2002.
51. Gilkes, K. W. R., Sands, H. S., Batchelder, D. N., Milne, W. I. and Robertson, J., Direct observation of sp^3 bonding in tetrahedral amorphous carbon UV Raman spectroscopy, *Journal of Non-Crystalline Solids*, **227–230**, 1998, 612–616.
52. Ferrari, A. C. and Robertson, J., Resonant Raman spectroscopy of disordered, amorphous, and diamondlike carbon, *Physical Review B*, **64**, 2001, 075414.
53. Sun, Z., Shi, J. R., Tay, B. K. and Lau, S. P., UV Raman characteristics of nanocrystalline diamond films with different grain size, *Diamond and Related Materials*, **9**, 2000, 1979–1983.
54. Nemanich, R. J., Solin, S. A. and Martin, R. M., Light scattering study of boron nitride microcrystals, *Physical Review B*, **23** (12), 1981, 6348–6356.
55. Richter, H., Wang, Z. P. and Ley, L., The one phonon Raman spectrum in microcrystalline silicon, *Solid State Communications*, **39**, 1981, 625–629.
56. Yoshikawa, M., Mori, Y., Obata, H., Maegawa, M., Katagiri, G., Ishida, H. and Ishitani, A., Raman scattering from nanometer-sized diamond, *Applied Physics Letters*, **67**, 1995, 694–696.
57. Gupta, B. K. and Bhushan, B., Fullerene particles as an additive to liquid lubricants and greases for low friction and wear, *Lubrication Engineering*, **50** (7), 1994, 524–528.
58. Abe, M., Kataura, H., Kira, H., Kodama, T., Suzuki, S., Achiba, Y., Kato, K., Takata, M., Fujiwara, A., Matsuda, K. and Maniwa, Y., Structural transformation from single-wall to double-wall carbon nanotube bundles, *Physical Review B*, **68**, 2003, 041405.
59. Matsumoto, N., Joly-Pottuz, L., Kinoshita, H. and Ohmae, N., Application of onion-like carbon to micro and nanotribology, *Diamond and Related Materials*, **16**, 2007, 1227–1230.
60. Retzko, I. and Unger, W. E. S., Analysis of carbon materials by X-ray photoelectron spectroscopy and X-ray absorption spectroscopy, *Advanced Engineering Materials*, **5** (7), 2003, 519–522.
61. Terrones, H. and Terrones, M., The transformation of polyhedral particles into graphitic onions, *Journal of Physical and Chemical Solids*, **58** (11), 1997, 1789–1796.
62. Howe, J. M. and Oleshko, V. P., Application of valence electron energy-loss spectroscopy and plasmon energy mapping for determining material properties at the nanoscale, *Journal of Electron Microscopy*, **53** (4), 2004, 339–351.
63. Lee, E. H., Hembree, D. M., Rao, G. R. and Mansur, L. K., Raman scattering from ion-implanted diamond, graphite, and polymers, *Physical Review B*, **48** (21), 1993, 15540–15551.
64. Richter, A., Ries, R., Smith, R., Henkel, M. and Wolf, B., Nanoindetation of diamond, graphite and fullerene films, *Diamond and Related Materials*, **9** (2), 2000, 170–184.
65. Xu, S., Tay, B. K., Tan, H. S., Zhong, L. and Tu, Y. Q., Properties of carbon ion deposited tetrahedral amorphous carbon films as a function of ion energy, *Journal of Applied Physics*, **79** (9), 1996, 7234–7240.
66. Bersani, D., Lottici, P. P. and Montenero, A., Micro-Raman investigation of iron oxide films and powders produced by sol-gel syntheses, *Journal of Raman Spectroscopy*, **30**, 1999, 335–360.
67. Sousa, M. H., Tourinho, A. and Rubim, J. C., Use of Raman micro-spectroscopy in the characterization of $M^{II}Fe_2O_4$ (M = Fe, Zn) electric double layer ferrofluids, *Journal of Raman Spectroscopy*, **31**, 2000, 185–191.
68. Joly-Pottuz, L., Matsumoto, N., Kinoshita, H., Vacher, B., Belin, M., Montagnac, G., Martin, J. M. and Ohmae, N., Diamond-derived carbon onions as lubricant additives, *Tribology International*, **41** (2), 2008, 69–78.
69. Ito, K., Martin, J. M., Minfray, C. and Kato, K., Low-friction tribofilm formed by the reaction of ZDDP on iron oxide, *Tribology International*, **39**(12), 2006, 1538–1544.

70. Yuansheng J and Shenghua Li, Superlubricity of *in-situ* generated protective layer on worn metal surfaces in presence of $Mg_6Si_4O_{10}(OH)_8$, in *Superlubricity* (Eds A. Erdemir and J.M. Martin), Elsevier, 2007, pp.447–471.
71. Nakagawa, H., Kibi, S., Tagawa, M., Umeno, M. and Ohmae, N., Microtribological properties of ultrathin C_{60} films grown by molecular beam epitaxy, *Wear*, **238**, 2000, 45–47.
72. Iijima, S., Helical microtubules of graphitic carbon, *Nature*, **354**, 1991, 56–58.
73. Tans, S. J., Verschueren, A. R. M. and Dekker, C., Room-temperature transistor based on a single carbon nanotube, *Nature*, **393**, 1998, 49–52.
74. Martel, R., Schmidt, T., Shea, H. R., Hertel, T. and Avouris, P., Single- and multi walled carbon nanotube field-effect transistors, *Applied Physics Letters*, **73** (17), 1998, 2447–2449.
75. de Heer, W. A., Chatelain, A. and Ugarte, D., A carbon nanotube field-emission electron source, *Science*, **270**, 1999, 1179–1180.
76. Dillon, A. C., Jones, K. M., Bekkedahl, T. A., Kiong, C. K., Bethune, D. S. and Bethune, M. J., Storage of hydrogen in single-walled carbon nanotubes, *Nature*, **386**, 1997, 377–379.
77. Liu, C., Fan, Y. Y., Cong, H. T., Cheng, H. M. and Dresselhaus, M. S., Hydrogen storage in single-wall carbon nanotubes at room temperature, *Science*, **286**, 1999, 1127–1129.
78. Baughman, R. H., Cui, C., Zakhidov, A., Iqbal, Z., Barisci, J., Spinks, G., Wallace, G., Mazzoldi, De-Rossi, D., Rinzler, A., Jaschinski, O., Roth, S. and Kertesz, M., Carbon nanotube actuators, *Science*, **284**, 1999, 1340–1344.
79. Zhang, P., Lammert , P. E. and Crespi, V. H., Plastic deformations of carbon nanotubes, *Physical Review Letters*, **81**, 1998, 5346–5349.
80. Song, Y. S. and Youn, J. R., Influence of dispersion states of carbon nanotubes on physical properties of epoxy nanocomposites, *Carbon*, **43** (7), 2005, 1378–1385.
81. Yu, A., Hu, H., Bekyarova, E. Itkis, M. E., Gao, J., Zhao, B.,and Haddon, R. C., Incorporation of highly dispersed single-walled carbon nanotubes in a polyimide matrix, *Composites Science and Technology*, **66** (9), 2006, 1190–1197.
82. Yeh, M.-K., Tai, N.-H. and Liu, J.-H., Mechanical behavior of phenolic-based composites reinforced with multi walled carbon nanotubes, *Carbon*, **44**, 2006, 1–9.
83. Iijima, S. and Ichihashi, T., Single-shell carbon nanotubes of 1-nm diameter, *Nature*, **363**, 1993, 603–605.
84. Journet, C., Maser, W. K., Bernier, P., Loiseau, A., Lamy-de-la-Chapelle, M., Lefrant, S., Deniard, P., Lee, R. and Fisher, J. E., Large-scale production of single-walled carbon nanotubes by the electric-arc technique, *Nature*, **388**, 1997, 756–758.
85. Despres, J. F., Daguerre, E. and Lafdi, K., Flexibility of graphene layers in carbon nanotubes, *Carbon*, **33** (1), 1995, 87–89.
86. Laplaze, D., Bernier, P., Maser, W. K., Flamant, G., Guillard, T. and Loiseau, A., Carbon nanotubes: the solar approach, *Carbon*, **36** (5–6), 1998, 685–688.
87. Schittenhelm, H., Geohegan, D. B. and Jellison, G. E., Synthesis and characterization of single-wall carbon nanotube-amorphous diamond thin-film composites, *Applied Physics Letters*, **81**, 2001, 2097–2099.
88. Cai, H., Yan, F. and Xue, Q., Investigation of tribological properties of polyimide/carbon nanotube nanocomposites, *Materials Science and Engineering A*, **364**, 2003, 94–100.
89. Wang, L. Y., Tu, J. P., Chen, W. X., Wang, Y. C., Liu, X. K., Olk, C., Cheng, D. H. and Zhang, X. B., Friction and wear behavior of electroless Ni-based CNT composite coatings, *Wear*, **254** (12), 2003, 1289–1293.
90. Chen, W. X., Tu, J. P., Wang, L. Y., Gan, H. Y., Xu, Z. D. and Zhang, X. B., Tribological application of carbon nanotubes in a metal-based composite coating and composites, *Carbon*, **41** (2), 2003, 215–222.
91. Lim, D.-S., An, J.-W. and Lee, H. J., Effect of carbon nanotube addition on the tribological behavior of carbon/carbon composites, *Wear*, **252** (5–6), 2002, 512–517.
92. Lim, D. S., You, D. H., Choi, H. J., Lim, S. H., and Jang, H., Effect of CNT distribution on tribological behavior of alumina-CNT composites, *Wear*, **259** (1–6), 2005, 539–544.
93. Sun, J., Gao, L. and Jin, X., Reinforcement of alumina matrix with multi walled carbon nanotubes, *Ceramics International*, **31** (6), 2005, 893–896.
94. Ohmae, N., Martin, J. M. and Mori, S., *Microtribology and Nanotribology*, ASME Press, 2005.
95. Miyoshi, K., Street, K. W., Wal, R. L. V., Andrews, R. and Sayir, A., Solid lubrication by multi walled carbon nanotubes in air and in vacuum, *Tribology Letters*, **19** (3), 2005, 191–201.
96. Buldum, A. and Lu, J. P., Atomic scale sliding and rolling of carbon nanotubes, *Physical Review Letters*, **83** (24), 1999, 5050–5053.

97. Ni, B. and Sinnott, S. B., Tribological properties of carbon nanotubes bundles predicted from atomistic simulations, *Surface Science*, **487** (1–3), 2001, 87–96.
98. Mieno, T., Sakurai, A. and Inoue, H., J × B arc jet fullerene producer with a revolver type automatic material injector, *Fullerene Science Technology*, **4** (5), 1996, 913–923.
99. Mieno, T., Matsumoto, N. and Takeguchi, M., Efficient production of single walled carbon nanotubes by J × B gas-arc method, *Japanese Journal of Applied Physics*, **43** (12A), 2004, L1527–L1529.
100. Zhao, X., Inoue, S., Jinno, M., Suzuki, T. and Ando, Y. Macroscopic oriented web of single walled carbon nanotubes. *Chemical Physics Letters*, **373**, 2003, 266–271.
101. Bacsa, R. R., Laurent, C., Peigney, A., Bacsa, W. S., Vaugien, T. and Rousset, A., High specific surface area carbon nanotubes from catalytic chemical vapor deposition process, *Chemical Physics Letters*, **323**, 2000, 566–571.
102. Kinoshita, H., Kume, I., Sakai, H., Tagawa, M. and Ohmae, N., High growth rate of vertically aligned carbon nanotubes using a plasma shield in microwave plasma-enhanced chemical vapor deposition, *Carbon*, **42**, 2004, 2735–2777.
103. Bower, C., Zhu, W., Jin, S. and Zhou, O., Plasma-induced alignment of carbon nanotubes, *Applied Physics Letters*, **77** (6), 2000, 830–832.
104. Choi, Y. C., Shin, Y. M., Lee, Y. H., Lee, B. S., Park, G. S., Choi, W. B., Lee, N. S. and Kim, J. M., Controlling the diameter, growth rate, and density of vertically aligned carbon nanotubes synthesized by microwave plasma-enhanced chemical vapour deposition, *Applied Physics Letters*, **76** (17), 2000, 2367–2369.
105. Ducati, C., Alexandrou, I., Chhowalla, M., Amarantuga, G. A. J. and Robertson, J., Temperature selective growth of carbon nanotubes by chemical vapor deposition, *Journal of Applied Physics*, **92** (6), 2002, 3299–3303.
106. Rols, S., Structure et dynamique des nanotubes de carbone. Une étude par diffusion élastique et inélastique des neutrons (in French), PhD Thesis, University of Montpellier 2, 2000.
107. Anglaret, E., Etudes spectroscopiques des nanotubes: absorption optique et diffusion Raman (in French), CNRS Thematic School 'Ecole Nanotubes', Aussois, France, 2003.
108. Kataura, H., Kumazawa, Y., Maniwa, Y., Umezu, I., Suzuki, S., Ohtsuka, Y. and Achiba, Y., Optical properties of single walled carbon nanotubes, *Synthetic Metals*, **103**, 1999, 2555–2558.
109. Joly-Pottuz, L., Dassenoy, F., Vacher, B., Martin, J. M. and Mieno, T., Ultralow friction and wear behaviour of Ni/Y-based single walled carbon nanotubes (SWNTs), *Tribology International*, **37** (11–12), 2004, 1013–1018.
110. Gavillet, J., Loiseau, A., Ducastelle, F., Thair, S., Bernier, P., Stephan, O., Thibault, J. and Charlier, J. C., Microscopic mechanisms for the catalyst assisted growth of single walled carbon nanotubes, *Carbon*, **40** (10), 2002, 1649–1663.
111. Massaldi, T.B. (Ed.), *Binary of Phase Alloys Diagrams*, ASM International, 1990.
112. Gavillet, J., Loiseau, A., Journet, C., Willaime, F., Ducastelle, F. and Charlier, J.-C., Root-growth mechanism for single walled carbon nanotubes, *Physical Review Letters*, **87** (27), 2001, 275504.
113. Loiseau, A., Gavillet, J., Ducastelle, F., Thibault, J., Stephan, O., Bernier, P. and Thair, S., Nucleation and growth of SWNT: TEM studies of the role of the catalyst, *Compte Rendu Physique*, **4**, 2003, 975–991.
114. Gavillet, J., Thibault, J., Stephan, O., Amara, H., Loiseau, A., Bichara, C., Gaspard, J. P. and Ducastelle, F., Nucleation and growth of single walled nanotubes: the role of metallic catalysts, *Journal of Nanoscience and Nanotechnology*, **4** (4), 2004, 346–359.
115. Cau, M., Dorval, N., Cao, B., Attal-Trétout, B., Cochon, J. L., Loiseau, A., Farhat, S. and Scott, C. D., Spatial evolutions of Co and Ni atoms during single walled carbon nanotubes formation: measurements and modelling, *Journal of Nanoscience and Nanotechnology*, **6** (5), 2006, 1298–1308.
116. Qiu, S., Zhou, Z., Dong, J. and Chen, G., Preparation of Ni nanoparticles and evaluation of their tribological performance as potential additives in oils, *Journal of Tribology*, **123**, 2001, 441–443.
117. Martin, J. M., Antiwear mechanisms of zinc dithiophosphate: a chemical hardness approach, *Tribology Letters*, **6**, 1999, 1–8.
118. Chang, Y. Y., Wang, D. Y. and Wu, W. T., Catalysis effects of metal doping on wear properties of diamond-like carbon films deposited by a cathodic-arc activated deposition process, *Thin Solid Films*, **420**, 2002, 241–247.
119. Wei, Q., Narayan, R. J., Narayan, J., Sankar, J. and Sharma, A. K., (1998), Improvement of wear resistance of pulsed laser deposited diamond-like carbon films through incorporation of metals, *Materials Science and Engineering*, **B53**, 1998, 262–266.
120. Novoselov, K. S., Keim, A. K., Morozov, S. V., Jiang, D., Zhang, Y., Dubonos, S. V. Grigorieva, I. V. and Firsov, A. A., Electric field effect in atomically thin carbon films, *Science*, **306**, 2004, 666–669.

121. Horiuchi, S., Gotou, T., Fujiwara, M., Asaka, T., Yokosawa, T. and Matsui, Y., Single graphene sheet detected in a carbon nanofilm, *Applied Physics Letters*, **84** (13), 2004, 2403–2405.
122. Iijima, S., Yudasaka, M., Yamada, R., Bandow, S., Suenaga, K., Kokai, F. and Takahashi, K., Nano-aggregates of single walled graphitic carbon nano-horns, *Chemical Physics Letters*, **309**, 1999, 165–170.
123. Kataura, H., Maniwa, Y., Kodama, T., Kikuchi, K., Hirahara, K., Suenaga, K., Iijima, S., Suzuki, S., Achiba, Y. and Krätschmer, W., High-yield fullerene encapsulation in single walled carbon nanotubes, *Synthetic Metals*, **121**, 2001, 1195–1196.

4

Reverse Micelles and Encapsulated Nanoparticle Approaches

Jean Louis Mansot and Jean Michel Martin

4.1 Introduction

In the boundary lubrication regime, the antiwear and friction-reduction processes are associated to the build-up of a tribologic film resulting from the chemical reaction between oil additives and sliding surfaces in contact [1–3]. The case of zinc dialkyl dithiophosphate (ZDDP) additive has been extensively studied by analytical transmission electron microscopy (ATEM) and the morphology, nature, structure and tribologic properties of the tribochemical film were obtained [4–6]. The Figures 4.1 and 4.2 summarize the main results.

Complementary analytical studies of the surface films using Auger electron spectroscopy (AES), X-ray photoelectron spectroscopy (XPS) and X-ray absorption near edges structures (XANES) [7, 8] lead to an accurate description of the antiwear film mainly composed of amorphous Fe/Zn poly(thio)phosphate embedding some nanocrystallites of ZnO and ZnS. The chemical reaction involved in the build-up of the tribo-film were recently interpreted with new theoretical approaches (hard and soft acid–base theory) [9, 10]. The action process of antiwear additives such as ZDDP presents two disadvantages:

- As shown on the wear rate and electrical contact resistance versus number of cycles diagram (Figure 4.1), the antiwear action of ZDDP requires an induction period for the build-up of the tribo-chemical film. During this period the wear rate is as high as additive free oil.
- The development and use of new materials for various mechanical applications presenting weak chemical reactivity with conventional (ZDDP) additives do not allow the build-up of the antiwear film.

Nanolubricants Edited by J. M. Martin and N. Ohmae

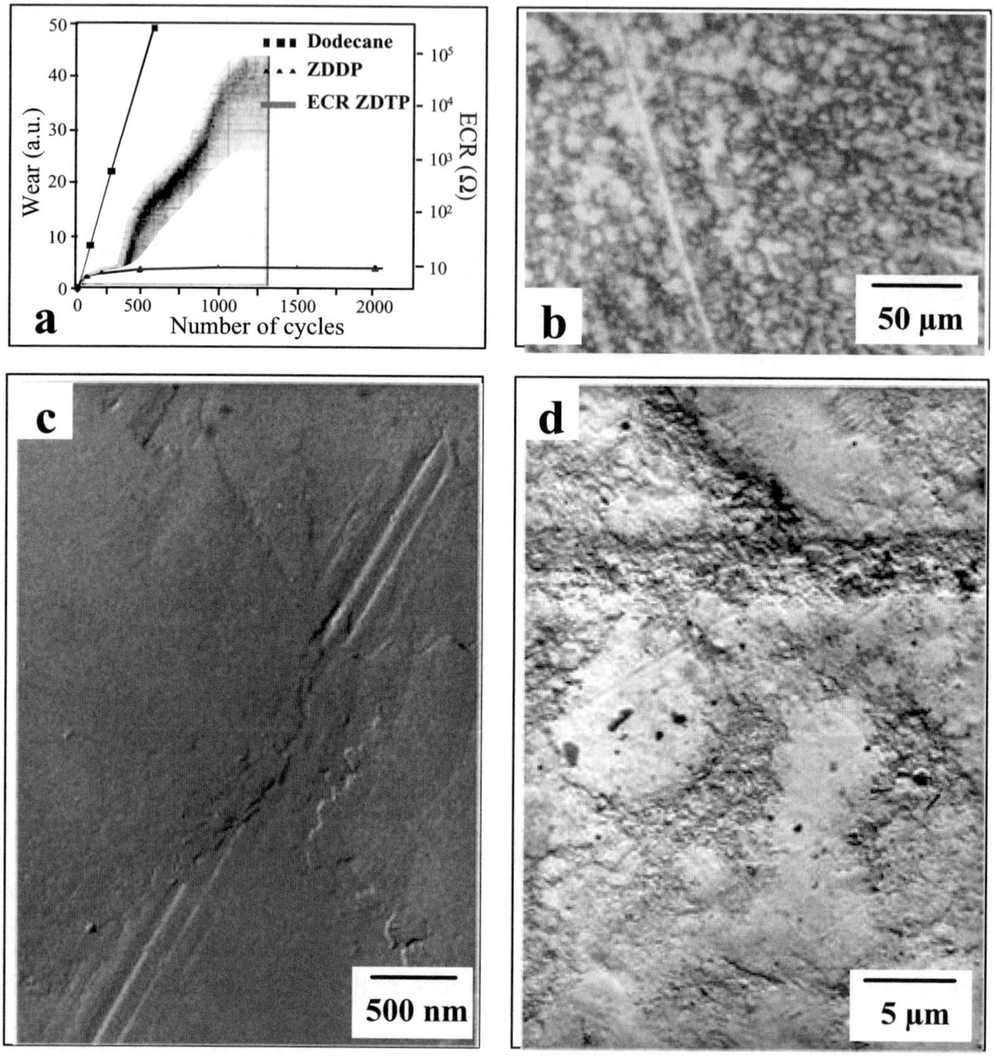

Figure 4.1 (a) Evolution of the electrical contact resistance (ECR) and wear rate as a function of the number of cycles (100C6 steel plane on cast iron plane test in the presence of 1 % ZDDP in dodecane solution at a temperature of 80 °C). It can be seen that the establishment of the antiwear film needs an induction period where the wear rate is as high as with pure oil base (dodecane). The build-up of the insulating film is shown by the increase of the ECR, which can be correlated to the decrease in the wear rate. (b) Optical microscopy image of the worn surface at the end of the test showing the presence of the insulating antiwear film (white and grey areas). (c) and (d) TEM micrographs recorded on a shadowed platinum carbon replica showing the morphology of the film with the presence of platelets on the surface, avoiding direct metal/metal contacts. These platelets (≈ 50 nm thick) correspond to the white areas seen in the optical micrograph.

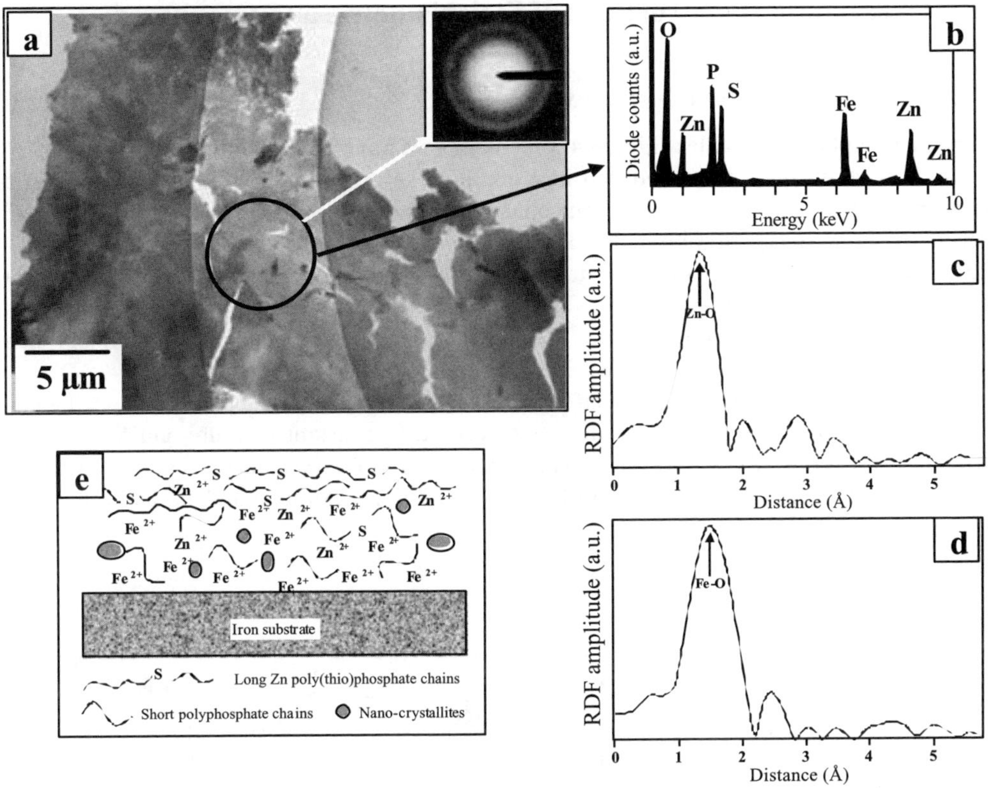

Figure 4.2 (a) TEM image recorded onto an anti-wear film particle and corresponding electron diffraction pattern. (b) X-ray emission spectrum (EDXS) analysis. The antiwear film is composed of iron and elements present in the ZDDP additive (O, P, S, Zn). (c) and (d) Fe and Zn radial distribution functions (RDFs) uncorrected from phase shift, extracted from EXAFS experiments on ZDDP antiwear film particles. The electron diffraction pattern and the presence of one major peak in each RDF corresponding to oxygen first-neighbours shell point out the amorphous structure of the antiwear film. (e) Schematic structure of the ZDDP tribo-film according to the multitechnique approaches [8]

In addition, the new constraining environmental European standards aiming to reduce the sulfur and phosphorus contents in lubricating oil, which could lead to the replacement of ZDDP antiwear additives.

This chapter is concerned with the development of an alternative way to build up the protective tribologic film between sliding surfaces that does not imply a chemical reaction of the additive with the substrates in contact. The approach is based on the use of nano-sized particles of tribologic film precursors consisting of reverse micelles or encapsulated mineral particles. First an overview of the various systems that could be used will be presented. Some examples of each case will then be deeply analysed (characterization of the colloidal system, tribologic properties, antiwear film structure and morphology and antiwear action mechanisms).

4.2 Overview of the Structures of Stoichiometric and Overbased Soap Additives

4.2.1 Dynamic Organic Micelles

When amphiphilic molecules (Figure 4.3) containing a polar (hydrophilic) group and a nonpolar (lypophilic) part are introduced in a nonpolar solvent (mineral oil for example) at a concentration superior to the critical micellar concentration (CMC) they will aggregate into reverse micelles in equilibrium with single molecules present in the solvent at the CMC [11–13]. The reverse micelles will associate in their cores with the polar groups and will present their nonpolar part to the nonpolar solvent. The molecular interactions responsible for aggregation are of a physical nature (van der Waals, Coulombian, hydrogen bonding etc.) and such micelles are dynamic and continuously exchanging molecules.

Depending on the solvent, molecular concentration, temperature and pressure, micelles can adopt various shapes and sizes. In the same solvent, at constant pressure and temperature from the lowest (just above the CMC) to the highest concentrations, micelles will evolve from a spherical shape to cylindrical, worm-like micelles (long flexible cylinders) and a lamellar assembly [12,13].

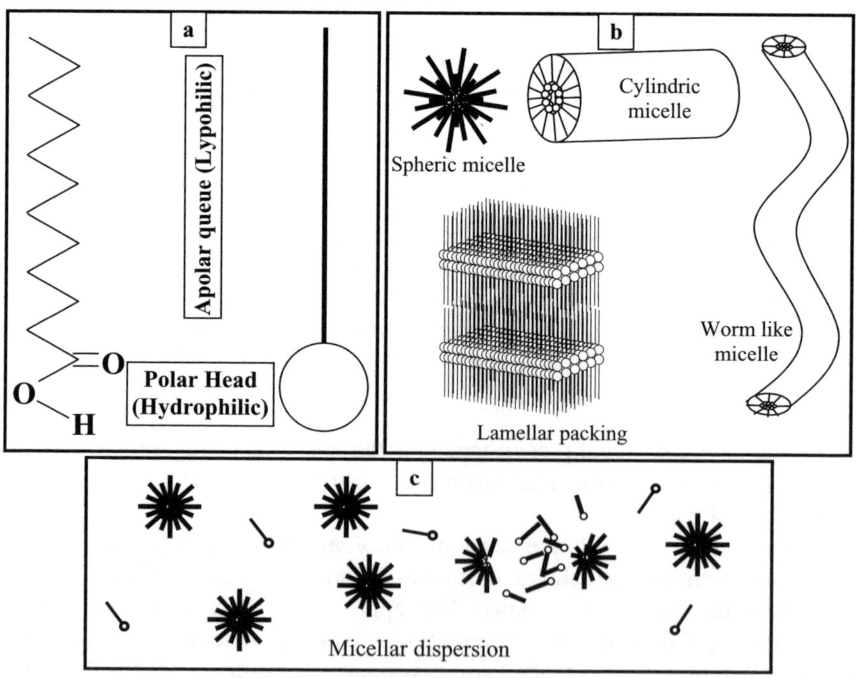

Figure 4.3 (a) Typical amphiphilic molecule and classical simplified representation. (b) Various types of molecular aggregation. (c) Schematic representation of the dynamic character of the micellar aggregates in equilibrium with single molecules dispersed at the CMC in the solvent

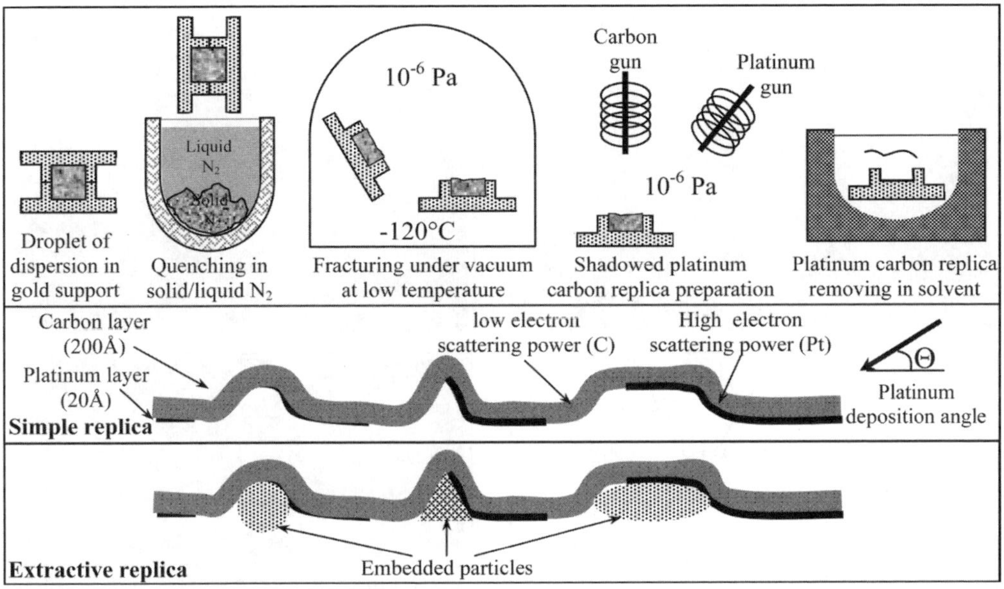

Figure 4.4 Freeze fracturing and etching technique (FFET) used to visualize micelles in their medium. After rapid freezing (quenching) into solid/liquid nitrogen to avoid calefaction, a small droplet of dispersion is then fractured under vacuum and a platinum shadowed replica is prepared. The areas coated by platinum will scatter the transmitted electron more than the Carbon-coated ones. The angle used for platinum deposition (shadowing angle Θ) permits the shape of the fracture to be revealed. The fracture propagates through minimum energy interfaces (solvent/particles, etc.) The replica can be cleaned with selective solvent to maintain in it some particles (extractive replica)

Various organic and organometallic compounds are subjected to build reverse micelles in nonpolar media: organic acids, organic esters, soaps and sequenced copolymers with alternating polar and nonpolar parts. Characterization of the CMC, the sizes, shapes, number of aggregations of such micelles can be achieved using quasi elastic light scattering (QELS). In some cases, transmission electron microscopy (TEM) coupled to a specific cryo-preparation method, the freeze fracturing and etching technique (FFET)[14], allows the visualization of micelle shape, size and dispersion in the continuous phase (Figure 4.4) [15].

The case of micelles of sequenced copolymer referred to as E1500 (Ugine Khulmann Trade Mark) is presented in Figure 4.5 [15–17]. In the 0.1 g/cm^3 E1500 in dodecane dispersion two phases coexist, spherical aggregates (diameters 6 to 8 nm) and lamellar packings ($\approx$ 6nm thickness) which can be induced by the freezing process. The complementary QELS studies [15–17] carried out an E1500 dispersion in the concentration range 10^{-4} g/cm^3 $< c <$ 10^{-1} g/cm^3 confirm the existence of 6 nm average diameter aggregates for $c > 10^{-3}$ g/cm^3. They reveal the existence of a CMC (10^{-3}g/cm^3) and the fact that micelles are on average constituted by three molecules. The analyses of particle interactions lead to a micelle structure consisting of a compact POPG (glycol polyoxipropylene) core ($\approx$ 3 nm diameter) surrounded by a soft shell ($\approx$1.5 nm thick) made of aliphatic chains.

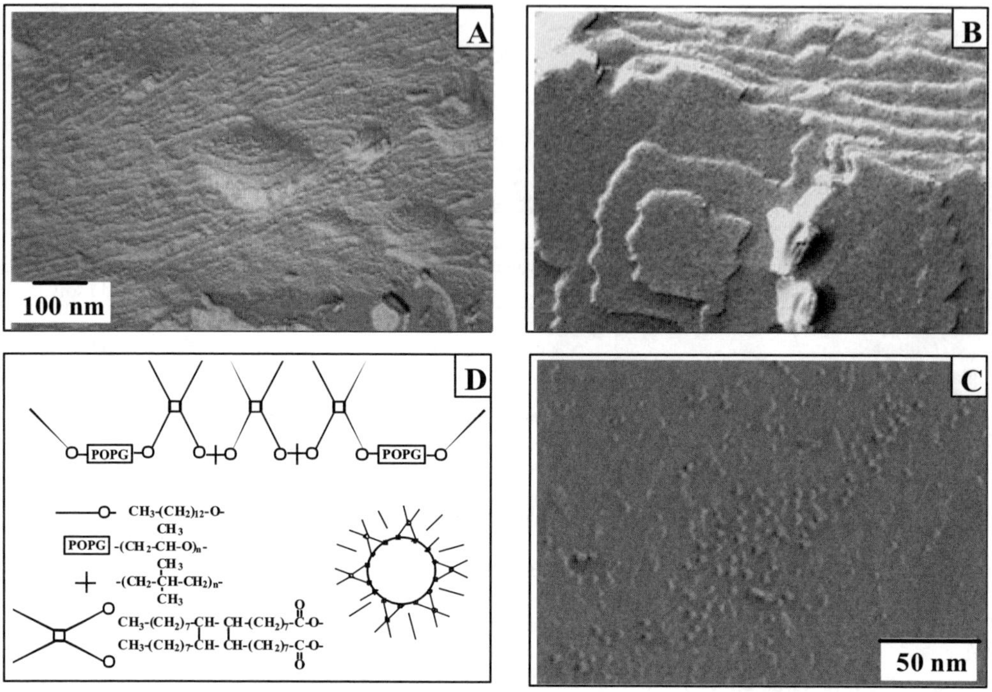

Figure 4.5 Transmission electron micrographs recorded on carbon platinum shadowed replica prepared by the freeze fracturing technique applied to the amphiphilic polymer dispersion in dodecane. (A) Aligned spherical micelles (10 % dispersion in dodecane). (B) Molecular lamellar packing induced by the preparation method (10 % dispersion in dodecane). (C) Well-dispersed micelles (1 % dispersion in dodecane). Their diameter of 6 nm corresponds to light scattering results. (D) Molecular formula of the studied sequenced copolymer (E1500 was used as the viscosity index improver additive). The average molecular mass (in weight) is 4300. According to the experimental results, an average number of three polymer molecules are associated in the micelles. The hydrophilic POPG (glycol polyoxipropylene) parts constitute the compact core

4.2.2 Dynamic Soap Micelles

As amphiphilic organic molecules, soap molecules, resulting from the reaction between organic acids and mineral bases (metal hydroxide, metal carbonate, etc), aggregate into reverse micelles when they are introduced into nonpolar media such as mineral oils or nonpolar solvents. The hydrophilic (mineral) part will agglomerate in the core of the micelles and the nonpolar queues will constitute the surrounding shell. Such assemblies are in dynamic equilibrium with single molecules dispersed at the CMC in the solvent.

Depending on the metal atom valence and on the synthesis process, various types of soap molecules can be obtained. In the case of stoichiometric additives, also called neutral soap, soap molecules syntheses will correspond to the following reactions [17, 18]:

Monovalent ions : $RCOOH + MeOH \rightarrow RCOOMe + H_2O$

Divalent ions: $2\ RCOOH + Me(OH)_2 \rightarrow (RCOO)_2\ Me + 2H_2O$

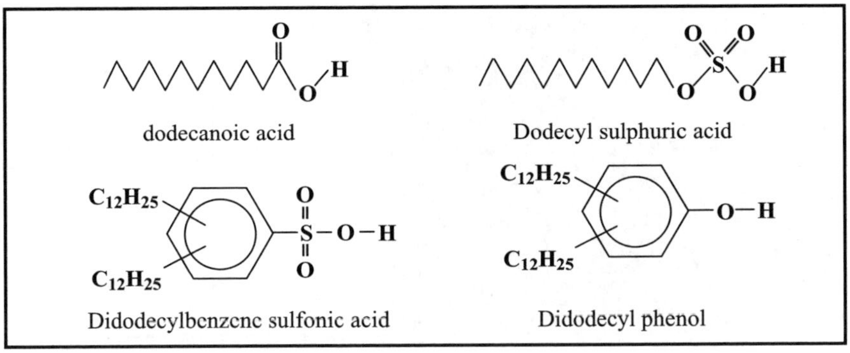

Figure 4.6 Formulae of four conventional organic compounds used for the synthesis of amphiphilic metal salts. In the case of alkyl aryl sulfonic acids and alkyl aryl phenols, the alkyl groups (propylene tetramer) are often ramified and their location on the aromatic ring depends on the synthesis process.

In the formulae, Me represents the metal atom. The organic acids RCOOH, where R is a hydrocarbon chain, can be replaced by alkyl sulfuric acids $ROSO_3H$), alkyl aryl sulfonic acids (RSO_3H), and alkyl phenols (ROH) (see Figure 4.6). In the case of divalent ions, nonstochiometric additives, also called overbased salts, can be obtained by the following reaction [19–21]:

$$2RCOOH + (1 + x)Me(OH)_2 + \underline{x}CO_2 \rightarrow (RCOO)_2Me, xMeCO_3 + (2 + x)H_2O$$

In the case of lead soaps, for $x = 1$ mass spectroscopy studies allows the following molecular formula to be determined [21]:

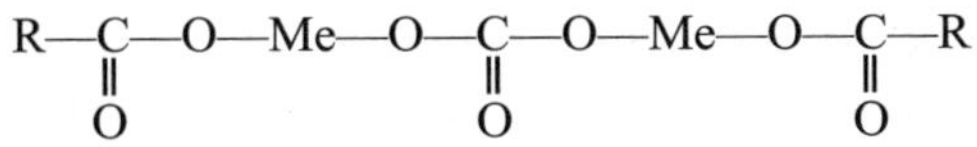

The micellar assemblies obtained in this case were studied by means of ATEM, small-angle X-ray Scattering (SAXS), QELS and extended X-ray absorption fine structures (EXAFS) [22–24]. The deduced micelle structure is shown in Figure 4.7(B) and (C).

The TEM micrograph recorded on aggregates deposited from a low concentration dispersion of the soap on a thin carbon grid is presented in Figure 4.7(A). The organized structure parameters of the single micelle (present in the liquid dispersion) and the structure obtained after coalescence of some micelles during the sample preparation process for TEM are deduced from the results of the multitechnique approach.

4.2.3 Encapsulated Nano-Sized Particles, also Called 'Overbased Reverse Micelles'

In this case the so-called 'reverse micelles' are static assemblies. The chemical processes [19, 20] lead to a mineral core mainly consisting of amorphous understoichiometric carbonate or borate surrounded by the amphiphilic molecules chemically bonded to the mineral core.

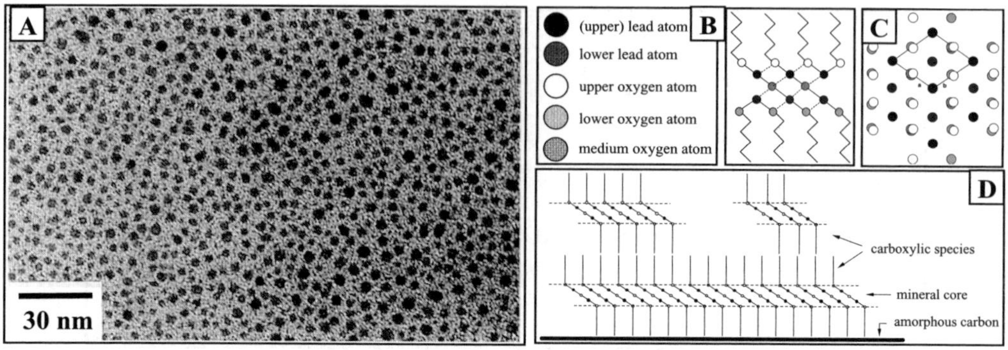

Figure 4.7 Overbased lead octanoate structure. The dynamic micelles (B) have a well-ordered core (C) (deduced from EXAFS) with a diameter of 1 nm (deduced from light scattering and cryo microscopy). Deposited on a carbon support (A), micelles will rearrange during the solvent evaporation and lead to lamellar aggregates (D)

Well-known examples of such aggregates are constituted by the overbased metal (calcium or magnesium) alkyl benzene sulfonates (OMeABS) used as corrosion inhibitor additives in lubricants for diesel engines [25]. Many techniques, such as small-angle neutron scattering (SANS) [26, 27], SAXS [28], chromatography/mass spectroscopy nuclear magnetic resonance (CP/MAS-NMR) [29], QELS [30], ATEM [31, 32] and EXAFS [33–35], and more recently XPS and time-of-flight secondary ion mass spectrometry (TOF SIMS)[36], were used in order to determine the size, the nature, the structure and the reactivity of such compounds.

The stability of such aggregates was first pointed out by QELS experiments that reveal the absence of CMC [30]. This was confirmed by dialysis experiments through ultrathin latex membranes used to purify these nano-sized particles before their study by TEM, AFM and tribological experiments. For the dialysis experiment a Soxhlet apparatus (Figure 4.8) is used where the dispersion of overbased particles in mineral oil is introduced in a latex socket. Light petroleum is introduced in the boiler of the system as the extraction solvent (boiling point of 36 °C).

After 24 hours of extraction, the solvent in the boiler is brown due to the presence of mineral oil and high molecular weight hydrocarbons generated during the synthesis of the OMeABS. Inside the latex socket, the dispersion of nano-sized particles is clear. After evaporation of petroleum ether a white powder is obtained corresponding to the purified particles. These particles are easily re-dispersed in hydrocarbons.

This experiment demonstrates the static state of the OMeABS, commonly called 'overbased micelles'. Effectively in the case of true dynamic micelles in equilibrium with single molecules, the continuous extraction with pure solvent would have completely extracted the molecules and the rebuild micelles would then be present in the boiler.

The TEM micrographs presented in Figure 4.9 show typical results obtained by means of the freeze fracturing technique applied to overbased calcium alkyl benzene sulfonate (OCABS) dispersion in dodecane. The OCABS aggregates, clearly identified by X-ray analysis on the extractive replica, appear as approximately spherical particles dispersed in the apolar medium [4, 35].

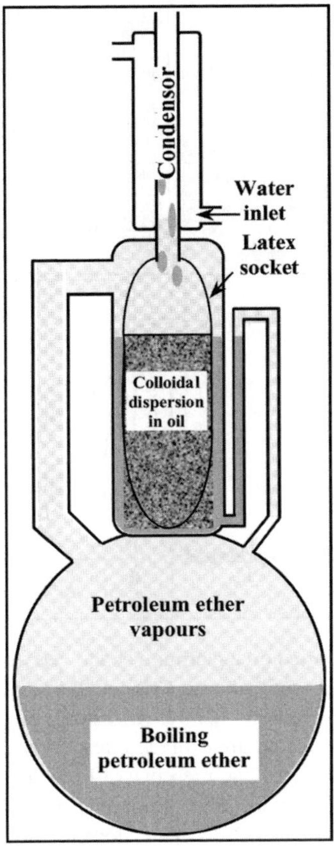

Figure 4.8 Soxhlet apparatus. Pure solvent is at the level of the condensor for efficient extraction. The solvent in the boiler progressively enriches in impurities

EXAFS results on the various OCABS dispersions point out the amorphous structure of the mineral calcium-rich part of the particles. The complementary TOf SIMS analyses [36] clearly established the presence of the understoichiometric calcium carbonate amorphous phase stabilized by residual calcium hydroxide.

4.3 Behaviour of the Micelles at the Solid–Liquid Interface

At the liquid–solid interface the behaviour of the aggregates will be different depending on their dynamic (true micelles) or static (encapsulated nano-sized particles) character. The dynamic reverse micelles (Figure 4.10) will disaggregate to form molecular adsorbed layers presenting a brush conformation. The molecules interact with the solid surface by their polar heads, often via Coulombian interactions or hydrogen bonding, sometimes by chemisorption.

The adsorption kinetics can be followed by ellipsometric measurements and the morphology of the film can be characterized by TEM coupled to a carbon platinum shadowed replica and atomic force microscopy (AFM) [14, 37]. For molecular concentration under the CMC,

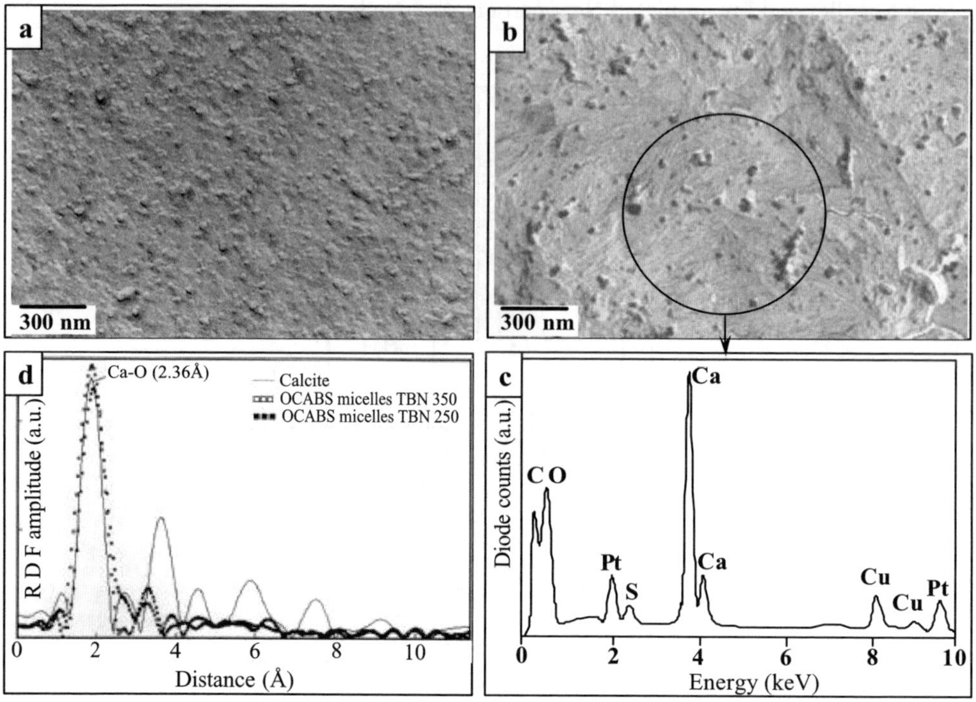

Figure 4.9 TEM micrographs recorded on a platinum shadowed replica prepared by FFET. (a) Sample prepared on 5 % dispersion of overbased calcium didodecylbenzene sulfonate in dodecane. Spherical nanoparticles of 10 nm diameter are easily visible. (b) Extractive replica prepared on the same dispersion. (c) X-ray analysis of the selected area (black circle) reveals the composition of the particles (Ca, S, O) corresponding to the sulfonate. The presence of Pt is due to the shadowing layer and copper to the copper support grid. (d) Radial distribution functions (from EXAFS studies) uncorrected from phase shifts obtained on calcite, and two OCABS dispersions in oil. The presence of a single peak corresponding to the first Ca–O distance points out the amorphous structure of the mineral part of the OCABS particles

Figure 4.10 (A) At the solid–liquid interface, reverse micelles will disaggregate and molecules will adsorb on to the solid surface as schematically drawn. (B) Adsorption kinetic of E1500 copolymer in dodecane at concentrations lower than the CMC (10^{-3} g/cm^3) and above the CMC (10^{-2} g/cm^3). The adsorption kinetics are different depending on the polymer concentration. At a concentration lower than the CMC, the adsorption phenomenon is governed by single molecule diffusion in the solvent. At a concentration greater than the CMC the establishment of the adsorbed film starts very fast, due to the existence of the micelles responsible for the first step of rapid feeding of the interface

the evolution of the fraction coverage as a function of time is mainly controlled by molecular diffusion in the solvent. At molecular concentrations greater than the CMC the kinetics often corresponds to a high-affinity type characterized by the rapid establishment of the first fractions of adsorbed film due to the feeding of the interface by micelle disassembly (Figure 4.10).

In the case of short amphiphilic molecules (fatty acids and the corresponding soap) the monomolecular adsorbed layer will adopt the well-known brush conformation and the corresponding thickness will be of the order of the molecular length or slightly less if the molecules are tilted away from the perpendicular to the surface.

As an alternative technique to adsorption experiments, the behaviour at the liquid–solid interface can be advantageously studied by dipping supports (thin carbon layer, mica sheet) into a low concentration dispersion of micelles or encapsulated nano-sized particles in pentane or other light solvent (hexane, toluene, etc.). After removing the solvent, ATEM or AFM can be used to study the films deposited on the chosen surfaces.

In the case of the dynamic strontium octanoate micelles (Figures 4.11 and 4.12) electron energy loss spectroscopy (EELS) and derived energy filtered transmission electron microscopy (EFTEM) studies reveal that the compound is homogeneously distributed at the carbon layer

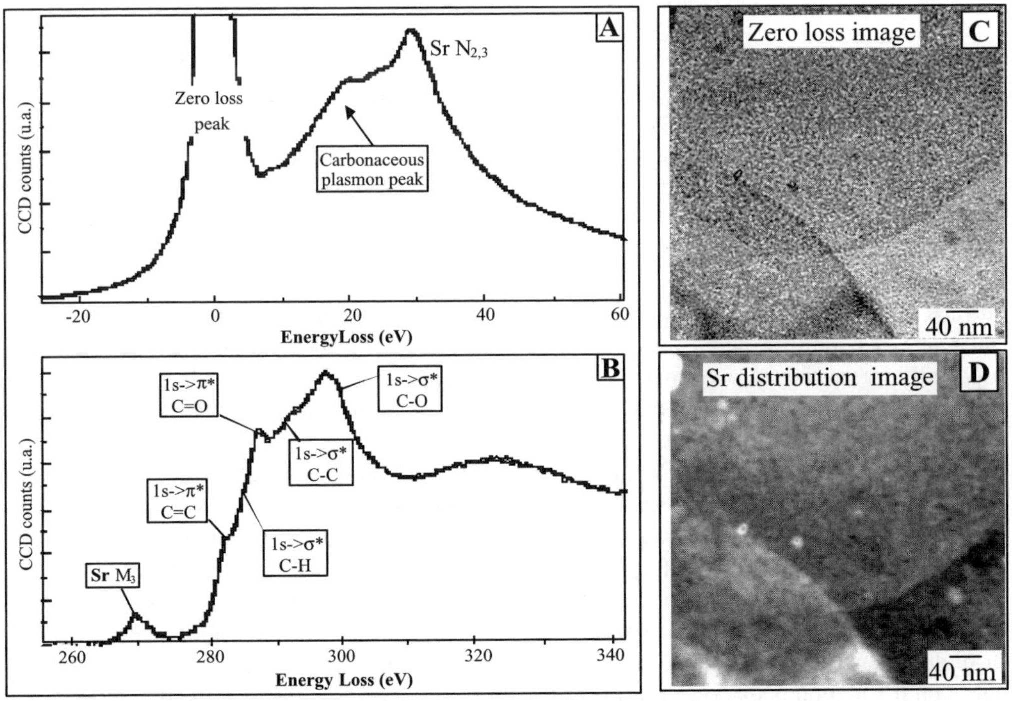

Figure 4.11 EFTEM micrographs and EEL spectra recorded on to a direct deposit of strontium octanoate on the thin carbon layer. (A) EEL spectrum collected on the sample in the low-loss region showing the presence of strontium. (B) EEL spectrum at the carbon K-edge, pointing out the presence of carboxylate groups (C = O and C–O characteristic near edge structures). (C and D) Zero loss and 32 eV energy filtered images showing the homogeneous spreading of strontium soap on the carbon layer

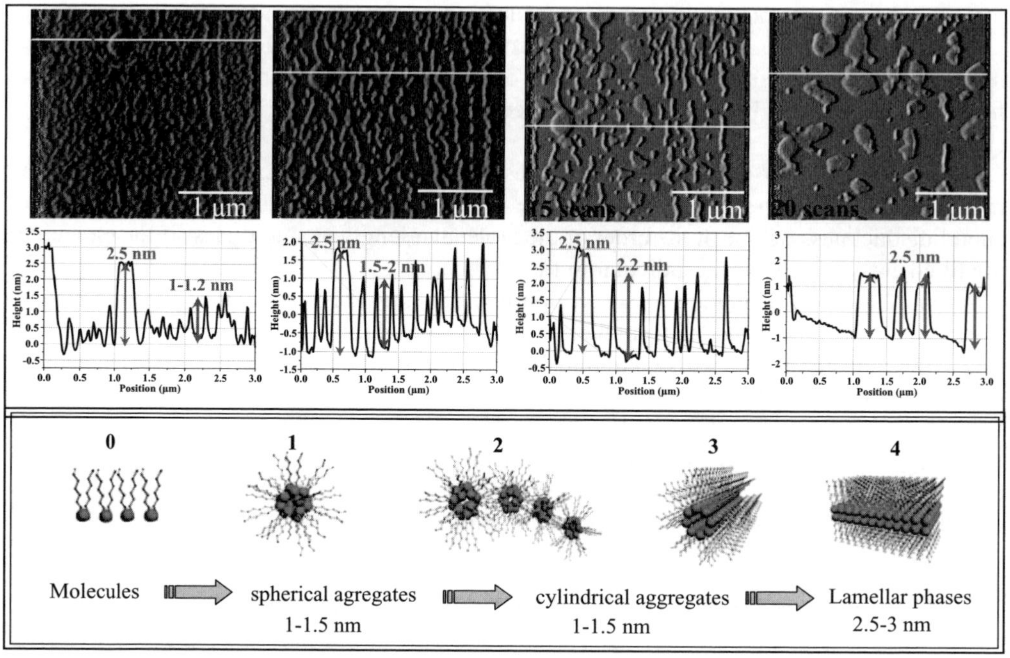

Figure 4.12 AFM topographic images and profiles corresponding to the white lines recorded after 1, 7, 15 and 20 scans on a strontium soap film deposited on to a freshly cleaved mica surface. After the first scan the surface is already perturbed; molecules have been displaced by the silicon tip and form small islands of 1 to 1.2 nm (length of the soap molecule). During the various scans molecule displacements continue and spherical, cylindrical and lamellar aggregates are successively obtained

surface [38]. AFM experiments carried out on Sr octanoate deposited on to freshly cleaved mica surfaces confirm the molecular spreading of soap molecules on the surfaces [39]. It is especially interesting to note that under the stress applied by the AFM silicon tip, the soap molecules are progressively displaced and successively aggregate into spherical, cylindrical and lamellar assemblies [39,40].

Due to their static character, the aggregates of OCABS (encapsulated nano-sized particles) are supposed to physically adsorb on to the surfaces, mainly by van der Waals interactions via their organic shell (nonpolar queues). The film deposits lead to a behaviour very different from that observed in the case of dynamic micelles. Typical TEM results are shown in Figures 4.13 and 4.14. Depending on the initial concentration of the OCABS dispersion in pentane, the deposited films appear as a bidimensional hexagonal close-packed lattice (Figure 4.13) or highly dispersed individual particles (Figure 4.14) [35]. The average film thickness measured on the platinum shadowed replica of 15 nm corresponds to the average diameter of the micelles.

In Figure 4.14(D), the combined EELS filtered images at 22 eV (carbonaceous species plasmon) and 36 eV (calcium $M_{2,3}$ core edges), representing respectively the carbon and calcium distributions on a single particle of 7 nm diameter, confirm the theoretical structure of the particle: an organic shell surrounds the inner mineral core, which is rich in calcium.

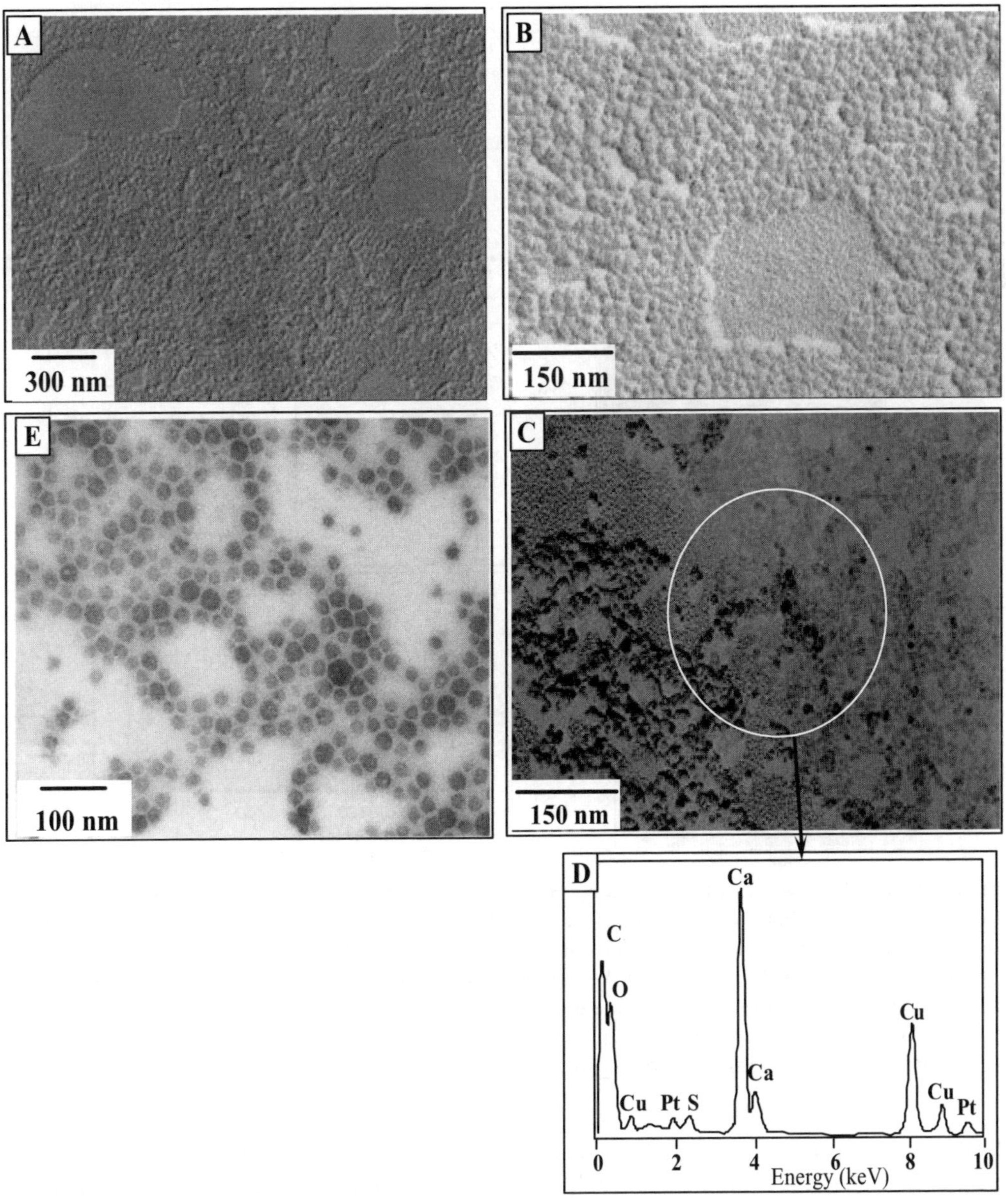

Figure 4.13 TEM micrographs recorded on OCABS colloids deposited on to thin carbon layers. (A) and (B) Platinum shadowed samples (shadowing angle of 45 °) revealing the shape and size of the particles (core and shell). (C) Partially platinum shadowed sample (shadowing angle 30 °). In the white circle shadowing and X-ray analysis (D) Dark grey areas can be identified as the mineral core of OCABS. (E) Simple deposit showing, in dark grey, the mineral cores of the colloidal particles. These pictures confirm that the individual particle character is maintained in these compact films. The distances separating the mineral cores (≈ 3.5 nm) correspond to twice the thickness of the organic shells

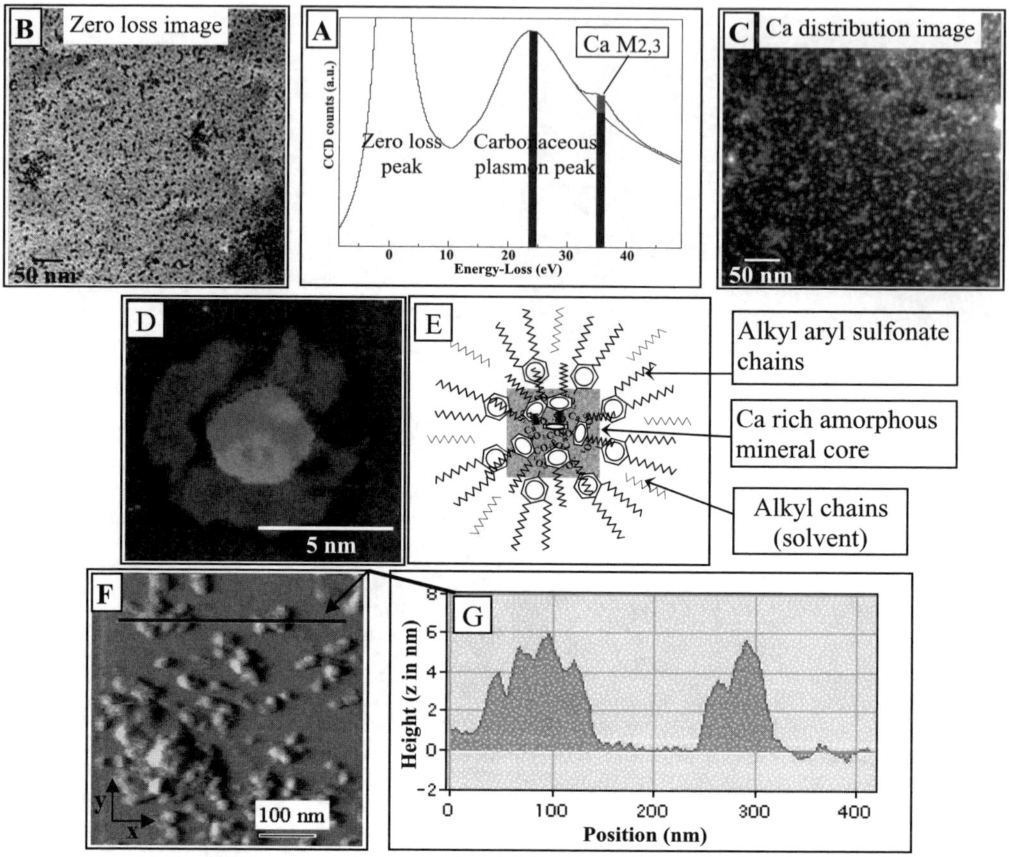

Figure 4.14 (A) Low-loss EEL spectrum, (B and C) zero loss and 36 eV energy filtered images. (D) Combined 24 eV (carbonaceous plasmon) and 36 eV (calcium $M_{2,3}$ edges) energy filtered image on an isolated particle showing the calcium-rich core (light grey) surrounded by the carbon-rich organic shell (medium grey). (E) Idealized model of the particle structure according to the experimental results. (F) AFM topographic image recorded on the same OCABS particles deposited on to freshly cleaved mica. (G) Height profile corresponding to the grey line in the picture allows the particle diameter of 4 to 6 nm to be deduced (the particle size in the x, y image is convolved with the tip radius)

In the case presented, the organic shell has a thickness of ≈ 2 nm close to the theoretical dodecylbenzene sulfonyl radical length [31,35]. The AFM experiments carried out on OCABS deposits on freshly cleaved mica surfaces confirm the individual and stable character of the OCABS nano-sized particles.

4.4 Tribologic Properties of Colloidal Systems

In order to describe the tribologic properties and associated mechanisms of the various types of colloids previously mentioned attention will be focused on representative examples of each type of system described in Table 4.1.

Table 4.1 Characteristics of the colloidal additives chosen as examples

System type	Compound	Formula	Aggregate shape and size
Reverse micelles (dynamic aggregates)	Strontium octanoate	$(C_7H_{15}COO)_2Sr$	Sphere Diameter = 2.7 nm
Encapsulated mineral particles (static aggregates)	Overbased calcium octanoate	$(C_7H_{15}COO)_2Ca,(CaCO_3)_x$	Sphere Diameter = 3 nm
Encapsulated mineral particles (static aggregates)	Overbased calcium alkyl aryl sulfonates	$(C_{24}H_{49}C_6H_4SO_3)_2Ca,(CaCO_3)x$	Sphere Diameter = 5 nm

For simplicity the colloidal additives are introduced alone in a model mineral oil base of low viscosity, normal dodecane, and the tribologic properties are evaluated using an alternative sphere on a plane tribometer. The main parameters of the experiments are summarized in Table 4.2.

4.4.1 Friction Reduction Properties of Micelles Related to Their Structure:

Depending on the nature and the static or dynamic character of the colloidal additives, the tribological behaviour at the level of friction will be different. Figure 4.15 summarizes results obtained on 1 % colloidal additive dispersions in dodecane.

It can be noticed that, in all cases, friction is reduced as soon as the tribologic tests start and remains stable up to the end of the tests. The friction coefficients obtained for various octanoate salts are summarized in Table 4.3. The friction coefficient decreases from octanoic acid (0.17) down to 0.11 for overbased calcium octanoate and 0.1 for OCABS. Assuming a dependency of the friction coefficient on the adsorbed film thickness already established by many authors in the case of organic acids, esters, hydrocarbons, etc. [41, 42], this variation of the friction coefficient is attributed to the increasing size of the adsorbed film

Table 4.2 Experimental parameters of the sphere/plane tribologic tests

Materials	Ball: 52100 AISI steel Hv = 850 Plane: 52100 AISI steel Hv = 850
Roughness (R_a)	Ball: 25 nm Plane: 10 nm
Normal load	10 N
Maximum pressure (Hertz)	1 GPa
Mean pressure	0.65 GPa
Sliding speed	3 mm/s
Temperature	25 °C

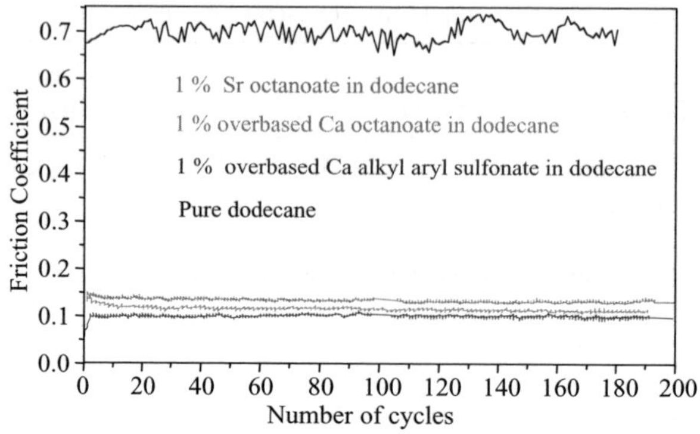

Figure 4.15 Evolution of friction coefficients as a function of the number of cycles for three types of colloidal additives and pure dodecane. The friction reduction is efficient as soon as the tests start

resulting from the size of the molecules and behaviour of the aggregates at the liquid solid interface.

In the case of organic acids or Sr, Ba and Zn soaps, the adsorbed film is a monomolecular film with a thickness close to the molecular length. In the case of overbased calcium compounds the adsorbed film consists of nano-sized particles with a thickness of the order of the particle diameter including the organic shell.

4.4.2 Antiwear Action Mechanisms of Colloidal Systems

Figure 4.16 summarizes wear results obtained for the various dispersions of colloidal additives in dodecane [3]. Their performances are compared to 1 % ZDDP in dodecane lubricant as the reference.

The main result is that for each type of colloidal additive, the antiwear action starts at the beginning of the test without any induction period. This behaviour, quite different from that

Table 4.3 Friction coefficients obtained to various octanoate salts

Compound	Aggregate type	Aggregates or molecular estimated size(nm)	Friction coeffficient
Octanoic acid	Dynamic micelles	0.9	0.17
Sr octanoate	Dynamic micelles	1.1	0.13
Barium octanoate	Dynamic micelles	1.1	0.13
Zn octanoate	Dynamic micelles	1.1	0.13
Overbased calcium octanoate	Static aggregates	3	0.11
OCABS	Static aggregates	5	0.1

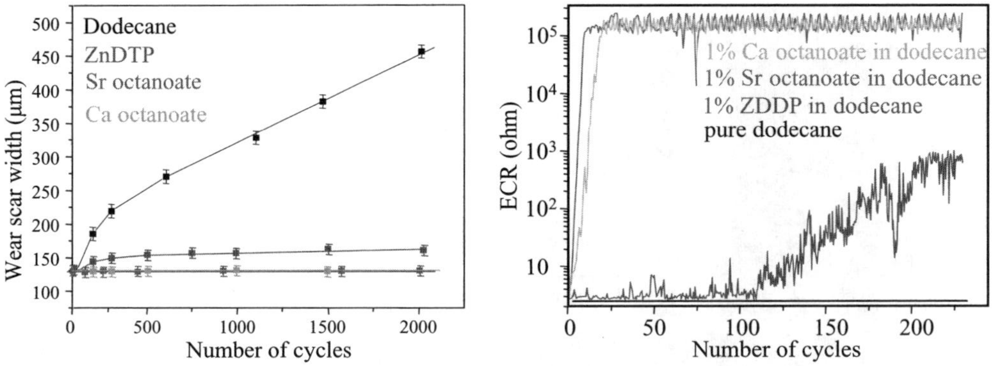

Figure 4.16 Evolution of the wear scar width (wear) and electrical contact resistance as a function of the number of cycles for tests carried out in the presence of various lubricants. The nonsignificant wear rate recorded in the case of Sr and Ca octanoate dispersions is attributed to the immediate build-up-of an anti wear film (increase in the ECR in the first cycles)

of ZDDP, is also pointed out by the rapid increase (2 to 3 cycles) in the electrical contact resistance, revealing the fast build-up of the insulating tribologic film between the surfaces.

The wear, estimated from the measurement of the wear trace width, is not significant in the case of colloids, even at the end of the test. Figure 4.17 shows photonic micrographs and wear trace profiles (optical measurements) obtained on the worn surfaces after 2000 cycles in the presence of ZDDP, strontium and calcium octanoate dispersions in dodecane as lubricants.

In each case a tribological film is formed. Wear trace profiles reveal that in the case of colloidal additives, the antiwear films are established on the surfaces and have a thickness of the order of 250 nm. Optical micrographs presented in Figure 4.17(G) and 4.18(A) and (B) show that the scratches of the initial surfaces, due to the polishing procedure for sample preparation, are still present under the antiwear film. This clearly demonstrates the exceptional protective efficiency of such additives compared to conventional ones.

The second interesting fact is the weak adhesion of these tribologic films on to the surfaces in the case of strontium and calcium compounds, pointed out by the extraction of these layers either by ultrasonic cleaning (Figure 4.18(A) and (B)) or slight stresses applied using the tip of the mechanical profilometer (Figure 4.17 (C)). This will constitute an advantage in the study of their structure and nature by ATEM.

In order to complete the investigation of the friction reduction and antiwear processes of the colloidal additives, intrinsic tribologic properties of the antiwear films are shown as an example in Figure 4.19. The antiwear film is first established in the presence of the lubricant. After 1000 cycles the test is stopped and the lubricant is removed. The sphere and the plane are separated and carefully rinsed with light petroleum in order to eliminate all lubricant traces. After evaporation of the solvent, the sphere is applied to the surface on the wear trace and the test is restarted on the dry antiwear film. The test is stopped again and a new sphere is used to test the surface outside the wear trace.

The rapid build-up of the insulating antiwear film is classically obtained in the presence of the lubricant. The friction, after cleaning, outside the wear trace points out the presence of an adsorbed film of colloidal additive (starting friction coefficient is around 0.18), which is

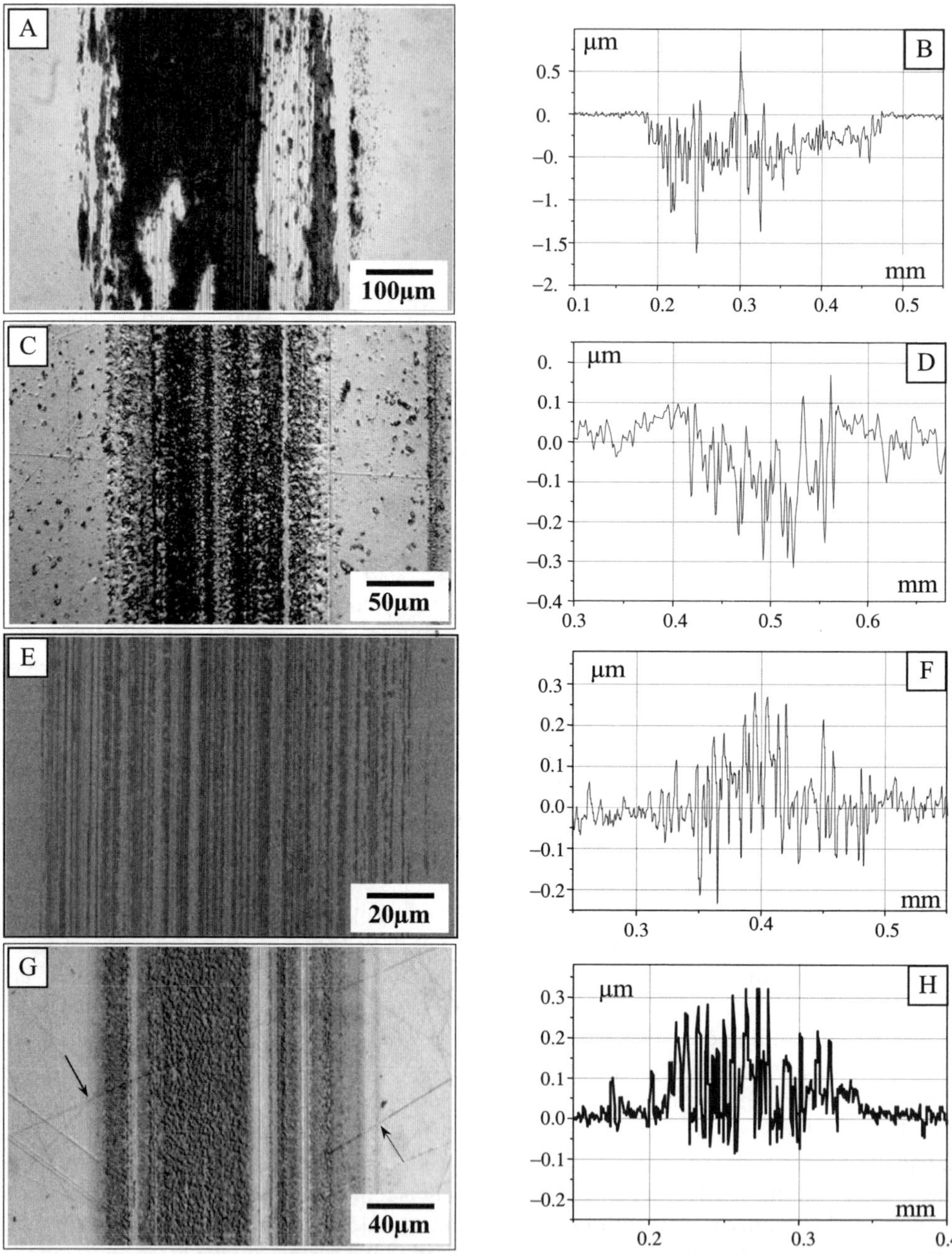

Figure 4.17 Optical micrographs and surface profiles collected on wear scars obtained after 500 cycles in the presence of pure dodecane (A and B) and 1 % ZDDP solution in dodecane (C and D). Optical micrographs and surface profiles recorded on wear scars obtained after 2000 cycles in the presence of a 1 % dispersion in dodecane of strontium octanoate (E and F) and overbased calcium octanoate (G and H). It can be seen that the tribologic film in the presence of octanoate additives is built on the initial surface. Scratches on the initial surface are easily seen (arrows in G)

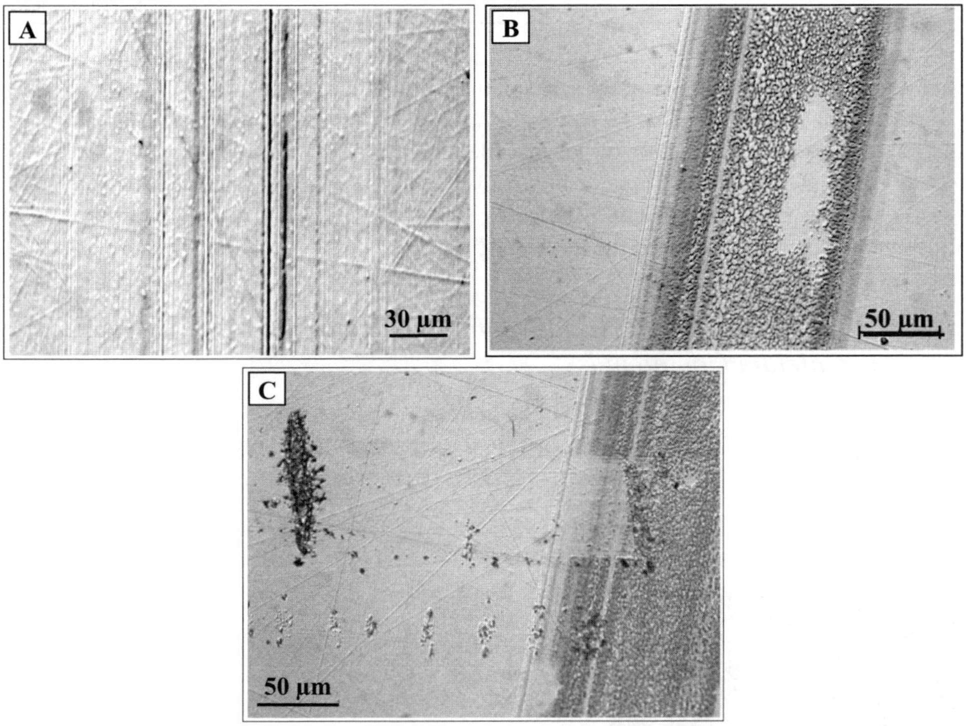

Figure 4.18 Optical micrographs collected on wear traces obtained in the presence of (A) 1 % Sr octanoate in dodecane and (B) 1 % overbased Ca octanoate in dodecane after ultrasonic cleaning in pentane at the end of tests of 2000 cycles. The total elimination of the antiwear film, in the case of Sr additive, shows that adhesion of the Sr antiwear film is weaker than the Ca one. The presence of polishing scratches on the initial surfaces under the film points out the exceptional efficiency of the additives. (C) Ca octanoate antiwear film extraction by the stress applied using the tip of a mechanical profilometer

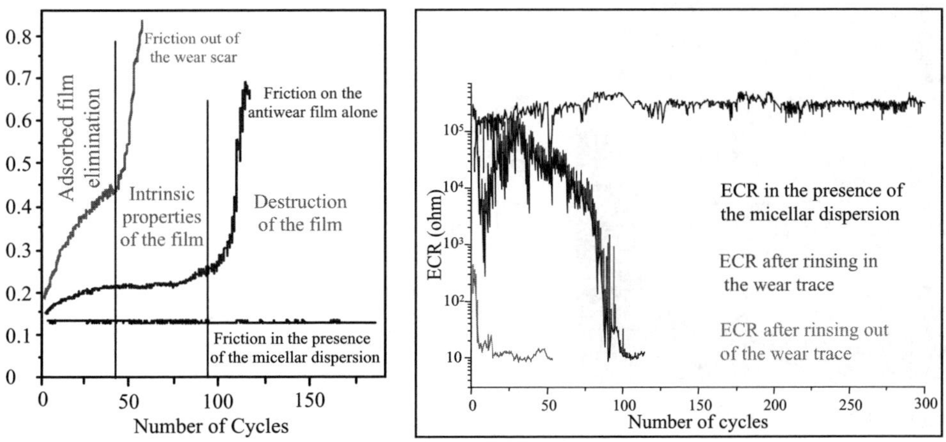

Figure 4.19 Friction tests carried out in the presence of the colloidal (Sr octanoate) dispersion (black curve) and after 1000 cycles, and pentane rinsing of the ball and the plane in the wear scar (medium grey curve) and out of the wear scar (light grey curve)

destroyed after 40 cycles. In the wear scar (on the tribologic film) the friction coefficient starts from 0.14, evolved during 30 to 40 cycles to raise a stable value of 0.21. This appears to be due to the progressive destruction of the adsorbed film present on the surface of the tribo-film. The value of 0.21 remains stable during 70 cycles and is attributed to the intrinsic friction properties of the film. Then friction increases up to the steel-on-steel coefficient. This step corresponds to the breakdown of the antiwear film.

4.4.3 Nature and Structure of Antiwear Films Obtained with Strontium and Calcium Compounds

In order to complete the understanding of the action mechanisms of colloidal additives the nature and structure of antiwear films obtained in the presence of two representative systems (Sr octanoate dynamic micelles and OCABS particles) are investigated and results are presented in Figures 4.20 and 4.21 [38, 43–45].

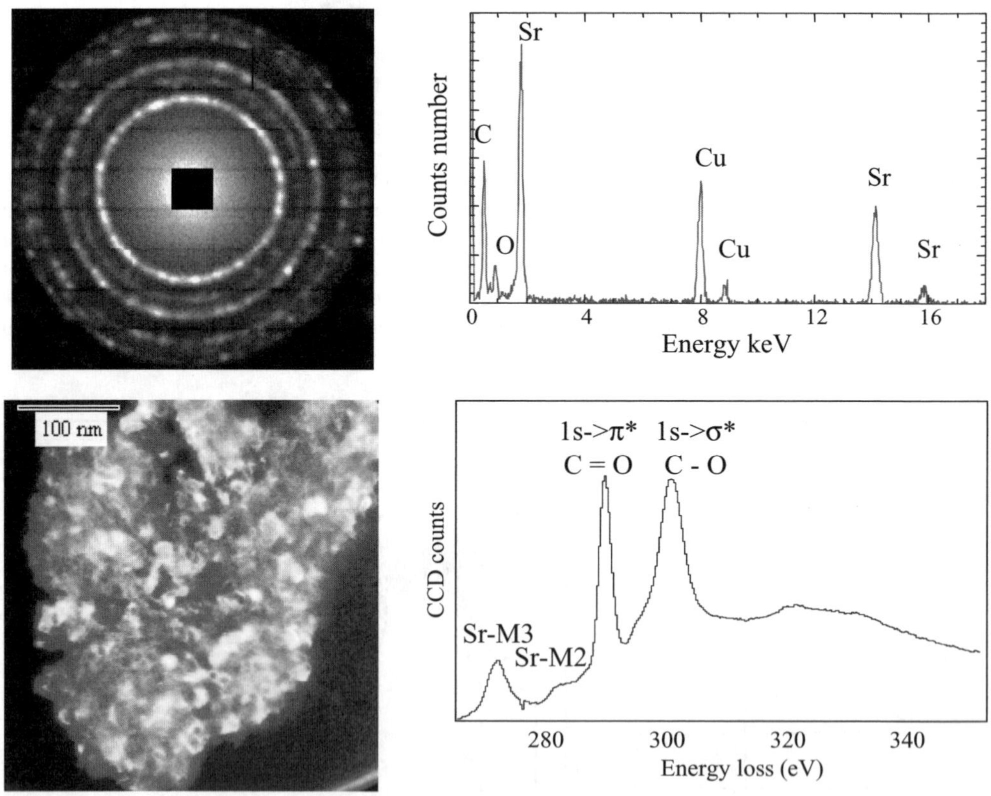

Figure 4.20 Dark-field electron micrograph and selected area electron diffraction pattern recorded on a tribological film particle obtained in the presence of Sr octanoate micelles. The film is made of polycrystalline strontianite and the EDX spectrum does not evidence the presence of iron, underlining the high antiwear properties. The EEL spectrum shows that hydrocarbon chains are no longer present in the film

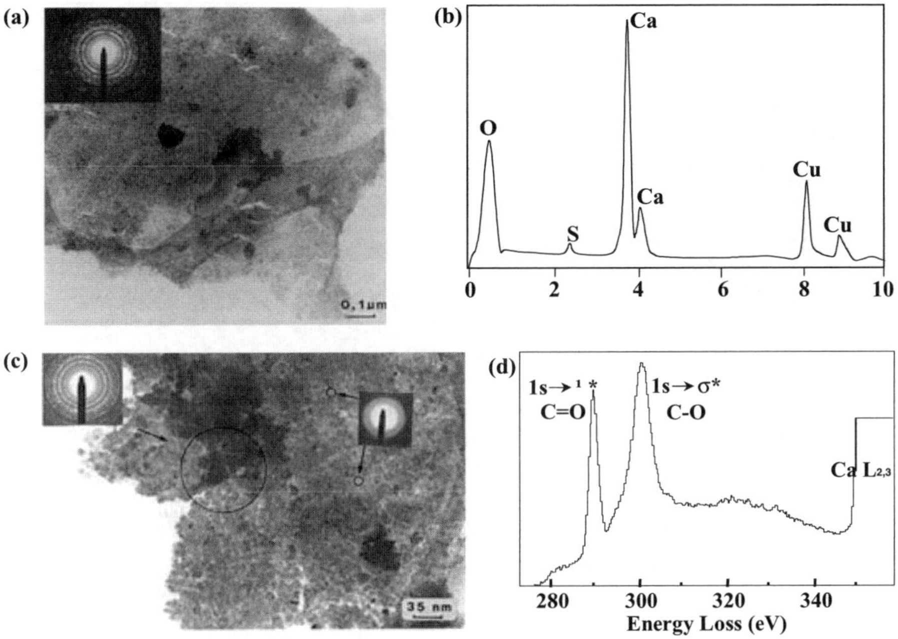

Figure 4.21 (A and B) Transmission electron micrographs and electron diffraction patterns recorded on tribologic film particles obtained in the presence of OCABS in a dodecane dispersion.(C, D) Corresponding EDX and EELS spectra. The film is composed of nanocrystallites of calcite (diffraction) with an average size of 10 nm aggregated by means of ill-ordered areas (diffuse diffraction). The EDX spectrum reveals that iron is not detected. The EEL spectrum at the carbon K-edge points out that no more hydrocarbon chains are present in the film

In the two cases, the antiwear films are polycrystalline, as clearly pointed out by electron diffraction patterns. Indexations of these last ones reveal that microcrystals are mainly carbonate phases, respectively strontianite and calcite. In the case of the OCABS film, electron microdiffraction patterns carried out between well-identified nanocrystallites demonstrate the existence of an amorphous phase responsible for the cohesion of the film. This phase, not identified by ATEM, probably corresponds to (amorphous) calcium oxide species recently detected by TOF SIMS analyses [46].

EEL spectra demonstrate that hydrocarbon chains are no longer present in the film, implying their elimination during the friction process. EDXS spectra show that iron is not present in the films, confirming that the film building does not involve any chemical reaction with the substrates. The absence of iron in the film also points out the high protective efficiency against wear of such additives.

A significant difference in behaviour of the two additives is evidenced by the size of the microcrystals present in the respective films. In the case of OCABS the microcrystals have dimension ($\approx$ 6 to 7 nm) close to the mineral core of the initial particles (5 nm). The crystals present in the strontianite film are much larger ($\approx$ 30 nm) than the initial micelle cores (1 nm) and involve a more complex recrystallization process in this last case.

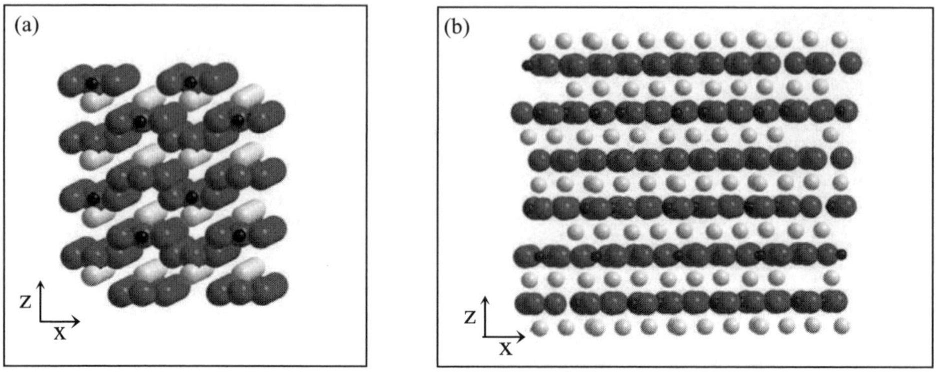

Figure 4.22 Views of (A) strontianite and (B) calcite structures showing the 'lamellar packing' inducing well-known anisotropic optical and mechanical properties. Directions of easy shearing (parallel to the x, y planes) are easily seen and confer to these structures good tribological behaviour

4.4.4 Associated Antifriction and Antiwear Actions in Tribological Behaviour of Colloidal Additives

The examples chosen to illustrate the tribologic behaviour allow the various actions of such colloidal additives to be described.

Friction reduction is mainly attributed to the adsorption film of the additive (organic molecule, soap molecules or encapsulated particles). The length of the organic shell (organic chains) is a first-order parameter according to Zisman experiments (where longer chains give lower friction).

The exceptional wear reduction in the wear test is associated with the quasi-instantaneous build-up of a stable tribologic film from the adsorbed additive layer. The polycrystalline carbonate nature of the film results from the elimination of organic parts (oxidization) and the crystallization of ill-organized cores of the micelles or encapsulated particles under the physical conditions of the sliding contact ($P \approx 1$ GPa, shear stress $\approx 10^5$ s^{-1}, oxidative species, flash temperatures, etc.).

It is important to note that the carbonate phases produced in the sliding contact present a lamellar structure (Figure 4.22) well known to confer good intrinsic tribological properties to the film (preferential shearing planes) [39, 44]. Sphere-on-plane friction experiments carried out on microcrystalline strontianite films deposited on the steel plane by the burnishing method [39] lead in air or in the presence of pure dodecane to a friction coefficient of 0.21, in excellent agreement with intrinsic properties of strontianite tribo-film (Figure 4.19). The friction coefficient on the burnished film decreases down to 0.13 in the presence of 1 % Sr octanoate dispersion, pointing out the friction reduction action of the adsorbed layer of soap molecules.

The action of the colloidal additives studied can be described as follows:

Step 1. Adsorption of the additive on the substrate surfaces.

Step 2. Loss of hydrocarbon chains and crystallization of the mineral part of the colloids under the physicochemical conditions of the sliding contact ($P \approx 1$ GPa, shear stress $\approx 10^5$ s^{-1}, flash temperature, oxidizing species, etc.). These transformations do not imply chemical reaction with surfaces components

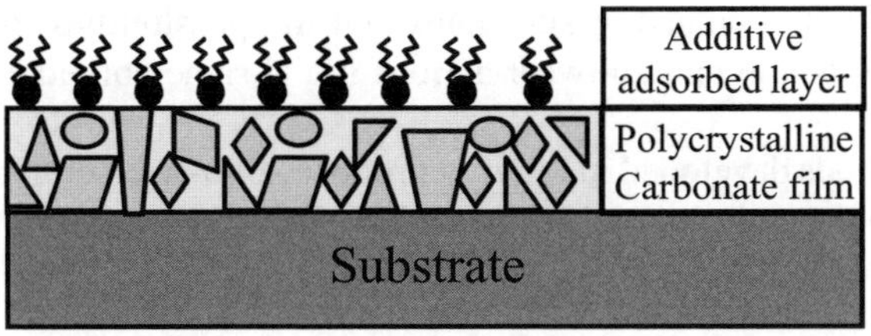

Figure 4.23 Schematic composite structure of the tribochemical film. The antiwear action is associated with the polycrystalline film (nanocrystallized metal carbonate aggregated via short extent ill-organized phase). The adsorbed layer of additive (molecules or aggregates) participates with the regeneration of the tribo-film and is responsible for friction reduction

Step 3. The tribofilm is worn instead of metal substrates (substitution wear) and regenerated from additive molecules or aggregates adsorbed on its surface (Figure 4.23).

The efficiency of the colloidal additives is due to the composite structure of the tribo-film (Figure 4.23). The friction reduction is associated with additive adsorbed layer. The antiwear action works via substitution wear of the tribochemical film, which is continuously regenerated from the additive adsorbed layer. The consumption of additive (apart from the initial adsorption on the whole surface) is selectively located in the sliding contact.

Such additives, whose action does not involve chemical reaction with substrates, are supposed to be efficient on nonreactive materials. Figure 4.24 presents the first results recorded on monocrystalline alumina. It can be seen that friction is heavily reduced and wear is not significant as it for metallic surfaces.

4.5 Conclusion and Perspectives

Colloidal physics and chemistry offer interesting ways of research in the reduction of friction and wear applied to various metallurgical contexts. The development of additives that do

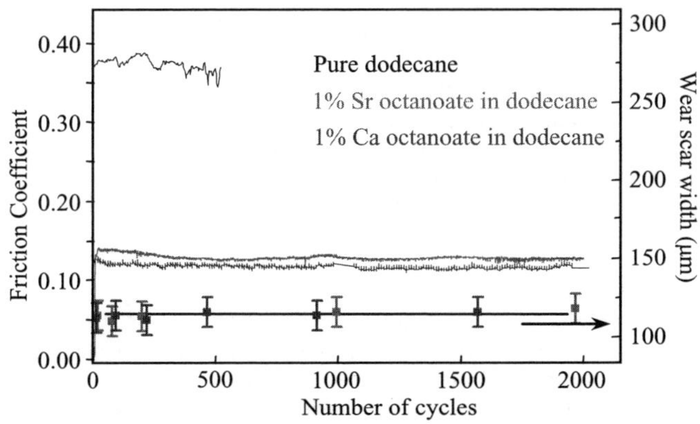

Figure 4.24 Tribologic results recorded during a sphere/plane test using monocrystalline alumina as the substrate. The friction coefficient (continuous curves) is stabilized at values of 0.13 (Sr) and 0.12 (Ca) as soon as the test starts. The wear estimated by measurement of the wear scar is not significant

not involve reaction with contacting substrates and the possibilities to associate corrosion inhibition, friction reduction and antiwear actions in the same colloidal particle are of great interest.

The chemical methods developed in order to produce nano-sized dynamic or static particles will allow the tribologist to design stable dispersion in oil of selected compound precursors of an efficient tribochemical film.

References

1. Martin, J.M. Contribution à la tribologie: etude du mécanisme d'action d'un additif anti-usure en regime de lubrification limite. Aspects chimiques dans le cas des organodithiophosphates métalliques, Thèse d'État ° 7827, Université Claude Bernard, Lyon, 1978.
2. Martin, J.M., Mansot, J.L. Berbezier, I. and Dexpert, H. The nature and origin of wear particles from boundary lubrication with ZDDP, Wear, **93**(2), 1984, 117–126.
3. Martin, J.M., Mansot, J.L., Berbezier, I. and Belin, M. Microstructural aspects of lubricated mild wear with zinc dithiophosphate, *Wear*, **107**, 1986, 355–366.
4. Martin, J.M., Belin, M. and Mansot, J.L. Friction induced amorphization with ZDDP. An EXAFS study, ASLE Transaction, **49**(4), 1986, 523–531.
5. Belin, M., Martin, J.M. and Mansot, J.L. Role of iron in the amorphization process in friction-induced phosphate glasses, *Journal de Physique*, **48**(C9), 1987, 1147–1153.
6. Bell, J.C., Delargy, K.M., and Seeney, A.M. The removal of substrate material through thick zinc dithiophosphate antiwear films, in *Wear Particles: from the Cradle to the Grave* (eds D. Dowson *et al.* Tribology Series 21 Elsevier), Amsterdam. 1992, pp. 387–396.
7. Martin, J.M., Grossiord, C., Lemogne, T. Bec, S. and Tonck, A, The two layer structure of ZnDTP tribofilm. Part 1: AES, XPS and XANES analyses, *Tribology. International*, **31**(10), 2001, 627–644.
8. Spikes, H., The history and mechanisms of ZDDP., Tribology. Letters, **17**(3), 2004, 469–489.
9. Martin, J.M., Lubricant additives and the chemistry of rubbing surfaces: metal dithiophosphates triboreaction films revisited, Japan. *Journal of Tribology*, **42**, 1997, 9.
10. Martin, J.M., Antiwear mechanisms of zinc dithiophosphate: a chemical hardness approach, *Tribology, Lettres*, **6**, 8, 1999, 1–8.
11. Langevin, D., in *Reverse Micelles*, (eds P.L. Luisi and B.E. Straub), Plenum, New York, 1982, p. 287.
12. Mittal, K.L., and Mukerjee, P. The wide word of micelles, in *Micellization, Solubilization and Microemulsions* vol. 1, (ed. K.L. Mittal), Plenum, New York, 1977, pp. 1–22.
13. Eicke, H.F., Micelles in apolar media, in *Micellization, Solubilization and Microemulsions*, vol. 1(ed. K.L. Mittal), Plenum, New York, 1977, pp. 429–444.
14. Benedetti, E.L., and Favaerd, P., *Freeze Etching Techniques and Applications*, Société Française de Microscopie Electronique, 1973.
15. Mansot, J.L., Aspects microscopiques de l'action des réducteurs de frottement en lubrification limite, These Docteur Ingénieur, Ecole Centrale de Lyon, 1982.
16. Mansot, J.L., Martin, J.M. and Candau, S.J. Amphiphilic complex ester as lubricant additive. Part one: structure of paraffinic solutions, Colloids and Surfaces, **7**, 1983, 301.
17. Belin, M., Martin, J.M., and Mansot, J.L. Mixed lubrication with a complex ester as a friction modifier, *ASLE Transactions*, **27**(4), p.398-404, 1985.
18. Martin, J.M., Mansot, Hurier and Marotel, Savons de calcium possèdant une réserve de basicité élevée , International Patent ° 85/19393, 1985.
19. Wittle, J.R., Method of preparing overbased calcium sulfonates, US Patent 4,427, 559, 1984.
20. Belle, C. *et al.*, Mécanismes des reactions polyphasiques: cinétique de formation de carbonate de calcium colloidal en milieu apolaire, *J. Chimica. Physique*, **87**, 1990, 93–104.
21. Hudson, L.K., Eastoe, J. and Dowding, P.J., Nanotechnology in action: overbased nanodetergent as lubricant oil additives, *Advances in Colloéd. and Interface Science* **123–126**, 2006, 425–431.
22. Wery-Venturini, J., Contribution à l'étude de la structure microscopique de micelles inverses d'octanoate de plomb, Thèse de Doctorat, Université de Nantes, 1992.

23. Wery, J., and Mansot, J.L. Quantitative study of irradiation damage in organometallic colloidal particles, *Mcroscopy. Microanalysis*. Microstrucure, 1993, **41**.
24. Mansot, J.L., Wery, J. and Lagarde, P., Local structure analysis of the mineral core of reverse micelles in dispersion in hydrocarbons, *Colloids and Surfaces*, **90**, 1994.167–182.
25. Marsh, J.F., Colloidal lubricant additives, *Chemistry and Industry*, **20**, 1977, 470–473.
26. Markovic, I., Ottewill, R.H., Cebula, D.J., Field, I. and Marsh, J.F. Small angle neutron scattering studies on non aqueous dispersions of calcium carbonate. Part 1: Guinier approach, *Colloid and Polymer. Science* **262**, 1984, 648.
27. Markovic, I., Ottewill, R.H., Cebula, D.J., Field, I. and Marsh, J.F. Small angle neutron scattering studies on nonaqueous dispersions of calcium carbonate. Part 2, *Colloid and Polymer. Science.*, **264**, 1986, 265–276.
28. Giasson, S., Espinat, D., Palermo, T., Ober, R., Pessah M. and Morizur, M. F. Small angle X-ray scattering (SAXS) on calcium sulfonate dispersions: effects of friction on microstructure, *Journal of Colloid and Interface Science*. **153**(2), 1992, 355–367.
29. Arndt, E.R., and Kreuz, K.L., Characterization of calcium alkylaryl sulfonates containing encapsulated solids, *Journal of Colloid and Interface Science* **123**, 1988, 230–237.
30. Makloufhi, R., Mansot, J.L. Hirsh, E. Wery, J. Candau, S.J., Thomas, J.P. and Rolland, J.P., Electron microscopy and light scattering study of organometallic inverse micellar systems, *Colloid and Polymer Science* **273**, 1995, 242.
31. Martin, J.M., Mansot, J.L., and Hallouis, M. Energy filtered electron microscopy (EFEM) of overbased calcium alkylaryl sulfonate micelles , *Ultramicroscopy*, **30**, 1990, 321–328,
32. Martin, J.M., Mansot, J.L., Hallouis, M. and Tenailleau, H. High resolution electron spectroscopic imaging (ESI) of reverse micelles, *Microscopy*, Microanalysis, Microstructure, **2**, 1990, 1–8.
33. Martin, J.M., Belin, M., and Mansot, J.L. EXAFS of calcium in overbased micelles, *Journal de Physique*, **12**(47), 1986, 887–890.
34. Mansot, J.L., Martin, J.M., Dexpert, H., Faure, D., Hoornaert, P. and Gallo, R. Local structure analysis in overbased reverse micelles, *Physica* B,**158**, 1989, 237.
35. Mansot, J.L., Hallouis, M. and Martin, J.M., Colloidal antiwear additives. Part One: structural study of overbased calcium alkylbenzene sulfonate micelles, *Colloids and Surfaces A*, **71**, 1993, 123–134.
36. Cisaire, L., Martin, J.M., Lemogne T. and Gresser, E. Chemical analysis of overbased calcium sulfonate detergents by coupling XPS, Tof-SIMS, XANES and EFTEM, *Colloids and Surface. A: Physicochem. Eng ineening. Aspects*, **238**, 2004, 151–158.
37. Mansot, J.L., Martin, J.M., Gagnaire A., and Joseph, J. Shear properties of partial monolayers, *Wear*, **84**, 1983 115–118.
38. Mantsot, J.L., Golabkan, V., Romana, L., and Cesaire, Th. Chemical and physical characterization by EELS of strontium hexanoate reverse micelles and strontium carbonate nanophase produced during tribological experiments, *Journal. of Microscopy*, **210**, 2003, 110.
39. Golabkan, V. Approche analytique des mécanismes d'action anti-frottement et anti-usure de nouveaux additives colloïdaux, Université des Antilles et de la Guyane, 2005.
40. Bilas, P., Golabkan V., Romana, L., Kraus, B., Bercion, Y. and Mantsot, J.L. Micellar phase transitions induced during nanoscale friction experiment, in 8th Inter-American Congress of *Electron Microscopy*, Cuba, 2005.
41. Bowden, F. P., and Tabor, D., *The Friction and Lubrication of Solids*, Clarendon Press, Oxford, 1950.
42. Zisman, W.A., Durability and wettability properties of monomolecular films on solids, in *Friction and wear* (ed. R. Davies), Elesevier, Amsterdam, 1959, pp. 110–148.
43. Mansot, J.L, Hallouis M., and Martin, J.M. Colloidal antiwear additives – Part two: tribological behaviour of colloidal additives in mild wear regime, *Colloids and Surfaces A*,**75**, 1993, 25–31.
44. Mansot, J.L., Golabkan, V., Romana, L., Bilas, Ph. Alleman, E. and Bercion, Tribological and physicochemical characterization of strontium colloidal additives in mild wear regime, *Colloids and Surfaces*. À, **243**, 2004, 67.
45. Giasson, S., Palermoa, T., Buffeteau, T., Desbat B. and Turlet, J. M., Study of boundary film formation with overbased calcium sulfonate by PM-IRRAS spectroscopy, *Thin Solid Films*, **252**, 1994, 111–119.
46. Kubo, T., Fijiwara, S., Nanao H., Minami, I., and Mori, S. Tof-SIMS analysis of boundary films derived from calcium sulfonates, Tribology. Letters., **23**(2), 2006, 171–176.

5

Nanolubricants Made of Metals

Weimin Liu and Xiaobo Wang

5.1 Introduction

Metals are the most widely used tribological materials. Rigid metals or alloys such as steel are used to prepare tribo-pairs for rolling or sliding bearings, while soft metals such as Au, Ag, Cu, Zn, Pb, Bi, Sn, In and their alloys are used as solid lubricants in the form of surface films or coatings. The lubricating behaviour of these soft metals arises from the face-centered structure, which has many more slip-planes and thus exhibits low shear strength along the sliding direction. This lubricating behaviour has a similarity to that of a fluid lubricant with very high viscosity. Lubricating films composed of soft metals may show somewhat self-repairing properties in low friction speed, but for most circumstances they have the following limitations: (a) they have no such behaviour under a higher sliding speed and/or load, (b) they have no capacity to remove friction heat, (c) relubrication is not possible in most instances and (d) application techniques are often complicated. A proposed route to solve these questions is adding these soft metals in the form of powders into a carrier material such as lubricating oils to form a colloidal or partially colloidal dispersion, or into a polymer host to form a self-lubricating material. When the carrier material is oil, the metal powders may be carried to the friction surface by oils to form a metallic thin film under the shearing of the tribo-pairs, and exhibit lubricating behaviour without the above limitations. However, this route also has a technical difficulty since most lubricating oils have the requirement that the solid particles dispersed in them should not be larger than 0.5 μm and not precipitate in the process of usage, which cannot be achieved through traditional preparation technology of a metal powder.

Metal nanoparticles are metal particles in the size range of 1 to 100 nanometres. In the past twenty years, they have attracted extensive research interest, mainly due to the quantum effect, and their properties can be adjusted by the particle size as well as by the surface morphology. The characteristic of these materials that gives special properties, whether in the form of powders or thin films, is their size. When at least one dimension is in the nanometre range, metals will exhibit remarkable and unusual properties that are absent in their bulk phase,

Nanolubricants Edited by J. M. Martin and N. Ohmae

whether in the form of powders or thin films. A large amount of research concerning the preparation, structure characterization, properties and application of metal nanoparticles have been conducted.

In tribology, it is anticipated that metal nanoparticles are good lubricants when used as additives in lubricating oils for the following reasons [1–4]: (a) metal nanoparticles are so small that a stable colloidal dispersion in oils can be achieved through proper methods, which can avoid the undesired precipitation caused by gravitation, (b) with the formation of a stable well-proportioned dispersion, metal nanoparticles are more liable to be trapped in the rubbing surface and (c) metal nanoparticles are liable to deposit on the rubbing surface. Many studies have been conducted to prove the above conjectures, and some valuable results and conclusions have been achieved. In these works, traditional soft metal lubricants such as Cu, Ag, Pb, Bi, Sn, In and their alloys have been prepared into nanoparticulate materials, the methods to disperse these nanoparticles into lubricating oils to form a stable colloidal dispersion have been developed and the tribological behaviour and mechanism have been studied and discussed. This chapter will give a general summarization of these research works.

5.2 Nanolubricants Made of Coinage Metal Nanoparticles

It is well known that the elements of group 1B consist of gold, silver and copper. They are famous as three inert coinage metals. Copper is the first metal to be harnessed by man, being second only to iron in its usefulness down the ages, and has been called the cornerstone of civilization. Copper is an excellent conductor of heat and electricity. It can be worked into various forms such as pipes, wire, rods, sheets and coins. It can be dissolved in acids, but is resistant to atmospheric corrosion. There are many uses for copper. It is a very versatile metal that has increased in importance to technology and society in the last century. Some major uses are electrical wiring and components, roofing, plumbing, alloys and electroplating. Silver is probably the third metal to be discovered after gold and copper. Chemically, silver is similar to the heavy metals. It is sometimes classed as a noble metal. Silver is somewhat rare and considered a commercially precious metal with many uses. Silver has the highest electrical and thermal conductivity of all metals. It is used for dental fillings, to silver mirrors, as a catalyst for chemical reactions, in water purification, in special batteries and in solar cells. Gold is classed as a heavy, noble metal. It is the most familiar of the precious metals. Gold is considered to be one of the first metals used by humans. The major used of gold is for making jewelry, but it has important applications in electronics, for electrical contacts in making computer transistors and diode wires, and in dentistry. Gold leaf finds many uses in surgery, space vehicles and art works.

The properties and applications of metallic gold, silver and copper have been greatly expanded by decreasing their size in the nanometre range through chemical or physical methods. In the past tens years, gold, silver and copper of different size, shape and surface morphology have been synthesized via different methods. These materials have been found to have remarkable and unusual properties that are absent in the bulk phase, and have important applications in electronics, optics, catalysis, biomaterial and other high-technology fields. Among the three elements, the studies concerning gold nanoparticles were the most abundant, followed by silver and copper. The reason is that for most situations the chemical stability of the particles is crucial to avoid degradation processes such as partial oxidation or undesired sintering of particles, and gold is more stable. In fact, the lack of sufficient stability of many nanoparticle preparations has impeded the development of real-world applications of nanomaterials to some extent. As a result, a number of studies have been conducted concerning the preparation of stable and dispersible metal nanoparticles in the past few years.

Metals of group IB are all 'soft' metals with a faced cubic crystal structure, which are traditional solid lubricants. Generally, they are interposed as a film between sliding and/or rolling surfaces to reduce friction and wear. Their powders (mainly copper or silver for the relative lower costs), in the size range of micrometres, can be used as fillers of polymer materials to enhance the antiwear (AW) and friction-reduction properties, but are not able to be employed in oils because they cannot form a stable dispersion. This problem has been solved to some extent by evolving new methods to prepare stable and dispersible metal nanoparticles along with the advancement of nanotechnology.

5.2.1 Organic Compound Surface-Capped Copper Nanoparticles as Oil Additives

Nanoparticles of group IB metals have been synthesized through various methods. However, most of the resulting nanoparticles (especially naked metal nanoclusters) are inherently unstable and easy to aggregate or erode by circumstances that eventually lead to precipitation when added in oils. The poor stability, solubility and dispersibility of these nanoparticles in lubricating oils have limited their further applications as additives in formulated lubricants to a great extent. An effective solution is to cap these nanoparticles with a layer of judiciously selected organic molecules, which are composed of a polar group and a long alkyl chain. The polar group should have a strong chemosorbing ability in order to cause the surfactant to be chemisorbed on the surface of the inorganic nanocores, while the long alkyl chain should have a suitable length and structure to enable inorganic nanoparticles to be soluble in oils. The capping process could proceed after the noncapping metal nanoparticles have been prepared, but is more effective during the preparation process. The most popular method at present involves the reduction of relevant metal salts or the thermal decomposition of metal organometallic precursors in the presence of a suitable capping agent. The main reported capping agents are long-chain alkyl thiols [5,6], oleic acid [7] and amines [8].

A series of works about the preparation and tribological properties of copper nanoparticles with O, O'-dialkyldithiophosphates (DDP) as the capping agent have been reported. The achieved dialkyldithiophosphate surface-capped copper nanoparticles (DDP-capped Cu nanoparticles), prepared by means of a redox surface modification technique, can readily disperse in common nonpolar solvents and mineral oils to give a transparent solution, and their liquid paraffin (LP) solution remains unchanged for several months in ambient conditions or for 3 days at 140°C. This finding suggests that the organic compound surface-capped metal nanoparticles are capable of being good additives of lubricant oils.

A lot of studies have been conducted to evaluate the tribological properties of surface-capped copper nanoparticles. Zhou *et al.* [5] pointed out that the size and additive concentration of copper nanoparticles have a remarkable effect on their lubricating properties, based on the experimental results shown in Table 5.1. According to the results, DDP-capped Cu nanoparticles with grain sizes of 2 and 5 nm can improve the AW ability of LP effectively. However, DDP-capped Cu nanoparticles 12 nm in size decreased the AW ability of LP to some extent, while nonmodified copper nanoparticles (with a size of 40 nm) as additives in LP caused instant seizure under the same conditions. A rational explanation of the phenomenon

Table 5.1 Antiwear performance[a] of DDP-capped Cu nanoparticles. Reproduced by permission of Elsevier from Zhou *et al.* [5], © (1999) Elsevier

		Grain size of copper nanoparticles (nm)		
Sample	Liquid paraffin	2	5	12
AWS (mm^2)	0.448	0.261	0.393	0.602

[a] Lubricating performance of oils with or without additive is usually evaluated by running tribological tests on a four-ball tester. The tester consists of a steel ball rotating under load on a set of three similar stationary balls. The speed, load and temperature during the test are precisely controlled. After the wear test, round wear scars are formed on the lower three balls and their diameters are measured using a microscope. The diameter of the wear scar (WSD) in millimetres or the area of the wear scar (AWS) is a measure of the AW characteristics of the lubricants. In this case the tests were preformed under conditions of 400 N, 1450 rpm and 30 min.

might be that the smaller the size of copper nanoparticles, the lower is the melting point and the higher the reactivity of the nano-sized copper core. Subsequently, copper nanoparticles of a smaller size are more liable to interact with the surfaces of the friction pairs to form a surface protective film, which contributes primarily to the increased AW ability. In the case of DDP-modified Cu nanoparticles of larger sizes, the reaction activity of the nano-sized copper core decreased, and the interaction between the nano-sized copper core and the rubbing surface became weaker, which was not beneficial to the formation of the surface protective film. Ultimately, nonmodified copper nanoparticles of a very large size act as abrasives during the friction process and cause severe wear.

Li *et al.* [6] have studied the influence of additive concentration of DDP-modified Cu nanoparticles on the tribological properties. Table 5.2 shows the four-ball friction and wear

Table 5.2 Four-ball test results[a] of oil containing DDP-capped Cu nanoparticles at various additive concentrations. With kind permission of Springer Science and Business Media from Li *et al.* [6], © (2006) Springer Science and Business Media

Concentration (wt %)	P_B (N)	P_D (N)	WSD (mm)
0	460.6	1568.0	0.60
0.1	509.6	1568.0	0.35
0.2	588.0	1960.0	0.38
0.3	637.0	1960.0	0.30
0.4	656.6	1960.0	0.30
0.6	686.0	2450.0	0.33
0.8	705.6	2450.0	0.32
1.0	735.0	3087.0	0.31

[a]The WSD value achieved under test conditions of 392 N, 1450 r/min, 75°C and 60 min (referring to ASTM D4172) is a measurement of the AW characteristics, while P_B (last nonseizure load, the last load at which the measured scar diameter is not more than 5 % greater than the compensation value at that load) and P_D (weld point, the lowest applied load at which sliding surfaces seize and then weld), measured according to GB3142-82 (similar to ASTM D2783), are used to evaluate the EP performance of the lubricants.

Table 5.3 Tribological performances of DDP-capped Cu nanoparticles under different loads. Reproduced by permission of Editorial Office of Tribology from Zhou *et al.* [9], © (2003) Editorial Office of Tribology

	DDP-capped Cu nanoparticles	
P (N)	WSD (mm)	Friction coefficient
300	0.40	0.93
400	0.45	0.08
500	Seizure	
600	Seizure	
700	Seizure	
1000	0.95	0.08
1200	1.02	0.06
1500	0.94	0.05

Conditions: LP is used as the base oil, at a rotation speed of 1450 r/min, for 30 min.

test results of the commercial base stock (HVI WH150) containing DDP-capped Cu nanoparticles at different concentrations. It can be seen that copper nanoparticles can significantly improve the AW and extreme pressure (EP) properties of the base oil, even at very low additive concentrations. With increasing additive concentrations, the P_B and P_D values increased, indicating that better EP properties can be achieved at a higher additive content. The wear scar diameter (WSD) of the steel ball decreased dramatically at 0.1 wt % of the copper nanoparticle added, and a minimum WSD was achieved at the concentration of 0.5 wt %.

The influence of the applied load and rotating speed on the tribological properties of DDP-capped Cu nanoparticles has been investigated by Zhou *et al.* [9] and Yu *et al.* [10]. The tribological performance of DDP-capped Cu nanoparticles under different loads is shown in Table 5.3. It is observed that DDP-capped Cu nanoparticles exhibit good AW and friction-reduction properties when the applied load is below 400 N or in the range of 1000 to 1500 N. The WSD at higher loads has no obvious change with the lubrication of LP containing DDP-capped Cu nanoparticles. In the tested load range, the friction coefficient reduces as the applied load increases. When the applied load is 1500 N, the corresponding friction coefficient is only 0.05, indicating that copper nanoparticles can perform excellent tribological properties at higher applied loads. It is found that severe abrasion of the test steel balls appears after several minutes of friction in the applied load range of 500 to 900 N, which might be owing to the fact that the copper nanoparticles could not form an adequate lubricating deposited film in this condition.

The study on the influence of rotating speed shows that the friction-reduction properties of Cu nanoparticles increase with decreasing rotating speeds at lower applied loads. At higher applied loads, the rotating speed has little effect on the tribological properties of copper nanoparticles.

The tribological properties of DDP-capped Cu nanoparticles in different base oils were studied by Yang *et al.* [11] and Qiu *et al.* [12]. The research results shown in Table 5.4 indicate that copper nanoparticles exhibit excellent AW and friction-reducing properties in all kinds of oils [11, 12] These apparent performances have some difference due to the intrinsic diversity of oils. The EP properties of the copper nanoparticles are more obvious in

Table 5.4 Wear scar diameters of different base oils with/without 0.7 % DDP-capped copper nanoparticles (abbreviated to NC). Reproduced by permission of Yang X from Yang *et al.* [11], © (2006) X Yang

The category of base oil	WSD (mm)	P_B(N)	P_D(N)
500 SN mineral oil	0.57	431	1568
500 SN mineral oil + NC	0.38	833	1960
Dioctyl sebacate	0.95	431	1568
Dioctyl sebacate + NC	0.41	580	1568
PAO101	0.63	490	1568
PAO101+ NC	0.43	1068	2450
Rap oil	0.58	559	1960
Rap oil + NC	0.46	755	1960

poly-alpha-olefin (PAO) synthetic oil and 500 SN mineral oil, while AW performance is more obvious in dioctyl sebacate.

Improper usage of the lubricating additives would accelerate the invalidation of the non-ferruginous accessories relative to ferruginous tribo-pairs. However, as a kind of 'inert' solid lubricant in nature, the function mechanism of copper nanoparticulate additive differs distinctly from the conventional polar or nonpolar organic compound additives, which are based on the chemical reaction of the 'active' elements such as S and P with the rubbing surface to improve the AW or EP properties; i.e. exhibitions of the tribological properties do not rely on the kind of tribo-pair materials. Some research works [13, 14] have revealed that copper nanoparticles have good AW and friction-reduction properties for friction pairs of steel–steel, steel–iron, steel–copper and steel–aluminium to a certain extent.

5.2.2 Copper Nanoparticles Passivated by Carbon Film Used as Oil Additives

Tarasov and his coworkers [13] have developed an energetic method to prepare metal nanoparticles that can be also dispersed in oils. In this method, nano-sized copper powder is produced by the electric explosion of metallic wire (the 'EEW' process) in an inert or active gas. Various copper nanoparticles with different average grain sizes and surface protection films can be used to disperse nano-scale copper powder in oils steadily by varying the gas media, such as argon, carbon dioxide, nitrogen and argon/oxygen mixture. In addition, the surfactant was also used to improve the dispersion stability of these nanoparticles in oils. Figure 5.1 gives an example of TEM images of the as-prepared copper nanoparticles produced in argon via the 'EEW' process.

The researchers have also performed wear and friction tests by using friction specimens made by rubbing medium carbon steel against an as-quenched roller with a hardness of 46 to 48 HRC under the lubrication of a commercial motor oil with/without copper nanoparticles under the following test conditions: normal load 110 and 184 N, sliding speed 1.56 and 2.62 m/s, length of the sliding path 27 km. The average friction coefficient, score limit as well as wear, given with respect to copper nanopowder obtained in different gases at an additive concentration of 0.3 wt %, are presented in Table 5.5. It can be observed that there is a clear tendency to reduce the friction coefficient and decrease the mass loss of wear for all the samples, especially for samples 2, 4 and 5.

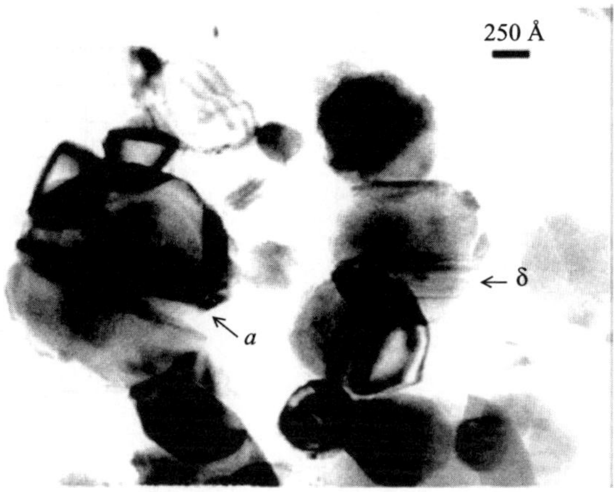

Figure 5.1 TEM image of copper nanoparticles produced in argon via the 'EEW' process. Reproduced by permission of Elsevier from Tarasov *et al.* [13], © (2002) Elsevier

Typical dependencies of the friction coefficient on wear path length (time) by the example of copper nanopowder additive obtained in carbon dioxide are shown in Figure 5.2. The data in Figure 5.2(a) to (c) provide the evidence of improved friction conditions where the nanopowder additive has been used. It can be concluded that copper nanopowder additive to SAE 30 motor oil reduces friction most effectively at higher loads and higher sliding speeds. Based on these results, the researchers demonstrated the feasibility that nanoparticulate copper can adhere preferentially to steel friction pairs and reduce the friction.

The copper nanopowder additive provides changes in worn surface topology and does not impair the lubrication characteristics of the motor oil. Local overheating caused by the direct contact of two surfaces initiates chemical deposition of copper on steel, providing a soft surface limited to the locality of the friction pairs.

5.3 Nanolubricants Made of Low Melting Point Metal Nanoparticles

It is known that many low melting point metal materials, such as indium, tin and bismuth, are good lubricating materials [15]. In particular, bismuth has attracted more interest as a potential thermoelectric and 'green' lubricant material [16]. It has been realized that soft metals

Table 5.5 Results of tribotechnical experiments. Reproduced by permission of Elsevier from Tarasov *et al.* [13], © (2002) Elsevier

Number	Lubricant	Moment of friction (N m)	Mass losses (g)
1	SAE 30	1.308	0.0058
2	SAE 30 + Cu (CO_2)	0.759	0.0028
3	SAE 30 + Cu (Ar)	1.283	0.0079
4	SAE 30 + Cu (Ar + O_2)	0.744	0.0023
5	SAE 30 + Cu (N_2)	0.543	0.0017

(The concentration of copper nanoparticles in oil is 0.3 wt %; rollers are made of steel.)

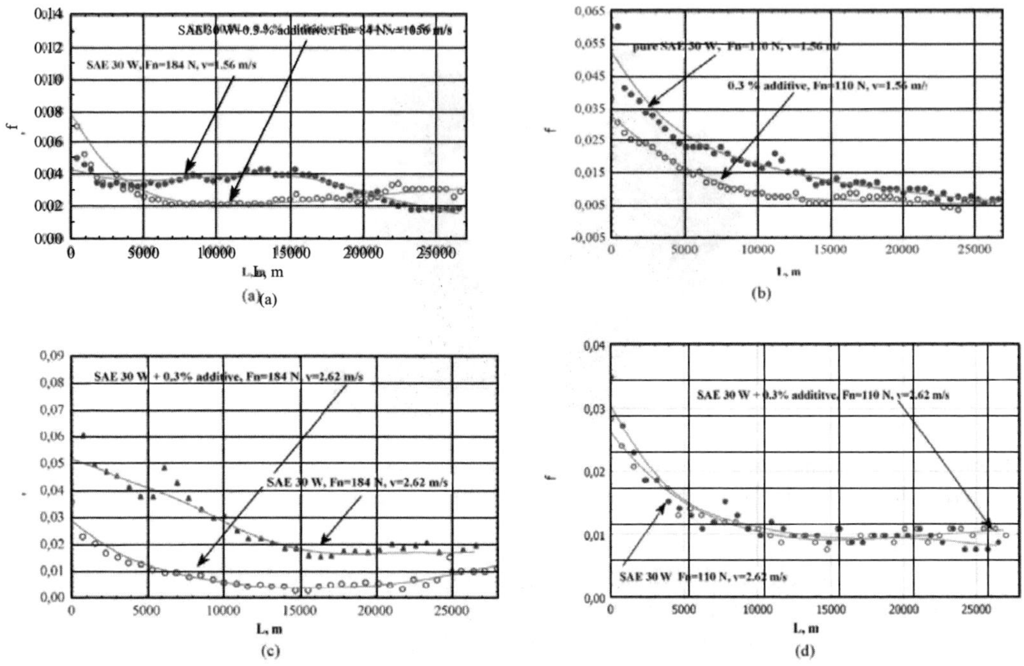

Figure 5.2 Wear path length dependencies of the friction coefficient. Reproduced by permission of Elsevier from Tarasov *et al.* [13], © (2002) Elsevier

with a low melting point on the friction surface can melt into liquid under the flash temperature caused by rubbing, and exhibit a low friction coefficient. Accordingly, it is expected that their corresponding nanoparticles should be excellent lubricants. It is reasonable to expect that the nanoparticles of low melting point metals, particularly in the form of a stable dispersion in oils, are able to move more easily into the rubbing surface and melt on it to form a liquid lubricating film, thus exhibiting a more effective lubricating performance. The problem is how to prepare these nanoparticles with an adequate stability dispersibility in lubricating oils. A simple solution strategy including a 'direct solution-dispersing method' or 'surfactant-assisted solution-dispersing method' has been proposed by Zhao and his coworkers, in which the inherent low melting point characteristic is ingeniously utilized [15, 17, 18]. Through these methods, indium, tin, bismuth and lead nanoparticles have been synthesized successfully. Their tribological behaviours as oil additives have been studied.

5.3.1 Nanolubricants of Indium, Tin and Bismuth via the Direct Solution-Dispersing Method

Zhao and his coworkers [15, 17, 19] present a simple preparation procedure of low melting point metal nanoparticles, which is very different from the conventional solution route. In this method, the bulk metal was melted in a hot organic liquid (which may be oils) and then turned into little liquid drops dispersed in oils by violent mechanical stirring. A dispersion

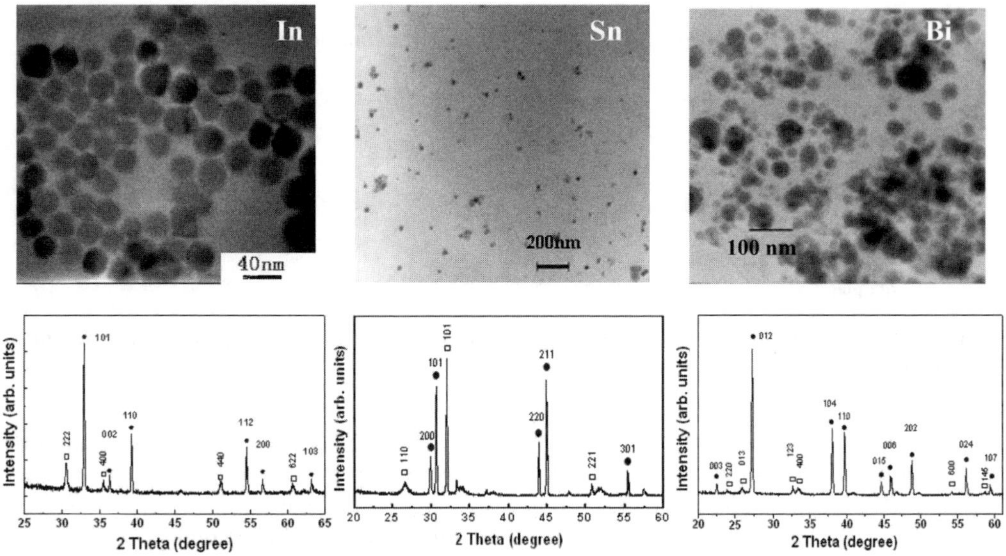

Figure 5.3 TEM images and the corresponding XRD patterns of the as-prepared indium, tin and bismuth nanoparticles. Reproduced by permission of American Chemical Society from Zhao *et al.* [15], © (2003) American Chemical Society, and of Elsevier from Zhao *et al.* [17, 18], © (2003 and 2004) Elsevier

composed of oils and little metal particles could be achieved after the synthetic system was fully stirred and cooled to room temperature. The preparation strategy allows these low melting point metal nanoparticles to be prepared from their bulk materials with ease and without using complicated apparatus and expensive agents, and is really a simple one-step method to prepare metallic nanoparticles.

Figure 5.3 presents TEM images and X-ray diffraction (XRD) patterns of the synthesized indium, tin and bismuth nanoparticles via the direct solution-dispersing method, in which the spherical In, Sn and Bi nanoparticles in the range of 30 to 50 nm can be observed. The size distribution of In nanoparticles is narrower than that of Sn and Bi nanoparticles. The XRD analysis reveals that both the pure metallic state and its corresponding oxide are present in the resulting product, indicating that a portion of metal has been oxidized during the dispersing process.

Based on these results, a possible pathway for the formation of indium, tin and bismuth nanoparticles via the direct solution-dispersing method has been proposed, as shown in Figure 5.4 [19]. At the beginning of the preparation process, large melted metal droplets are broken into small droplets in hot solvent by mechanical stirring, and the fresh surfaces of these metal droplets are oxidized to generate a metal oxide layer. The metal oxide layer can efficiently prevent the coagulation of smaller droplets and further oxidation of inner metal into little droplets. During the dispersion process, the larger droplets bear a shearing force larger than that of the smaller ones. Therefore, the larger droplets reduce faster than the smaller ones and, as a result, the size distribution can be focused down to one that is nearly monodisperse.

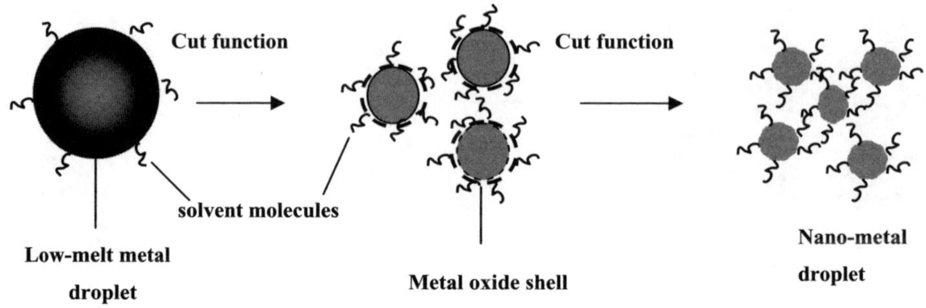

Figure 5.4 Schematic illustration of the formation process of metal nanoparticles by the direct solution-dispersing method

Soft metals such as indium, tin and bismuth are good solid lubricants, but they cannot be used as a lubricating oil additive because of their poor dispersibility in oils. However, their nanoparticles prepared by the direct solution-dispersing method have been found to have better oil solubility and can be directly used as lubricating additives. Their tribological properties have been evaluated on a four-ball tester and the results are presented as follows.

Figure 5.5 shows the variation of WSD with the additive concentration of the In, Sn and Bi nanoparticles in LP, which was used as the base oil [15, 19]. It can be seen that the WSD of pure LP is as high as 0.72 mm. However, the WSD of LP containing Sn nanoparticles are obviously decreased in all the tested additive concentration ranges, indicating that Sn nanoparticles can improve the AW property of the base oil very well. The WSD of LP containing Sn nanoparticles is only 77 % of that of pure LP, even at an additive concentration of 0.12 %, and reaches a minimum value when the additive concentration increases to 1 %. In and Bi nanoparticles also exhibit excellent AW performance in a broad additive concentration range of 0.06 to 1.0 %. They can efficiently improve the AW property of LP at a very low concentration and their AW performance is evidently improved when the additive concentration is increased. These results indicate that the as-prepared Sn, In and Bi nanoparticles have

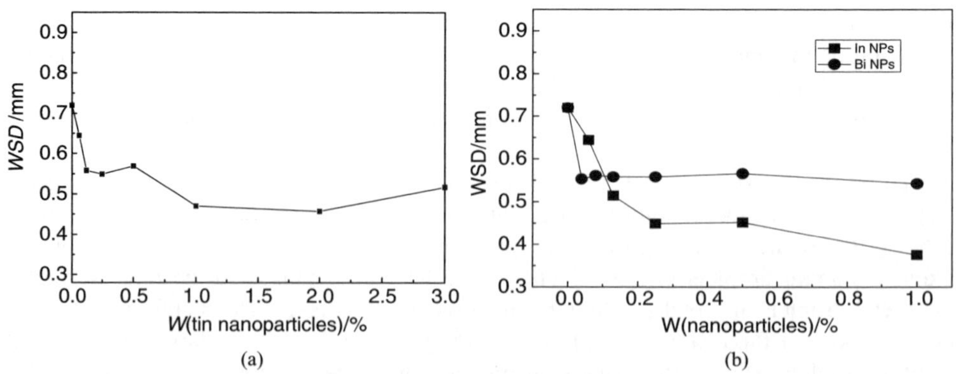

Figure 5.5 WSD as a function of the additive concentration of (a) Sn and (b) In and Bi nanoparticles in LP (under a rotation speed of 1450 rpm and load of 300 N). Reproduced by permission of American Chemical Society from Zhao *et al.* [15], (c) (2003) American Chemical Society

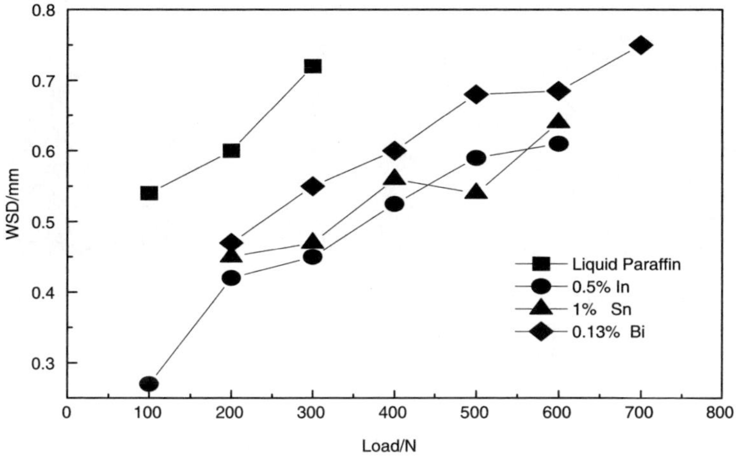

Figure 5.6 Relationship between WSD and the applied load with lubrication of LP and LP containing these different metal nanoparticles

excellent AW properties when used as oil additives. Comparatively, the AW property of indium nanoparticles is better than those of bismuth and tin nanoparticles.

The relationship between WSD and the applied load of LP and LP containing these soft metal nanoparticles is shown in Figure 5.6 [15, 17–19]. It can be seen that the WSD of LP containing different metal nanoparticles are much smaller than that of pure LP, especially for In nanoparticles. During the four-ball friction and wear test, seizure of the tribo-pairs appeared when the applied normal load was above 400 N for LP, but as high as 600 N for LP containing In or Sn nanoparticles and 700 N for Bi nanoparticles. In fact, all of these soft metal nanoparticles have an excellent load-carrying capacity when used as oil additives.

Figure 5.7 shows the variation of the friction coefficient with the additive concentration of the In, Sn and Bi nanoparticles in LP, which was used as the base oil [19]. It can be observed

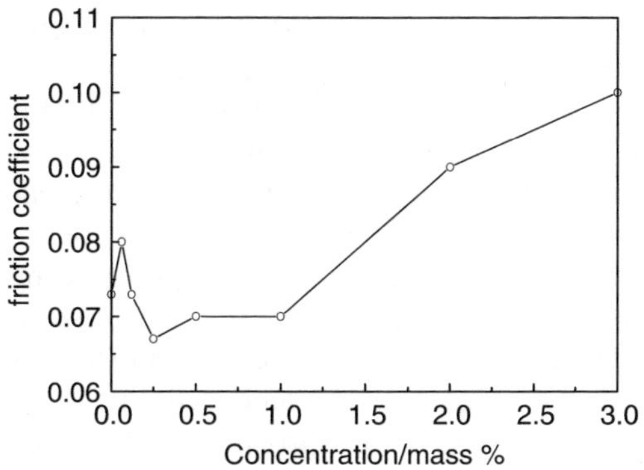

(a)

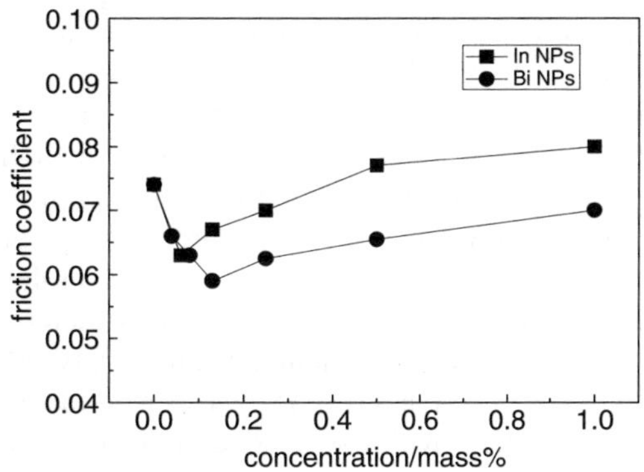

(b)

Figure 5.7 Friction coefficient as a function of additive concentration of (a) Sn and (b) In and Bi nanoparticles in LP under 300 N

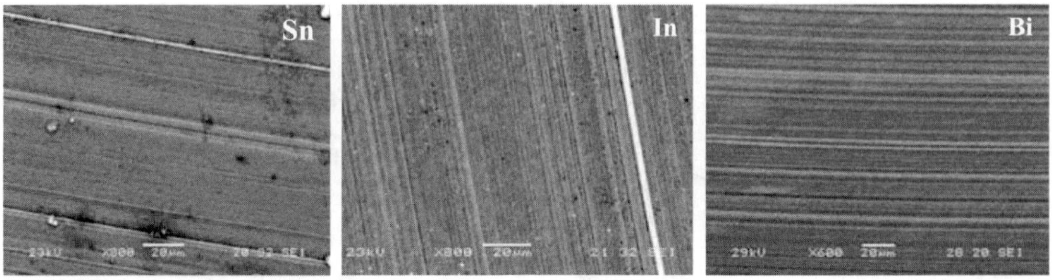

Figure 5.8 Morphologies of the wear surfaces under lubrication of LP containing 1 % Sn, 0.5 % In and 0.13 % Bi nanoparticles at an applied load of 300 N for 30 min. Reproduced by permission of Elsevier from Zhao *et al.* [18], © (2004) Elsevier

that the friction coefficient is sensitive to the addition concentration of these metal nanoparticles. At the concentration range of 0.25 to 1.00 % for the Sn nanoparticle, 0.06 to 0.25 % for the In nanoparticle and 0.04 to 1.00 % for the Bi nanoparticle, respectively, a low friction coefficient can be achieved.

SEM analysis is used to observe the worn surface of the steel ball after the friction test under the lubrication of LP with the as-prepared metal nanoparticles; these results are presented in Figure 5.8 [18, 19]. It can be observed that all wear scar surfaces are relatively smooth and the rubbed traces are shallow. The results imply that these metal nanoparticles can melt and form instantaneous metallic films between rubbing surfaces, which prevents direct contact of the rubbing surfaces, and they exhibit good lubricating properties.

5.3.2 Nanolubricants of Lead and Bismuth via the Surfactant-Assisted Solution-Dispersing Method

The direct solution-dispersing method has been found to be a simple route for the preparation of the low melting point metal nanoparticles such as tin, bismuth and indium, which is based on the low melting point properties to enable the formation and emulsification of liquid metal drops in hot oils and the simultaneous formation of an oxide layer on the outside of the liquid metal drops acting as the stabilizer. However, the spontaneous and simultaneous formation of the oxidation of the metals is somewhat uncontrollable, especially for conditions of longer preparation time or higher synthetic temperature that are required for metals with a relatively high melting point, which will lead to overoxidizing of the products. In order to resolve the problem, an improved method based on the *in situ* formation of a surfactant layer chemisorbed on the surface of metal drops has been proposed by Zhao and coworkers [19, 20]. In this method, larger metal droplets are dispersed to small droplets in solvent containing judiciously selected surfactants such as stearic acid (SA) or polyethylene glycol (PEG) by a stirring force. The fresh surfaces of the generated little metal droplets react with SA or PEG molecules simultaneously to form a closed-packed organic protective layer.

Figure 5.9 shows the TEM images and XRD patterns of Bi nanoparticles stabilized by SA (Bi-SA), Pb nanoparticles stabilized by SA (Pb-SA) and Pb nanoparticles stabilized by PEG (Pb-PEG) prepared by the surfactant-assisted solution-dispersing method [19, 20]. It can be

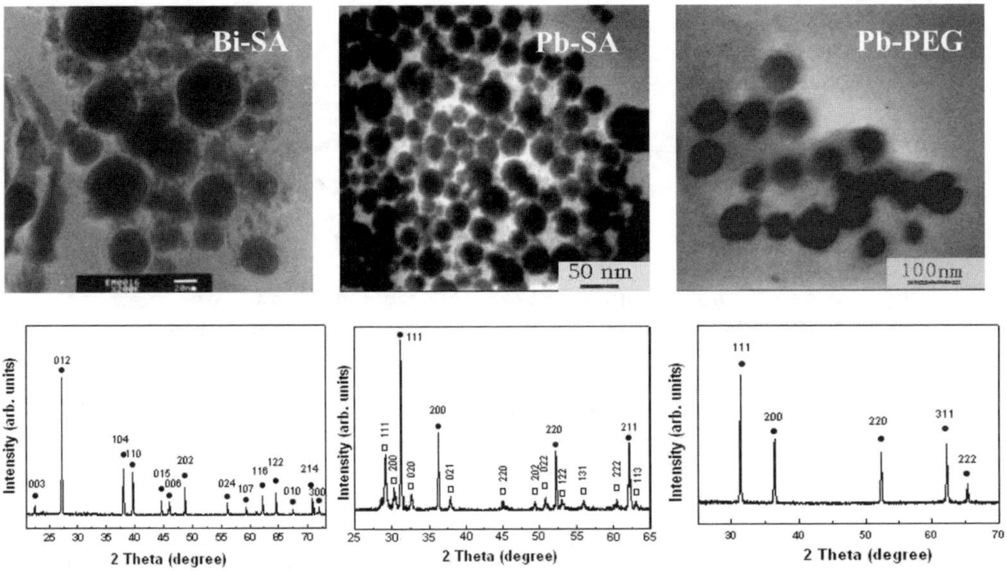

Figure 5.9 TEM images and XRD patterns of Bi nanoparticles stabilized by SA and Pb nanoparticles stabilized by SA and by PEG. With kind permission from Springer Science and Business Media from Zhao *et al.* [20], © (2004) Springer Science and Business Media

observed that the Bi-SA nanoparticles exhibit an irregular shape and a broad size distribution with an average particle diameter of 60 nm. The diffraction peaks in the corresponding XRD pattern can all be assigned to the cubic phase of metal Bi, and no diffraction peaks of the oxide can be found. The result shows that stearic acid can restrain well the bismuth from being oxidized in the formation process of the nanoparticles.

The as-prepared Pb-SA nanoparticles appear to be spherical in shape with an average particle diameter of 70 nm, and have a wide size distribution. In their corresponding XRD pattern, the prominent peaks at scattering angles (2θ) of 31.40, 36.42, 52.45 and 62.42 can be assigned to the scattering from the (111), (200), (220) and (311) crystal planes of cubic phase Pb, but there are some broad and weak peaks that match the crystal planes of PbO well. The Pb-PEG nanoparticles have a spherical shape with an average diameter of 40 nm, and also exhibit a wide size distribution. Their corresponding XRD pattern exhibits five narrow diffraction peaks, which can all be indexed to the cubic phase of metallic Pb, indicating the good crystallinity of lead nanoparticles. The result shows that PEG is better than SA in the prevention of lead oxide formation during the preparation process of lead nanoparticles in the surfactant-assisted solution-dispersing method.

The tribological behaviours of the as-prepared nanoparticles used as oil additives were studied via a series of four-ball friction and wear tests. Figure 5.10 shows the variation of WSD with the additive concentration of Bi-SA, Pb-SA and Pb-PEG nanoparticles in LP. In Figure 5.10(a) [19], it can be observed that the WSD of LP has reduced from 0.72 to 0.53 mm after adding Bi-SA nanoparticles at a very low additive concentration of 0.02 %, which suggests that the addition of Bi-SA nanoparticles can improve the AW properties of base oil. As the additive concentration is increased, the value of the WSD increases very slowly,

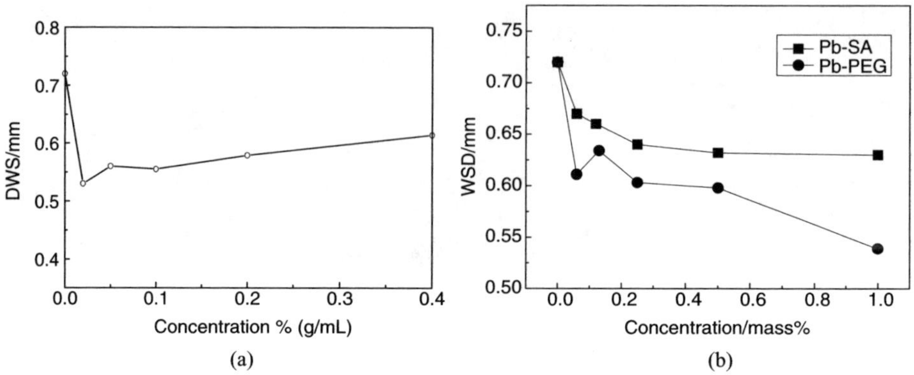

Figure 5.10 WSD as a function of the additive concentration of (a) Bi-SA, (b) Pb-SA and Pb-PEG nanoparticles prepared in LP under 300 N. With kind permission from Springer Science and Business Media from Zhao *et al.* [20], (c) (2004) Springer Science and Business Media

indicating that Bi-SA nanoparticles dispersed in base oils can have AW performance in a relatively broad additive concentration range.

Lead is a kind of conventional lubricant, so the application of Pb nanoparticles as lubricating additives is potentially interesting. It can be seen from Figure 5.10(b) [20] that the WSD values of LP containing Pb-SA and Pb-PEG nanoparticles are much lower than those of base oil in a broad concentration range (0.06 to 1.0 %), especially for Pb-PEG nanoparticles. They can improve the AW property of LP efficiently, even at a very low concentration, and it is evident that their AW performance can be improved by increasing the additive concentration, indicating that lead nanoparticles are good lubricating additives.

The relationship between the WSD of LP and LP plus Bi-SA, Pb-SA and Pb-PEG nanoparticles as a function of the applied load is shown in Figure 5.11 [19]. It can be seen that these

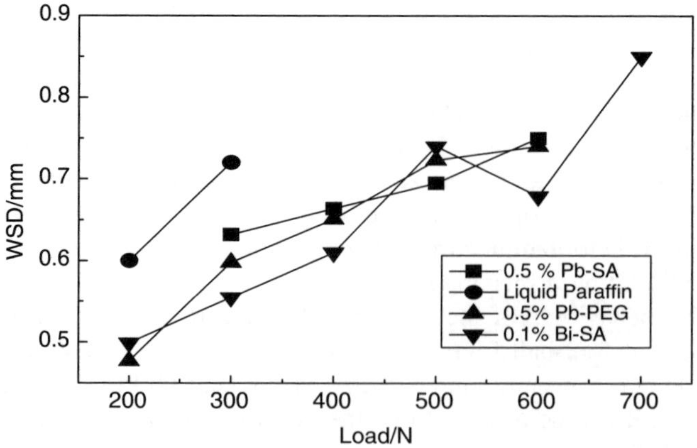

Figure 5.11 Relationship between the WSD and applied load with lubrication of LP and LP containing Bi-SA, Pb-SA and Pb-PEG nanoparticles at certain concentrations

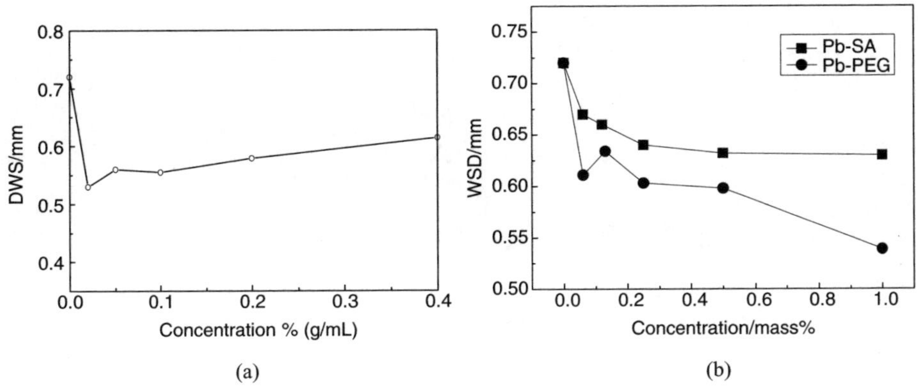

Figure 5.12 Friction coefficient as a function of the additive concentration of Bi-SA and Pb-PEG nanoparticles in LP under 300 N

metal nanoparticles prepared by the surfactant-assisted solution-dispersing method have excellent antiwear properties under different loads. At the load of 300 N, the WSD values of LP containing 0.1 % Bi-SA, 0.5 % Pb-SA and 0.5 % Pb-PEG nanoparticles are 0.56, 0.63 and 0.60 mm, respectively, which are smaller than those of pure LP. In the friction tests, seizure of the tribo-pairs appeared when the applied normal load is above 400 N for LP, but is as high as 600 N for LP containing Pb-SA and Pb-PEG nanoparticles and 700 N for Bi-SA nanoparticles, indicating that the load-carrying capacity of LP can be greatly improved by the addition of as-prepared metallic nanoparticles. At a load of 600 N, the WSD values of LP containing 0.1 % Bi-SA, 0.5 % Pb-SA and 0.5 % Pb-PEG nanoparticles are only 0.68, 0.75 and 0.74 mm, respectively, illustrating that these metal nanoparticles have excellent antiwear properties at high load.

The variations of the friction coefficient with the additive concentration of Bi-SA and Pb-PEG nanoparticles in LP are given in Figure 5.12 [19]. The friction coefficient is very sensitive to the addition concentration of the Bi-SA nanoparticles. To achieve the best friction-reduction property, a proper additive concentration of the Bi-SA nanoparticles is needed, which is in the range of 0.02 to 1.00 % under selected testing conditions. Comparably, the friction coefficient of LP can obviously be decreased by adding Pb-PEG nanoparticles over a wide concentration range, which suggests that the Pb-PEG nanoparticles could be used as a friction-reducing additive (friction modifier). When the concentration of Pb-PEG nanoparticles is about 0.5 %, the best friction-reducing property is obtained.

SEM analysis is also conducted to observe the worn surface of the steel ball after the friction test under the lubrication of LP with the as-prepared nanoparticles; the results are presented in Figure 5.13 [19, 20]. It can be observed that the wear scar surfaces are all relatively smooth and the rubbing traces are shallow, which is in good accordance with their better AW property.

In conclusion, nanoparticles of low melting point metals, including indium, bismuth, tin and lead, can be easily prepared via the solution-dispersing method. The simultaneously formed oxide layer or the *in situ* generated surfactant layer is retained to prevent the reconjunction of little metal drops during the preparation process and improved the dispersibility of the

Figure 5.13 The morphologies of the worn surfaces under lubrication of LP containing 0.1 % Bi-SA, 0.5 % Pb-SA and 0.5 % Pb-PEG nanoparticles at an applied load of 300 N and friction duration of 30 min. With kind permission from Springer Science and Business Media from Zhao *et al.* [20], © (2004) Springer Science and Business Media

resulting metal nanoparticles. A series of pilot studies on the tribological properties of these nanoparticles used as oil additives has proved their potential value as novel lubricants.

5.4 Nanolubricants Made of Low Melting Point Metal Alloy Nanoparticles

When two or more kinds of metals are melted together to form a metal alloy, the physical, electronic, mechanical and chemical properties will change considerably. Therefore, nanoparticles of metal alloys are expected to have more fascinating properties in electronics, catalysis and magnetism [21, 22] compared to nanoparticles of single metals. As a result, metal alloy nanoparticles have been the focus of intensive research in recent years [23–25].

In practice, low melting point metals are often used to prepare lubricating coatings in the form of metals alloys to achieve the most optimizing properties, and nanoparticles of alloyed metals are anticipated to have tunable excellent tribological properties as the composition of the alloys is varied. Until today, nanoparticles of In-Sn, Bi-In, Pb-Bi, Sn-Bi and Sn-Cd alloys that can be dispersed in oils have been synthesized and their tribological properties have been studied.

5.4.1 In-Sn, Bi-In and Pb-Bi Nanoparticles Prepared by the Direct Solution-Dispersing Method

Zhao and his coworkers [19, 26–28] have extended the direct solution-dispersing method to the synthesis of low melting point metallic alloy nanoparticles. Figure 5.14 shows the TEM images and their corresponding XRD patterns of the resulting In-Sn, Bi-In and Pb-Bi alloy nanoparticles [26–28]. From TEM images, it can be seen that both In-Sn and Pb-Bi alloy nanoparticles prepared by this method appear to have a spherical shape and have an average particle diameter of 60 and 50 nm, respectively. However, Bi-In alloy nanoparticles prepared by the same method have an evident dendritic morphology, with an average diameter of 5 nm and an elongated shape.

The XRD pattern of bulk In-Sn alloy displays nine strong peaks [27], which are assigned to the reflections for crystalline In_3Sn and $InSn_4$, respectively. Comparably, the XRD pattern of In-Sn alloy nanoparticles contains clearly distinguishable multiple peaks, which can be

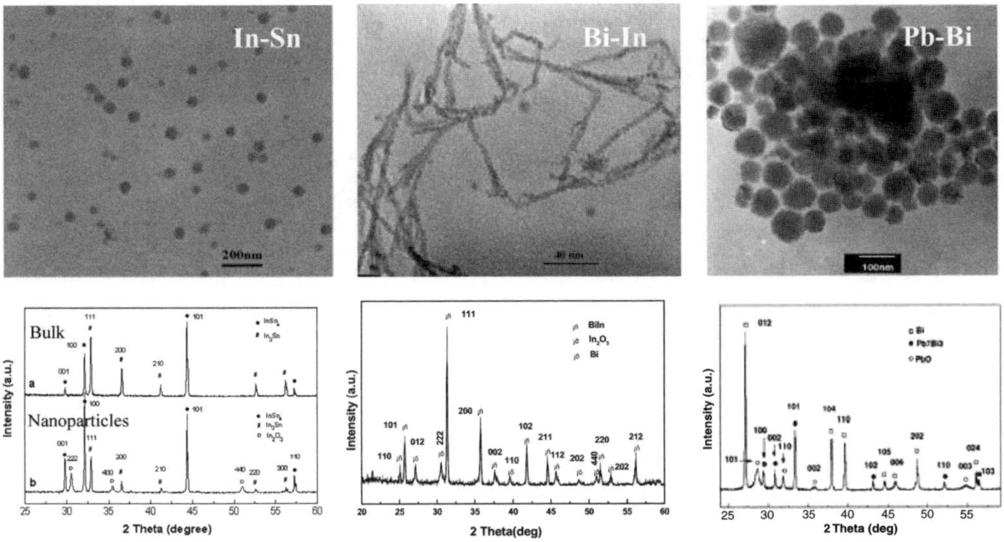

Figure 5.14 TEM images and their corresponding XRD patterns of In-Sn, Bi-In and Pb-Bi alloy nanoparticles. Reproduced by permission of The Royal Society of Chemistry and reprinted with permission from Zhao *et al.* [26], © (2004) American Chemical Society and from Zhao *et al.* [28], © (2006) Elsevier

indexed to crystalline *b*-In_3Sn, $InSn_4$ and In_2O_3. It is clear that In-Sn alloy nanoparticles have the same crystal structure as that of the bulk In-Sn alloy. The XRD pattern of Bi-In dendritic nanocrystals also contains multiple peaks [26], some of which correspond to the tetragonal phase of Bi-In alloy. There also appear weak additional peaks indexed to crystalline Bi and In_2O_3, respectively. The XRD pattern of Pb-Bi alloy nanoparticles displays multiple strong peaks [28] that can be easily assigned to the reflections for crystalline Pb_7Bi_3, Bi and PbO, respectively. This reveals the presence of the equilibrium phase Pb_7Bi_3 and Bi in the nanoparticles. The analysis of these XRD patterns reveals that the resulting products are all mixtures, and one component is one of two corresponding metal oxides; i.e. the oxidization must happen in the preparation process of metal alloy nanoparticles.

Figure 5.15 gives the relationship between the concentration of three kinds of bimetal alloy nanoparticles in base oil and their corresponding WSD values [19, 27]. For comparison, the results of monometallic nanoparticles as additives in oil are also provided. It can be seen that these bimetal alloy nanoparticles are all good lubricating additives, which can improve the AW property of the base oil over a wide concentration range. It is evident that, in the tested concentration range, the AW performance of In-Sn and Bi-In nanoparticles is much better than that of their corresponding monometallic nanoparticles, which might be attributed to their nano-alloy structures. However, this phenomenon is not obvious for Pb-Bi alloy nanoparticles.

Figure 5.16 shows the variations of WSD values with the lubrication of LP containing the as-prepared bimetal alloy nanoparticles under different applied loads [19, 27, 28]. The results of zinc dialkyl dithiophosphate (ZDDP, a commonly used commercial lubricating additive), Sn and In nanoparticles are also presented for comparison in Figure 5.16(a). It is clear that the In-Sn alloy nanoparticles displayed an excellent AW property under the applied load range

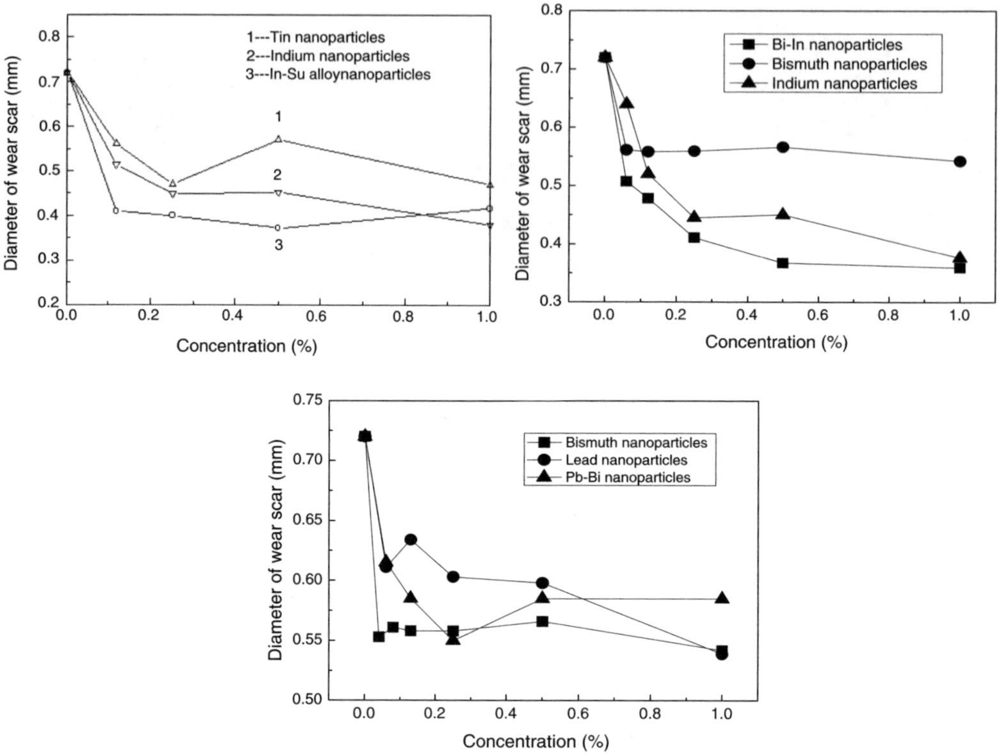

Figure 5.15 WSD as a function of the additive concentration of bimetal alloy nanoparticles and their corresponding monometallic nanoparticles in LP under 300 N. Reproduced by permission of The Royal Society of Chemistry and reprinted with permission from Zhao *et al.* [27] and Zhao [19]

of 300 to 500 N, in which a relatively lower WSD was observed as compared to those of LP containing 4 % ZDDP, 0.5 % indium nanoparticles and 1 % tin nanoparticles. When the applied load was increased to 600 N, the AW ability of In-Sn alloy nanoparticles was adjacent to ZDDP. A similar result can be observed for Bi-In and Pb-Bi alloy nanoparticles.

Figure 5.17 [19, 27, 28] gives the morphologies of the worn surfaces with lubrication of LP containing 0.5 % In-Sn, 0.5 % Bi-In and 0.25 % Pb-Bi alloy nanoparticles, which shows that the worn surfaces are all relatively smooth and the rubbing traces are shallow.

5.4.2 Sn-Bi and Sn-Cd Alloy Nanoparticles Prepared by the Ultrasonic-Assistant Solution-Dispersing Method

When liquids are irradiated with high-intensity ultrasound, the acoustic cavitations can lead to a collapse of bubbles, producing intense local heating (5000°C) and high pressures (1000 atm) in a very short lifetime (heating and cooling rates above 1010 K/s). The transient, localized hot spots can generate high-energy physical or chemical transitions and can be utilized to induce and accelerate reactions [29–33]. A variety of nano-sized metal and alloy particles have been prepared using the sonochemical method in the last few years [34–36].

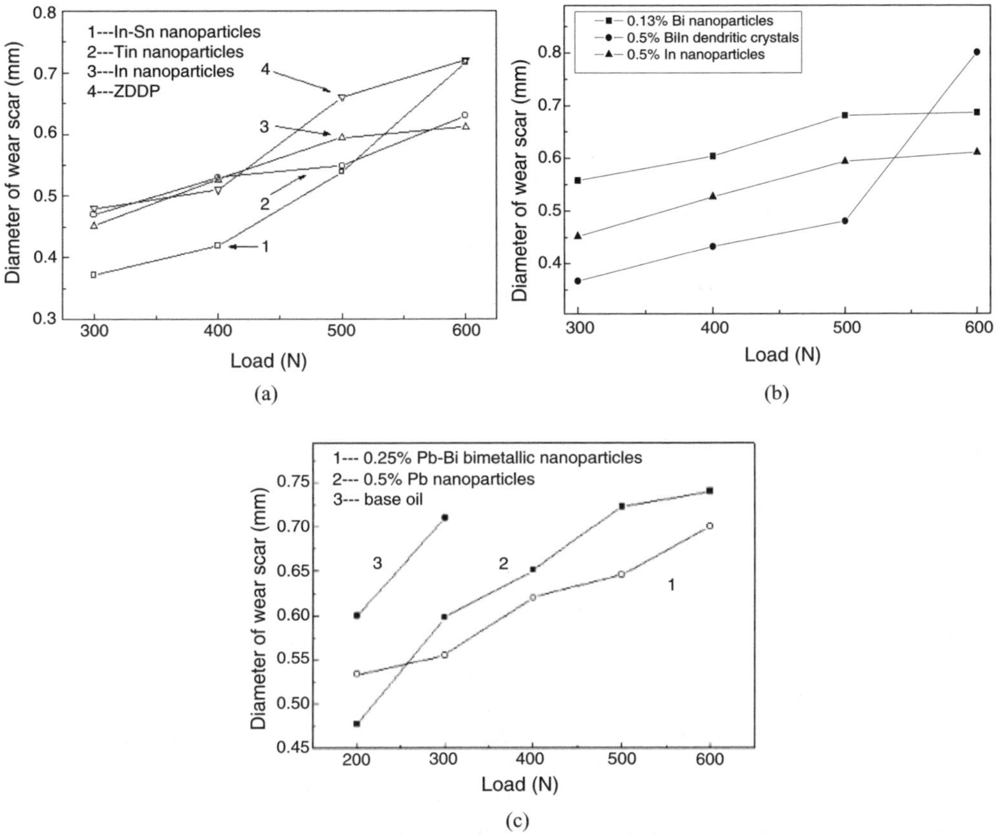

Figure 5.16 Relationships between the WSD value and applied load with lubrication LP containing 0.5 % In-Sn, 0.5 % Bi-In and 0.2 5% Pb-Bi alloy nanoparticles. Reproduced by permission of The Royal Society of Chemistry and reprinted with permission from Zhao *et al.* [28], © (2006) Elsevier

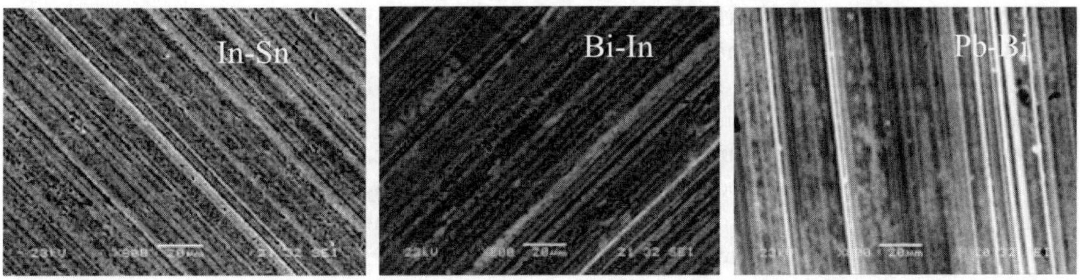

Figure 5.17 The morphologies of the worn surfaces with lubrication of LP containing 0.5 % In-Sn, 0.5 % Bi-In and 0.25 % Pb-Bi alloy nanoparticles with an applied load of 300 N for 30 min. Reproduced by permission of The Royal Society of Chemistry and reprinted with permission from Zhao *et al.* [28], © (2006) Elsevier

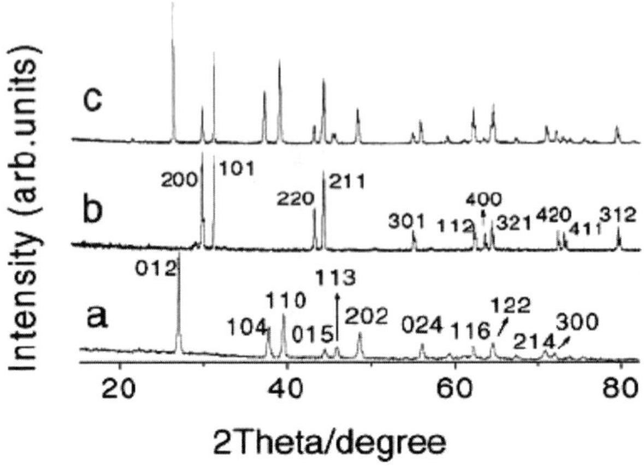

Figure 5.18 XRD pattern of (a) bismuth, (b) tin and (c) the as-produced Sn-Bi nanoparticles. Reproduced by permission of Elsevier from Chen *et al.* [37], © (2005) Elsevier

Chen and his coworkers [37, 38] reported a new sonochemical method used to prepare low melting point metal alloy nanoparticles for use as oil additives. In this method, bulk alloys of low melting point metals were put in oils and then directly dispersed into the oils by ultrasonification at room temperature. The XRD patterns of Sn-Bi and Sn-Cd nano-alloy materials prepared by this method are shown in Figures 5.18 [37] and 5.19 [38]. It can be clearly seen that the all of the prominent peaks of these Sn-Bi alloy nanoparticles in Figure 5.18 could be assigned to those of the metals bismuth or tin. Only one sharp and narrow endothermic peak at 140.8°C was found in its corresponding DTA curve, revealing that the powder consisted of Sn-Bi eutectic alloy nanoparticles and not simply of a mixture of tin and bismuth nanoparticles. Therefore the resulting Sn-Bi nanoparticles were the eutectic alloy consisting of the tetragonal phase of tin and the rhombohedral phase of bismuth. In addition, no other peaks attributed to metal oxides were found in pattern (c) of Figure 5.19, which indicated that no oxidation reactions took place in the process. A similar result was found in the XRD pattern of Sn-Bi nanoparticles.

Figure 5.20 shows the TEM images of as-prepared Sn-Bi and Sn-Cd nanoparticles [37, 38]. It is clear that these nanoparticles are well separated, mostly hexagonal and spherical, and the average particle diameter is in the range of 10 to 25 nm. A proposed growing mechanism of these bimetal alloy nanoparticles includes the following three steps. First, the bulk bimetal alloy melts in LP, which is mainly induced by the energy of collapse of the cavitations of bubbles. Second, the melted bimetal alloy is dispersed into unequal ultrafine liquid particles in the solvent by the results of sonication. Finally, the unequal particles are homogenized to become nano-sized particles by collision and cutting resulting from the process of ultrasonic irradiation. According to this mechanism, it can be assumed that ultrasonic power will have a significant impact on the size distribution of the bimetal alloy nanoparticles and that smaller alloy particles could be generated with increasing ultrasonic power. This has been proved by the experimental results shown in Table 5.6. These researchers also pointed out that the sonication time had little impact on the size of the nanoparticles.

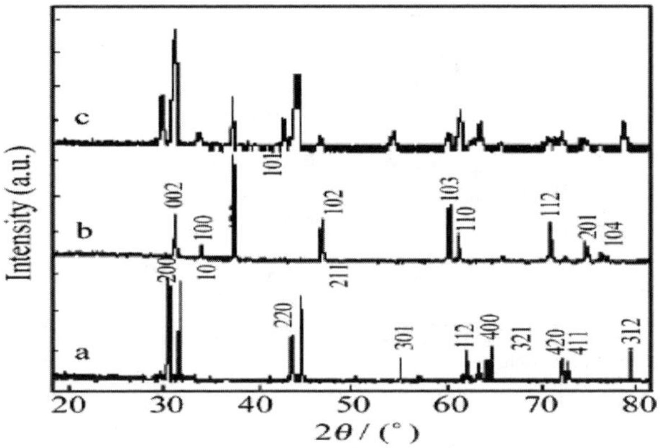

Figure 5.19 XRD pattern of (a) tin, (b) cadmium and (c) the as-produced Sn-Cd nanoparticles. Reproduced by permission of Elsevier and Editorial Office of Chemical Research from Chen *et al.* [38], © (2005) Elsevier and Editorial Office of Chemical Research

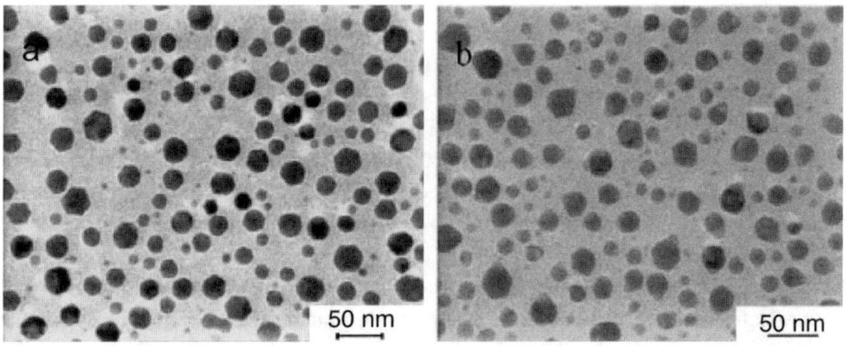

Figure 5.20 TEM images of (a) Sn-Bi and (b) Sn-Cd nanoparticles prepared at 1000 W/cm^2. Reproduced by permission of Elsevier from Chen *et al.* [37[, © (2005) Elsevier and of Elsevier and Editorial Office of Chemical Research from Chen *et al.* [38], © (2005) Elsevier and Editorial Office of Chemical Research

Table 5.6 The size distribution of Sn-Bi nanoparticles obtained at different ultrasonic power. Reproduced by permission of Elsevier from Chen *et al.* [37], © (2005) Elsevier

Ultrasonic power (W/cm^2)	Size range (nm)
700	60–80
800	45–60
900	25–40
1000	10–25

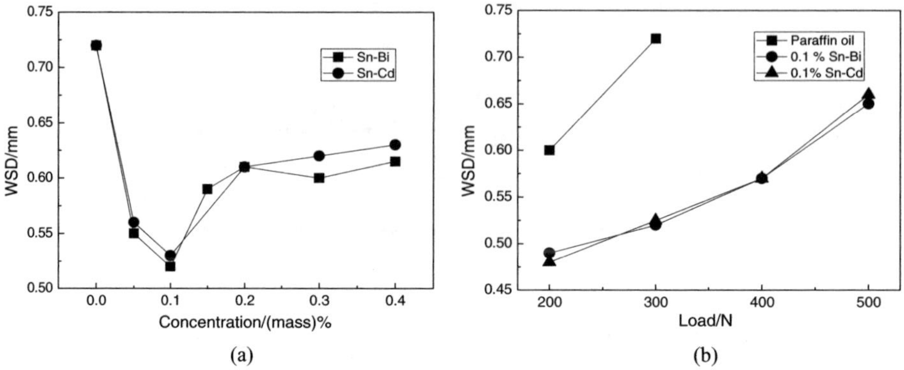

Figure 5.21 (a) Relationship of the WSD with the additive concentration of Sn-Bi and Sn-Cd nanoparticles under 300 N and (b) the WSD as a function of applied load with the lubrication of LP and LP containing 0.1 % Sn-Bi and 0.1% Sn-Cd nanoparticles. Reproduced by permission of Elsevier from Chen *et al* .[37], (c) (2005) Elsevier and of Elsevier and Editorial Office of Chemical Research from Chen *et al.* [38], (c) (2005) Elsevier and Editorial Office of Chemical Research

Nanoparticles of Sn-Bi and Sn-Cd alloys prepared by the ultrasonic-assistant solution-dispersing method are reported to have good oil solubility and can be used directly as lubricating oil additives. The tribological properties of Sn-Bi and Sn-Cd alloy nanoparticles tested in a four-ball tester are shown in Figure 5.21. The variation of WSD with the different concentrations of alloy nanoparticles in LP under 300 N is shown in Figure 5.21(a). Apparently the LP containing alloy nanoparticles gives a smaller WSD in a wide concentration range compared with the base oil, indicating that these bimetal alloy nanoparticles can improve the AW property of LP. At the concentration of 0.1%, both Sn-Bi and Sn-Cd nanolubricants present the best AW property. Figure 5.21(b) shows the variation of the WSD of LP containing 0.1 % alloy nanoparticles at different loads. It can be seen that both kinds of nanolubricants show a better AW property in the load range of 200 to 500 N than pure LP. These tribological data illustrate that these alloy nanoparticles can be used as lubricating oil additives with good tribological properties.

5.5 Mechanism of Metal Nanoparticles Used as Oil Additives

The research and application of nanoparticles as additives in lubricant oils have been ongoing for twenty years. However, the exact or real action mechanism of nanoparticulate oil additives is still unclear. Two hypotheses seem reasonable:

(a) 'Microrolling bearing' between tribo-pairs. This hypothesis [39] contends that when nanoparticles are dispersed equably in lubricant oils, each individual particle between the rubbing tribo-pairs can avoid the direct contact of rubbing surfaces and change the sliding friction of tribo-pairs into rolling friction, thus exhibiting excellent AW and EP performance. This process comes into existence when the normal load between the tribo-pairs is very low and the nanoparticles must be dispersed uniformly and keep their rigidity under local high temperature and high pressure in the microcontact zones [42].

(b) Boundary film formation by *in situ* formation of metal nanoparticles. This hypothesis [13, 41–43] contends that nanoparticles dispersed in lubricant oils can form a layer of depositing film that can decrease the shear strength of tribo-pairs, offset the microdamage and scratch the friction surface. This process can make the surface slick and even, thus favouring stress release, friction reduction, prevention of wear and even the self-repairing of a worn surface. The situation may be more suitable for soft metal nanoparticles such as copper, silver, lead, tin and bismuth.

Li *et al.* [6] have discussed the mechanism of DDP-capped copper and silver nanoparticles as oil additives. In their studies, the traditional electrical contact resistance (ECR) measuring technique, which has been applied to various studies of the contact of tribo-pairs in mixed lubrication or in boundary lubrication, has been used to investigate the *in situ* formation of the boundary film of metal nanoparticles. It is easy to understand that the ECR value between metallic tribo-pairs is very high under hydrodynamic lubrication, but very low (<0.1 ohm) when metallic contact occurs. Of course, the ECR value is larger than several thousands of ohms when a nonmetallic layer exists on top of specimens. For boundary lubrication of a metallic interface with AW/EP additives such as ZDDP blended in lubricating oils, antiwear efficiency relies on the reaction film formation, which acts also as an insulating barrier for electrical current. The ECR can therefore be used to monitor the formation process of the reaction film.

Figure 5.22(a) shows the ECR curve between tribo-pairs lubricated with a drop of pure LP. It is seen that the ECR value fluctuated acutely in the range from several to tens of kilo-ohms. The AW/EP mechanism at this stage may be the ‘wearing off the natural oxide layer, reoxidizing the surface, wearing off the oxide layer’ process, according to the general cognition of boundary lubrication. If a droplet of LP with ZDDP is added, an obvious increase in the ECR could be observed, indicating the generation of a tribo-film to some extent. A similar phenomenon appears when a droplet of LP containing copper nanoparticles is added. With the addition of ZDDP, the ECR value immediately increased from 20 to 30–50 kilo-ohms and became unstable. With the addition of copper nanoparticles, the ECR value immediately increased from 20 to 50–70 kilo-ohms respectively, and then became steady, indicating a thicker and firmer tribo-film was formed in the process of friction compared to that of ZDDP. This is in accordance with the good AW/EP properties of LP containing copper nanoparticles. In the case of silver nanoparticles, the ECR value tended to decrease at an early stage. The reason might be that silver nanoparticles are more inert than other metal nanoparticles, and can form a metallic silver tribo-film, which has good electrical conductivity.

The formation of tribo-films is further proved by surface analysis on the wear scar by means of SEM, EDS and XPS. It has been shown that the SEM images of the wear scar of steel balls, lubricated by LP with nanoparticulate additives, show a much smoother wear surface. The typical EDS and XPS analysis results usually indicate that the metallic elements in the nanoparticles were deposited and enriched on the worn surface. Figure 5.23 [6] gives the typical element distribution on a worn steel surface lubricated by oils containing copper nanoparticles.

Based on the above results, the researchers deduced that when metal nanoparticles are added into lubricant oils, a homogeneous and stable colloid solution is formed. When the lubricant film between tribo-pairs become thinner, mixed lubrication or boundary lubrication occurs, and the nanoparticles perhaps carry a proportion of the load and separate the two

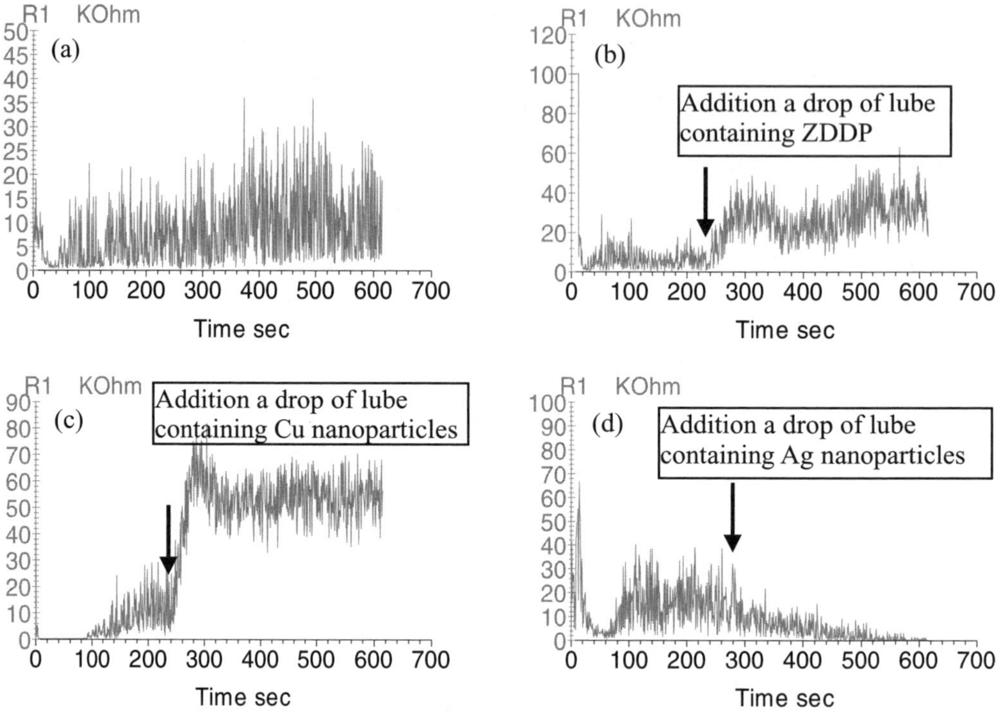

Figure 5.22 Variations of electrical contact resistance between tribo-pairs versus friction time (a) lubricated by pure LP, (b) lubricated by pure LP during the first 250 seconds, then adding a drop of LP containing 5 % zinc dialkyl dithiophosphate (ZDDP), (c) Cu nanoparticles and (d) Ag nanoparticles. Test condition: a steel ball reciprocally rubbing against a steel plate at a frequency of 15 Hz and a normal load of 5 N at 25°C with the lubrication of a drop of oil. With kind permission from Springer Science and Business Media from Li *et al.* [6], (c) (2006) Springer Science and Business Media

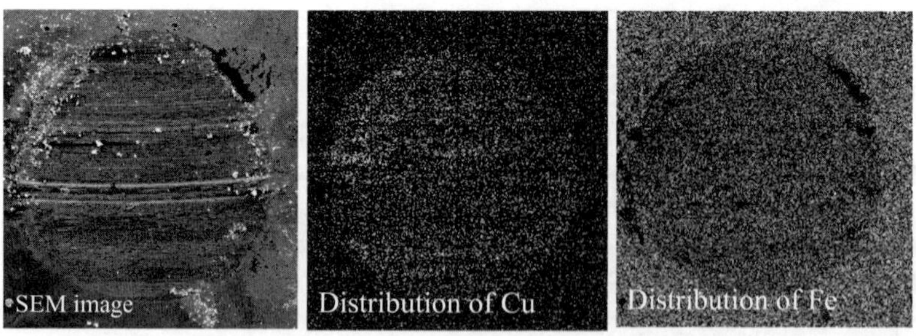

Figure 5.23 SEM micrographs and typical element distributions on a worn steel surface lubricated by 1 % wt Cu nanoparticles in liquid paraffin (four-ball, 1450 r/min, 392 N, 30 min). With kind permission from Springer Science and Business Media from Li *et al.* [6], © (2006) Springer Science and Business Media

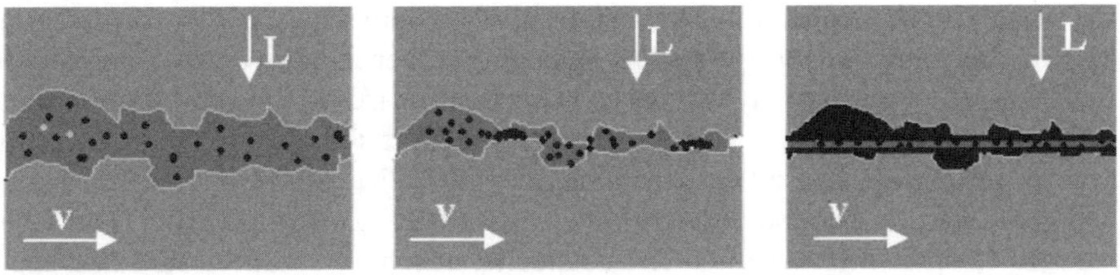

The thickness of lubricant film is decreasing | Mixed or boundary lubricanton | The formation of deposition film

Figure 5.24 Schematic representation of the mechanism of nanoparticles as a lubricant additive. With kind permission from Springer Science and Business Media from Li *et al.* [6], © (2006) Springer Science and Business Media

surfaces to prevent adhesion, benefiting the improvement of the antiwear property. When the shearing is strong, the core/shell structure of nanoparticles may be destroyed, the surface capping layer desorbed and decomposed, the inorganic core melted and welded on to the shearing surface, or the structure may react with the specimen to form a protective layer providing good AW and EP properties. This process is illustrated in Figure 5.24 [6].

Tarasov and his coworkers [13] have also studied and discussed the action mechanism of copper nanoparticles produced by the electric explosion of metallic wire (EEW) process used as an additive in SAE 30 motor oil. In the topology analysis of the corresponding worn surfaces shown in Figure 5.25, it was found that the worn surfaces looked very smooth under high-speed and low-load ($P = 110$ N, $v = 2.62$ m/s) tests conditions, and a rippled pattern was a characteristic of the worn surface obtained at a high load of 184 N irrespective of the sliding speed value. Optical microscopy showed copper agglomerates trapped on the worn surfaces. The microstructure of the copper agglomerates indicated by TEM images shows them to be heavily deformed.

It is pointed out that the addition of copper nanoparticles into lubricating oils can replace or at least cover the real wear particles with those of a more ductile metal, which would provide better lubrication as compared, for example, with steel wear particles. The softer metal would also probably decrease adhesion between the frictional surfaces, which is dependent on the particular metal used as the wear surface. It is also realized that the copper nanoparticles added

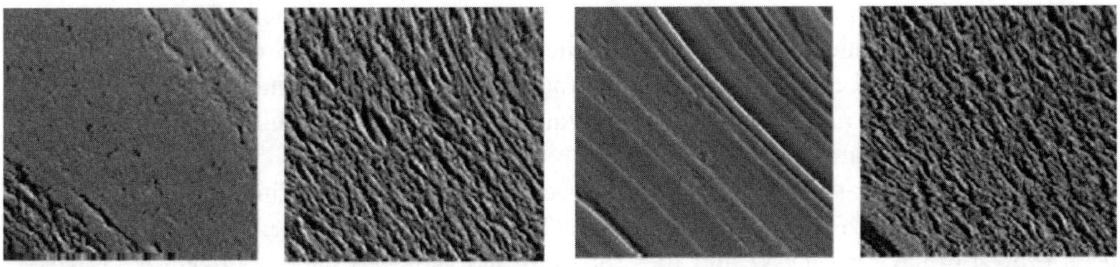

(a) 110 N; 1.56 m/s (b) 184 N, 1.56 m/s (c) 110 N, 2.62 m/s (d) 184 N, 2.62 m/s

Figure 5.25 SEM images of worn surfaces after friction tests with copper nanopowder additive (×1500). Reproduced by permission of Elsevier from Tarasov *et al.* [13], © (2004) Elsevier

in oils are trapped on the worn surfaces in the form of agglomerates, and have a deformation incompatibility between the nanocrystalline subsurface layer and the underlying base metal during the friction process. The layer behaves like a highly viscous fluid and friction-induced stresses effectively damp inside it. It is reasonable to suppose that the higher the temperature (sliding speed) of the subsurface layer, the more likely it is that such a layer will behave like a fluid and provide lower friction together with a smoother wear surface. This is what might have happened to specimens during a test conducted at a high speed of 2.62 m/s and 110 N, where a smooth wear path was formed. At a higher load of 184 N, plastic deformations were more severe, creating a rippled surface. Based on these discussions, the mechanism of copper nanoparticles from the EEW method can be summarized as: 'local overheating due to direct contact of two surfaces initiates chemical deposition of copper on steel, providing a soft surface limited to the locality of the friction pair to reduce friction and wear'.

It is anticipated that the nanoparticles of low melting point metals and their alloys have a similar mechanism. However, the reported results of EDS analysis on a worn surface reveal that no corresponding metallic elements of nanoparticles were in the rubbing surface [18, 19, 27, 28]. The reason might be that these metallic elements in nanoparticles have less of a tendency to attach on to the friction surface. However, it is reasonable to assume that the low melting point metal and alloy nanolubricants might be melted to form an instantaneous liquid metal film to prevent the direct contact of rubbing surfaces under boundary lubrication conditions, thus generating AW, friction-reducing and EP properties.

For the traditional organic compound AW/EP additives, the tribo-film formation is carried out through the reaction of 'active' elements with the surfaces of the metallic tribo-pair. In other words, the phenomenon of 'corrosive wear' will exist as long as these additives begin to operate on the prevention of severe wear. Thus, under boundary lubrication or mixed lubrication using organic compound additives, the mass loss of trio-pairs owing to 'corrosive wear' will be inevitable. However, the improvement of the friction-reducing and AW properties of the frictional system are mainly induced by the preferential depositing of metal nanoparticles and the *in situ* generation of metallic tribo-film when metal nanoparticles are used as additives. The depositing film is made up of metallic nanoparticles; i.e. the tribo-film formation need not consume the materials of tribo-pairs.

5.6 Perspective

Nowadays, there are much more increasing demands for the development of novel oil additives, mainly due to three factors. The first is the increased complexity, higher speeds and loads, and higher working temperatures that occur in modern machinery. The second is the increasing attention paid to environmental concerns. The third is the increasing use of more highly refined mineral oils and synthetic hydrocarbons for the lubrication of extreme conditions (high/low temperature, erosive circumstance, high vacuum), which usually have very little, or no, inherent antiwear performance.

Traditional organic sulfur and phosphorus compounds, such as zinc dialkyl (or diaryl) dithiophosphate (ZDDP), tricresyl phosphate (TCP) and sulfurised isobutylene (SIB), have been proved to have good AW and EP properties and achieved wide application. However, these AW/EP additives are chemically more or less reactive, and have lower resistance to heat and oxidation than most modern base oils, and cannot be used to formulate oils for high-temperature applications. The more powerful EP additives usually have some corrosive effect in contact with metals and are only used when necessary. In addition, 'active' elements such as sulfur, phosphorus or chlorine in their molecular form are hazardous to human health and plant growth. Large searches for substitutes for these organic compounds as AW/EP additives have been carried out over the past decades, yet no satisfactory results have been achieved until now.

With the rapid development of nanotechnology, nanolubricants made of metals have been proposed and studied. The reported experimental results have proved that metal nanoparticles can improved the friction-reducing, AW and EP properties when added in oils. Compared with traditional organic compound additives, they generate lubricating properties by means of providing *in situ* deposition of a microlubricating film limited to the locality of the friction pair without corrosive effect, and can be used at higher temperatures owing to their inherent character. They have the potential to become the new generation of AW/EP additives.

However, the investigations concerning nanolubricants made of metal have been inadequate until recently. The reported experimental results are very primative to some extent and some technique problems have not been solved: (a) the long-term stability of metal nanoparticles and oils containing metal nanoparticles has not been solved very well; (b) the influence of metal nanoparticles on the base oil has not been revealed; (c) interactions between metal nanoparticles and other additive integrants for formulated oils have not been fully studied. In-depth research of these technical obstacles would stimulate the application of the novel nanolubricants. Additionally, economic pressure is likely to be considered before the best additives available today are replaced by nanomaterials.

References

1. Zhang, Z. J., Xue, Q. J. and Zhang, J., *Wear*, **209**, 1997, 181.
2. Xu, T., Zhao, J. Z. and Xu, K., *Journal of Phyics D: Applied Physics*, **29**, 1996, 2932.
3. Qiu, S. Q., Zhou, Z. R., Dong, J. X. and Chen, G. X., *Journal of Tribology*, **123**, 2001, 441.
4. Zhou, J. F., Wu, Z. S., Zhang, Z. J., Liu, W. M. and Xue, Q. J., *Tribology Letters*, **8**, 2000, 213.
5. Zhou, J. F., Yang, J. J., Zhang, Z. J., Liu, W. M. and Xue, Q. J., *Materials Research Bulletin*, **34** (9), 1999, 1361.
6. Li, B., Wang, X. B., Liu, W. M. and Xue, Q. J., *Tribology Letters*, **22** (1), 2006, 78.
7. Wang, B., Shen, J. M., Zhu, H., Kang, X. H. and Zhe, L., *Journal of Beijing Jiaotong University*, **30** (3), 2006, 43.
8. Yu, H. L., Xu, Y., Wang, X. L., Shi, P. J. and Xu, B. S., *Journal of Academy of Armored Force Engineering*, **20** (5), 2006, 86–89.
9. Zhou, J. F., Zhang, Z. J., Wang, X. B., Liu, W. M. and Xue, Q. J., *Tribology*, **20** (2), 2000, 123.
10. Yu, H. L., Xu, Y., Liu, Q., Shi, P. J. and Xu, B. S., *China Surface Engineering*, **2**, 2005, 23.
11. Yang, X., Zhang, G. Z. and Bai, W. X., *Lubricating Oils*, **21** (4), 2006, 35.
12. Qiu, S. Q., Dong, J. X. and Chen, G. X. *Lubrication Engineering*, **133** (3), 1999, 14.
13. Tarasov, S., Klubaev, A., Belyaev, S., Lerner, S. and Tepper, F., *Wear*, **252**, 2002, 63.
14. Xia, Y. Q., Ding, J. Y., Ma, X. G., Yang, W. T., Liu, W. M. and Xue, Q. J., *Lubrication Engineering*, **131** (1), 1999, 40.
15. Zhao, Y. B., Zhang, Z. J. and Dang, H. X., *Journal of Physical Chemistry B*, **107**, 2003, 7574.
16. Rohr, O., *Industrial Lubrication Tribology*, **54**, 2002, 153.
17. Zhao, Y. B., Zhang, Z. J. and Dang, H. X., *Material Science Engineering A*, **359**, 2003, 405.
18. Zhao, Y. B., Zhang, Z. J. and Dang, H. X., *Materials Letters*, **58**, 2004, 790.
19. Zhao, Y. B., Preparation of metal, alloy, oxide and sulfide nanomaterials by solution dispersion and its tribological properties, PhD Dissertation, Lanzhou, Lanzhou Institute of Chemical Physics, Chinese Academy of Sciences, 2004.
20. Zhao, Y. B., Zhang, Z. J. and Dang, H. X., *Journal of Nanoparticle Research*, **6**, 2004, 47.
21. Mandal, M., Kundu, S., Sau, T. K., Yusuf, S. M. and Pal, T., *Chemical Materials*, **15**, 2003, 3710.

22. Dong, X. L., Zhang, Z. D., Zhao, X. G., Chuang, Y. C., Jin, S. R. and Sun, W. M., *Journal of Materials Research,* **14** (2), 1999, 398.
23. Huang, S. P. and Balbuena, P. B., *Journal of Physical Chemistry B,* **106**, 2002, 7225.
24. Park, K.W., Choi, J. H., Kwon, B. K., Lee, S. A., Sung, Y. E., Ha, H. Y., Hong, S. A., Kim, H. and Wieckowski, A., *Journal of Physical Chemistry B,* **106**, 2002, 1869.
25. Hodak, J. H., Henglein, A. and Hartland, G.V., *Journal of Physical Chemistry B,* **104**, 2000, 5053.
26. Zhao, Y. B., Zhang, Z. J., Liu, W. M., Dang, H. X. and Xue, Q. J., *Journal of American Chemical Society,* **126**, 2004, 6854.
27. Zhao, Y. B. Zhang, Z. J. and Dang, J., *Materials Chemistry,* **14**, 2004, 299.
28. Zhao, Y. B., Liu, J., Cao, L. Q., Wu, Z. S., Zhang, Z. J. and Dang, H. X., *Materials Chemistry and Physics,* **99**, 2006, 71.
29. Flint, E. B. and Suslick, K. S., *Science*, **253**, 1991, 1397.
30. Suslick, K. S., *Science*, **247**, 1990, 1439.
31. Kurikka, V. P. M. S., Aharon, G. and Ruslan, P., *Journal of Materials Chemistry*, **8** (3), 1998,769.
32. Kurikka, V. P. M. S., Aharon, G., Ruslan, P., Adam, R. and Janos, L., *Journal of Materials Research*, **15** (2), 2000, 332.
33. Hyeon, T., Fang, M., Cichowlas, A. A. and Suslick, K. S., *Materials Science Engineering A*, **204**, 1995, 186.
34. Li, Q. L., Li, H. L., Pol, V. G., Bruckental, I., Koltypin, Y., Calderon-Moreno, J., Nowik, I. and Gedanken, A., *New Journal of Chemistry*, **27**, 2003, 1194.
35. Zhu, J. J., Wang, H., Xu, S. and Chen, H. Y., *Langmuir,* **18**, 2002, 3306.
36. Li, H. L., Zhu, Y. C., Palchik, O., Koltypin, Y., Gedanken, A., Palchik, V., Slifkin, M. and Weiss, A., *Inorganic Chemistry*, **41**, 2002, 637.
37. Chen, H. J., Li, Z. W., Wu, Z. S. and Zhang, Z. J., *Journal of Alloys and Compounds,* **394**, 2005, 282.
38. Chen, H. J., Li, Z. W., Tao, X. J., Wu, Z. S. and Zhang, Z. J., *Chemical Research,* **16** (3), 2005, 34.
39. Yu, L. Y., Hao, C. C., Sui, L. N. and Cui, Z. L., *Journal of Materials Science and Engineering,* **22** (6), 2004, 901.
40. Alexeyev, N. M., Kuzmin, N. N., Trankovskaya, G. R. and Shuvalova, E. A., On the similarity of friction and wear processes at different scale levels, *Wear,* **156**, 1992, 251–261.
41. Tarasov, S. Y. and Kolubaev, A. V., Wear, 231, 1999, 228.
42. Makowski, R., *Tribologia,* **32** (4), 2001, 669.
43. Jiang, B. X., Chen, B. S. and Dong, J. X., *Lubrication Engineering,* **132** (3), 1999, 50.

6

Boron-Based Solid Nanolubricants and Lubrication Additives

Ali Erdemir

6.1 Introduction

Solid lubricants have been around for quite a long time and are used widely by industry in a large variety of moving mechanical systems [1]. During the last two decades, the popularity of these lubricants has increased significantly, mainly because their liquid counterparts could not alone meet the increasingly stringent conditions of more advanced mechanical systems operating under much harsher conditions involving very high or low temperatures, ultra-high vacuum, radiation, extreme contact pressures, very low or high speeds, etc. Moreover, much tighter emission control strategies and environmental mandates imposed by various government agencies in recent years have been discouraging the use of certain types of additives in these lubricants, mainly because of their adverse effects on the environment. Overall, it looks as though, in the near future, experts in the solid and liquid lubrication fields will have to work together to bridge the gap that exists between these two lubrication media and come up with novel solid-liquid formulations that can safely meet the increasingly harsher application conditions that will be demanded of future mechanical systems. In short, combining the potentials of solid and liquid lubricants in one package sounds like the most logical next step for achieving better performance and efficiency in future tribological systems.

For many decades, fine (micro-scale) powders of certain solid lubricants have been mixed with a range of greases and liquid lubricants to achieve better anti-friction and -wear properties in a variety of applications [2]. For example, molybdenum disulfide (MoS_2), graphite, hexagonal boron nitride (h-BN), boric acid (H_3BO_3), and poly-tetrafluoroethylene (PTFE) have all been mixed with greases to enhance lubrication. Colloidal dispersion of micro-powders of these solid lubricants in mineral and/or synthetic oils has also been tried to enhance their performance [3]. Key advantages of using solids with liquid and/or grease lubricants are summarized in Table 6.1. The reported results indicate that such combined uses of solid and

Nanolubricants Edited by J. M. Martin and N. Ohmae

Table 6.1 Key advantages that can be attained by blending solid lubricants with liquid lubricants in various tribological applications

Application environment and/or condition	Advantage
Vacuum	Long life (MoS_2 lubricates extremely well in high vacuum and does not evaporate or disappear in the long run like some of the liquids).
Pressure	Most solid lubricants have excellent endurance under extreme pressures; hence, may further increase the extreme pressure range of formulated oils and greases.
Temperature	Lubricating property of most solid lubricants is less sensitive to fluctuations in temperature. Some can function at very low and high temperatures. In contrast, liquid lubricants are very sensitive to changes in temperature and may not function at temperature extremes.
Electrical/Thermal Conductivity	Some of the solid lubricants may provide better electrical and thermal conductivity, whereas most liquid lubricants are insulating.
Radiation	Most solid lubricants are relatively insensitive to UV, gamma, or nuclear radiation, whereas liquids may undergo chemical/physical changes.
Wear	Solid lubricants can provide excellent durability and lubricity at very slow speeds, whereas liquids may break down if the pressure is extreme and severe wear or scuffing proceeds.
Friction	Extremely low friction coefficients are feasible with some solid lubricants, and when combined with liquids, such low friction may persist over much longer sliding distances or much heavier loads.
Storage/Aging	Solid lubricants can be stored for very long times without decay in their lubricity (MoS_2 is sensitive to humidity and oxygen).
Hygiene, Environmental Safety	Solid lubricants have better industrial hygiene than liquids due to little or no hazardous emissions. Some of them are naturally occurring, so even if they are released to the environment, there is no danger of contamination.
Compatibility with Tribological Surfaces	Solid lubricants are very compatible with all kinds of hard-to-lubricate surfaces like Al, Ti, stainless steels, and ceramics; hence, if they are present in liquid lubricants, effective lubrication of such surfaces becomes feasible.
Resistance to Aqueous and Chemically Aggressive Environments	Most solid lubricants are relatively insensitivity to aqueous environments, chemical solvents, certain acids and bases; hence, when added to liquid lubricants, they can increase their application range.

liquid lubricants have significant beneficial effects on the friction and wear performance of sliding surfaces [4–7]. With the use of solid lubricants in oils and greases, it may be possible to further reduce some of the extreme-pressure, anti-wear, and anti-friction additives from these oils and, hence, achieve better compliance with environmental mandates. In particular, heavy metals and sulfur- and phosphorous-bearing additives may be eliminated from current additive packages with the incorporation of such solid lubricants in these oils.

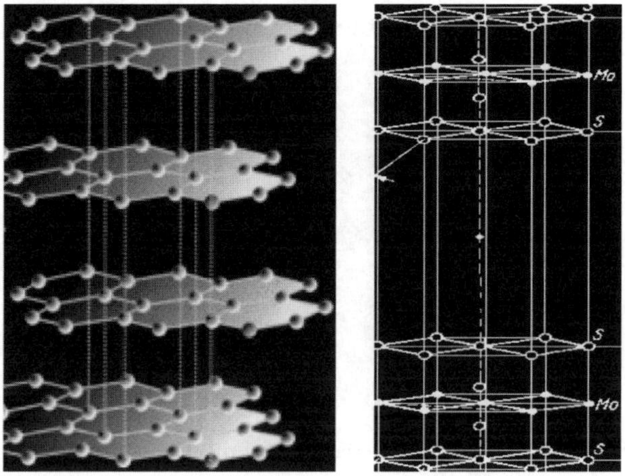

Figure 6.1 Schematic illustration of the layered lattice structures of (a) graphite and (b) molybdenum disulfide

6.1.1 Brief Overview of Lubrication Mechanisms of Solid Lubricants

Most of the solid lubricants mentioned above owe their low-friction characteristic primarily to a lamellar or layered crystal structure (see two of them in Figure 6.1 as typical examples). When present at a sliding contact interface, these solids shear easily along their atomic shear planes and thus provide low friction. Some of the solid lubricants do not have such layered crystal structures, but in applications, they too provide very low friction and wear. For example, certain soft metals (In, Pb, Ag, Sn, etc.), PTFE, a number of solid oxides and rare earth fluorides, diamond and diamondlike carbons, etc., can also provide fairly good lubrication despite the lack of a layered crystal structure like the ones shown in Figure 6.1 [1]. In fact, diamondlike carbon films are structurally amorphous but provide some of the lowest friction and wear coefficients among all other solid materials available today [8].

In general, the lubricity and durability of all solid lubricants (both lamellar and nonlamellar) are controlled by mechanisms that may involve interfilm sliding, intrafilm flow, and film/substrate or interface slip (with and/or without third body or transfer layer formation). In the case of lamellar solid lubricants (i.e., graphite, h-BN, H_3BO_3,and transition-metal dichalcogenides, like MoS_2 and WS_2), the atoms lying on each layer are closely packed and strongly bonded to each other, while the layers themselves are relatively far apart, and the forces that hold them together, e.g., van der Waals, are weak [1]. As illustrated in Figure 6.2 and revealed in Figure 6.3, when present on sliding surfaces, the atomic layers in these solids can align themselves parallel to the direction of relative motion and slide over one another with relative ease to provide low friction. Previous transmission electron microscopy (TEM) and X-ray diffraction (XRD) studies have also verified the formation of such preferred crystalline orientations on most other lamellar solid lubricants that were subjected to sliding tests [9,10].

While the interlayer shear mechanism is believed to be generally responsible for the low friction of most lamellar solid lubricants, extensive research by many scientists in previous years has also confirmed that a favorable layered crystal structure in itself is not sufficient

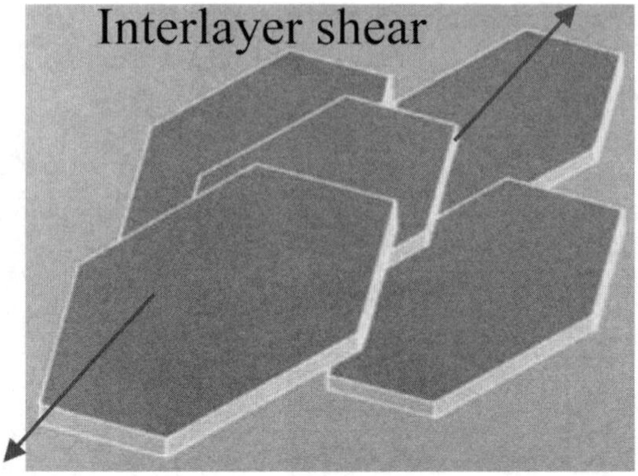

Figure 6.2 Illustration of interlayer shear mechanism of lamellar solids

for achieving ultra low friction or effective lubrication in most solid lubricants. These fundamental studies have indicated that the presence or absence of certain chemical adsorbates is needed for attaining easy shear in most solid lubricants that are in use today. Moisture seems to be very important for the lubricity of graphite, H_3BO_3, and h-BN, while MoS_2 and WS_2 seem to work best in non-humid and non-oxidizing environments [11–14]. The superior lubricity of graphite, H_3BO_3, and h-BN in humid test environments has been attributed to the weakening effects that water molecules have on the residual bonds between the atomic layers of these solid lubricants. As for the very poor lubricity of MoS_2, WS_2, and other dichalcogenides in humid environments, it has been suggested that water molecules react directly or indirectly with MoS_2 and, thus, alter the interatomic array and bonding, which increase friction [11].

Apart from the microscopic mechanisms described above, the electronic states of atoms lying in each layer can also play important roles in the lubricity of lamellar solids. Electronically, graphite and h-BN are similar (except for the differences between their bonding configurations). In the case of graphite, the π-bonding and π-antibonding bands overlap weakly at the Brillouin-zone boundary and makes graphite electrically conductive. In contrast, these bands are well-separated in the case of h-BN, causing it to be an insulator. In both cases,

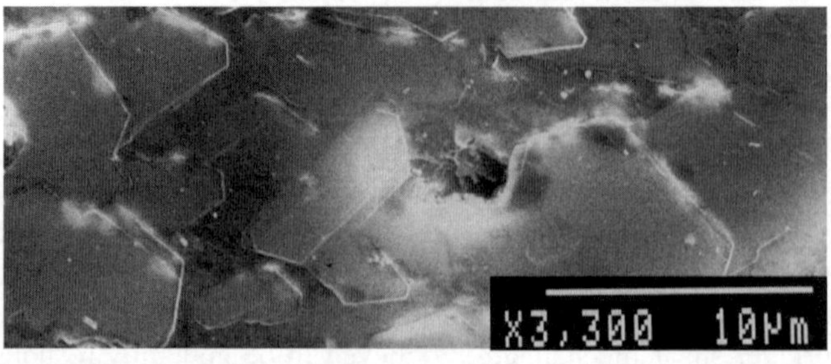

Figure 6.3 SEM micrograph revealing interlayer shear for a boric acid lubricant after sliding test in a pin-on-disk machine

if residual π-attractions between atomic layers are not eliminated or reduced, high friction and wear may result [14]. In graphite, π-bond interactions are generally reduced or eliminated by intercalation. As a result, the interlayer shear properties of graphite are markedly improved. However, attempts to find effective intercalation species for h-BN were mostly unsuccessful.

In the case of transition metal dichalcogenides like MoS_2, metal atoms are sandwiched between the chalcogen atoms in planar arrays of S-Mo-S (see Figure 6.1b). The interatomic bonding within the layers of these solids is strong and mainly of the covalent type, whereas the bonding between adjacent layers is rather weak and of the van der Waals type. Within the rigid two-dimensional layers of MoS_2 crystals, the S anions have a trigonal prismatic coordination around the Mo cations and, hence, screen them completely. The presence or absence of electrostatic attractions between the layers of dichalcogenides can affect the lubricity of these solids. For example, NbS_2, TiS_2, VS_2, TaS_2, etc., have a layered structure similar to that of MoS_2 but are not as lubricious [15]. Based on molecular orbital and valence bond theories, Jamison [15,16] proposed that the poor lubricating performance of these lamellar solids is primarily due to the following: within their layered crystal structures, a region of negative electrical charge not only concentrates above the chalcogen atoms of a given layer, but also extends well into the pockets between the chalcogen atoms of neighboring layers. Because the bottoms of the pockets are positively charged (due to the exposed ion cores of the surrounding atoms), an electrostatic attraction exists between the layers, making the layers of these solids shear with more difficulty. As for the excellent solid-lubricating capacities of MoS_2 and WS_2, the region of negative electrical charge is contained within the layers. Thus, the surfaces of the chalcogen atoms are positively charged, creating an electrostatic repulsion between the layers and making interlayer slippage much easier [15]. Relatively greater interlayer separation in MoS_2 and WS_2 crystals is thought to result from the same electrostatic repulsion between the successive atomic layers of these solid lubricants.

In most cases of sliding, the solid lubricants may chemically interact or react with rubbing surfaces, and hence they are chemically bonded and/or transferred to these surfaces. The formation of such bonded transfer layers on sliding surfaces is very important for the accommodation of sliding velocity and dissipation of frictional energy. Such a strong bonding can also make solid lubricants last very long and endure very high contact pressures in actual applications. If such a strong bonding is not feasible, then their wear live becomes very short. Overall, the friction coefficients of solid lubricants are typically in the range of 0.002 to 0.2, depending on test conditions and environments.

In the past, most solid lubricants have been applied on tribological surfaces as thin or thick solid films by means of burnishing, spraying, and smearing of their fine powders. In certain applications, fine powders of solid lubricants have been incorporated into an aerosol carrier and sprayed directly onto those surfaces to be lubricated. In other applications, such fine powders are strongly bonded to a surface by means of adhesives and/or epoxy resins to provide lubrication. As mentioned above, such fine powders can also be dispersed or mixed with oils and greases and then used to achieve improved lubrication. This particular practice will be the focus of the following sections. In general, previous studies have clearly demonstrated that the friction and wear performance of solid lubricants is strongly affected by both intrinsic (crystal-specific) and extrinsic (operating-environment-specific) factors. Therefore, no single solid lubricant can provide low friction and wear in all environments. Furthermore, not all layered solids are good solid lubricants. The type and magnitude of interlayer bonds are also important.

6.1.2 Recent Advances in Solid Lubrication

In most modern practices, solid lubricants are generally produced on tribological surfaces as strongly bonded thin solid films (ranging in thickness from 10 nm to 10 μm) by various physical and chemical vapor deposition (PVD and CVD) processes, such as sputtering, ion plating, and plasma-enhanced CVD [1,17]. These plasma-based processes can also result in very dense morphology, which increases durability. With these advances, much better lubricity and longer wear lives have already been achieved with numerous solid lubricants in a variety of tribological applications. In recent years, the design and manufacture of nanocomposite bulk materials and thin solid films incorporating nano-scale pockets or layers of solid lubricants have become very popular. Such materials have resulted in further improvements in the lubrication performance and durability of coated surfaces. Furthermore, the synthesis of adaptive or chameleon-type, self-lubricating coatings incorporating a variety of solid lubricants has improved the friction and wear performance of tribosystems operating under steep fluctuations in test temperature, environment, and other conditions [18].

Great strides have also been made in the production and increased use of unique nanostructures and nano-powders of various solid lubricants. In particular, graphite and WS_2 were prepared in the forms of hollow nano-tubes, nano-onions, etc., and then mixed with oils and greases to achieve very impressive friction and wear properties under boundary-lubricated sliding conditions [19–21]. Further details of recent accomplishments with these and other nano-materials in the lubrication field can be found in other chapters of this book. The main focus of this chapter is boron-based nanolubrication additives. Specifically, this chapter will concentrate primarily on the effectiveness of boric acid as a solid dispersion additive in oils. It will also provide a brief background on elemental boron and certain boron compounds (such as hexagonal boron nitride, h-BN), which are all used in the lubrication field for a range of purposes. Accordingly, in the following sections, boron and its compounds are reviewed briefly, then the next section is devoted to h-BN, and finally, the main characteristics of nano-boric acid powders and their lubrication properties are presented.

6.2 Brief Overview of Boron and its Self-Lubricating Compounds

Boron occupies a place next to carbon in the periodic chart, yet it does not have as rich a chemistry. Nevertheless, it is the main building block for many inorganic boron and organo-boron compounds (some of which are used by the additive industry as anti-oxidant, -bacterial, or -fungal agents). Solid boron exists in both the crystalline and amorphous forms. In the crystalline form, it is very resistant to chemical attack by strong acids, such as HF and HCl. Boron has a high melting point (above 2000 °C), and electrically, it is a semiconductor. Its band gap (i.e., 1.50 to 1.56 eV) is higher than that of silicon or germanium. Other electrical and physical properties of boron are close to those of metals and non-metals, and hence, boron is considered as a metalloid and/or semi-metallic element. Chemically, boron is very reactive and, hence, forms numerous covalently bonded organic/inorganic compounds (especially at elevated temperatures). It does not exist in nature in the elemental form but can be extracted from a variety of boron minerals such as borax ($Na_2B_4O_7{\cdot}10H_2O$), ulexite ($NaCaB_5O_9{\cdot}8H_2O$), colemanite ($Ca_2B_6O_{11}{\cdot}5H_2O$), kernite ($Na_2B_4O_6(OH)_2{\cdot}3H_2O$), and boric acid ($H_3BO_3$). Boron and its compounds are used widely by industry, especially in the glass-making and fiberglass industry, as well as in ceramics and enamels, textiles, detergents, medicine, pest control, fire retardants, and agricultural fertilizers.

In the field of tribology, certain boron compounds (such as boron carbide, boron nitride, various transition metal borides like TiB_2, and boric acid) have long been used to combat friction and wear in both the bulk and thin film forms [22–24]. Boron carbide is a superhard material (the third hardest after diamond and cubic boron nitride) and used as a thin hard coating by industry to control wear in various tribological applications. It also makes a very good armor material. Boron nitride is another very important boron compound and exists in two crystalline forms. In a cubic crystalline form, it is extremely hard (the second hardest after diamond) and has excellent thermal conductivity (like diamond), and hence, it is used as a super-abrasive in grinding media or applied as a thin solid film on metal-cutting and -forming tools to combat wear [25]. In the hexagonal form, BN is very soft and useful as a solid lubricant or lubrication additive in oils and greases [6,14].

Boric acid represents yet another interesting boron compound for tribological applications [23,24,26]. Structurally, it is a lamellar solid like h-BN and has excellent lubricity in moist environments. Besides these superhard and self-lubricating boron compounds, other types of organo-boron compounds and/or borate esters are used as anti-bacterial, anti-oxidant, and extreme-pressure additives in metalworking fluids and lubricating oils. In the following sub-sections, the general characteristics of solid lubricant h-BN are summarized first, and then boric acid is covered in much greater detail. The major emphasis will be on boric acid and its lubricity as a thin solid film first, and then as micro- and nano-colloidal forms in oils.

6.2.1 *Hexagonal Boron Nitride*

Hexagonal-BN is a lamellar solid just like graphite and has high refractory and lubricity qualities over a broad range of temperatures [1,14]. Its resistance to corrosive and oxidative degradation is also high. In fact, it is stable up to $\sim$1000 °C and, hence, is used extensively as a solid lubricant at elevated temperatures. Typical friction coefficients of h-BN in air are 0.2–0.3 up to about 700 °C. Its structure is similar to that of graphite, but unlike graphite, it is white (hence, occasionally referred to as 'white graphite'). Its atomic layers are made of two-dimensional arrays of boron and nitrogen atoms, and as in graphite, the bonding between nitrogen and boron atoms is covalent and hence very strong, while bonding between the layers is primarily van der Waals type and weak.

Hexagonal-BN is produced by reacting B_2O_3 with urea or ammonia at high temperatures. Depending on temperature and process conditions, turbostratic, quasi-turbostratic, mesographitic, and graphitic h-BN may also be produced. In terms of lubricity, the graphitic grade h-BN structures provide the best performance. In applications, h-BN can be used in powder form or compacted into dense solid structures by hot-pressing. It can also be incorporated into a range of composite structures to provide lubricity. It has been used as a release agent for many years in metalworking, especially at elevated temperatures. In high vacuum, h-BN loses its lubricity. In short, experimental studies to date on h-BN have not only verified that it is similar to graphite in its crystal structure but also in its lubrication behavior, thus calling it 'white graphite' seems appropriate.

In the past, h-BN has been tried as a colloidal dispersion additive in oils and greases. Recent tests by Kimura *et al.* showed that addition of h-BN powders into oil concentrations as little as 1 wt % reduces wear by an order of magnitude for bearing steels sliding against each other in line contacts [6]. At higher concentrations, the reduction in wear is even much greater. Similar improvements in anti-wear and -friction properties were reported for h-BN-containing greases by Denton and Fang [27]. However, there exists little or no experimental work dealing with the lubrication behavior of nano-h-BN structures dispersed in oils or greases (even though various types of nanotubes, nanohorns, etc., have been produced from h-BN in recent years) [28].

6.2.2 Boric Acid

Boric acid is made up of boron, hydrogen, and oxygen. It exists in two major crystalline forms: metaboric acid ($H_2O{\cdot}B_2O_3$ or HBO_2) and orthoboric acid ($3H_2O{\cdot}B_2O_3$ or H_3BO_3) [24]. Orthoboric acid is the most common form and may easily result from the hydration of boric oxide (B_2O_3). It also exists as a natural mineral known in mineralogy books as 'sassolite.' Metaboric acid crystallizes in three forms: orthorhombic or α-metaboric acid, monoclinic or ß-metaboric acid, and cubic or γ-metaboric acid. Among all these boric acid types, orthoboric and orthorhombic metaboric acids exhibit layered-crystal structures as shown in Figures 6.4 and 6.5, respectively. Figure 6.6 shows a bulk crystal of orthoboric acid or sassolite. Its layered triclinic crystal structure is stable up to about 170 °C. Beyond this temperature it begins to lose some of its OH content and transforms to metaboric acids. At or above 400 °C, it gradually turns into boron oxide or B_2O_3 (see diagram in Figure 6.7). Please note that in this chapter the term 'boric acid' is used to mean orthoboric acid, H_3BO_3, but not the other rare boric acids mentioned above.

Just like h-BN, graphite, and MoS_2, boric acid is a lamellar solid and has unique lubrication properties [24]. In its triclinic crystal structure, boron, oxygen, and hydrogen atoms are strongly bonded to one another to form extensive atomic layers that are parallel to the basal plane of the crystal (see Figure 6.4). There are four boric acid molecules in each unit cell. In each of these layers, as shown in Figure 6.8, each boron cation is bonded to three oxygen anions to form the triangular BO_3 groups. Bonding between boron and oxygen is predominantly covalent, but some ionic character may also exist. Hydrogen bonds link the planar BO_3

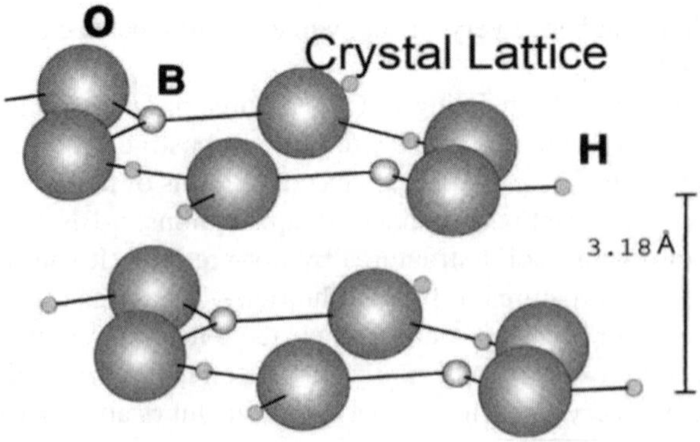

Figure 6.4 Layered lattice structure of boric acid (H_3BO_3)

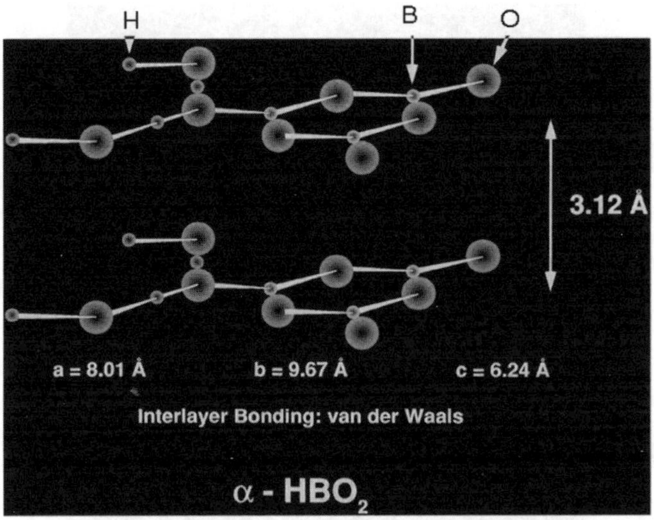

Figure 6.5 Layered lattice structure of orthorhombic metaboric acid (HBO_2)

groups together [29]. Because of the triclinic crystal structure, the *c* axis is inclined to the basal plane at an angle of 101 °, which causes shifting of alternate layers along the *c* axis. Bonding between the atoms lying on the same plane is of the covalent/ionic and hydrogen type. The layers are 0.318 nm apart and held together by weak van der Waals forces. Because of the ionic bond character, boric acid can dissolve in water and some other solvents. Hydrogen bonds hold the planar BO_3 groups together (see Figure 6.8). Overall, with its layered crystal structure, H_3BO_3 resembles other layered solids well known for their excellent lubrication capabilities (e.g., MoS_2, graphite, and h-BN).

Boric acid is a very mild, non-toxic acid. It is soluble in water (about 6 wt % at room temperature), and at high concentrations has a pH of about 5. Diluted water solutions of boric

Figure 6.6 A large single crystal of boric acid (also known as sassolite)

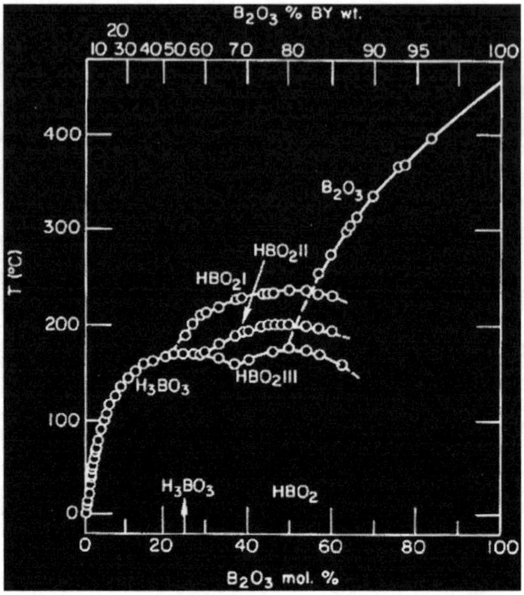

Figure 6.7 Boric acid/boron oxide phase diagram

acid are often used to wash eyes. Boric acid is also used as an antiseptic, insecticide, flame retardant, and food preservative. The Environmental Protection Agency has established that boric acid is benign and is not classified as a pollutant under the Clean Water Act. Material safety data sheets show no serious illnesses or carcinogenic effects resulting from exposure to boric acid solutions or aerosols.

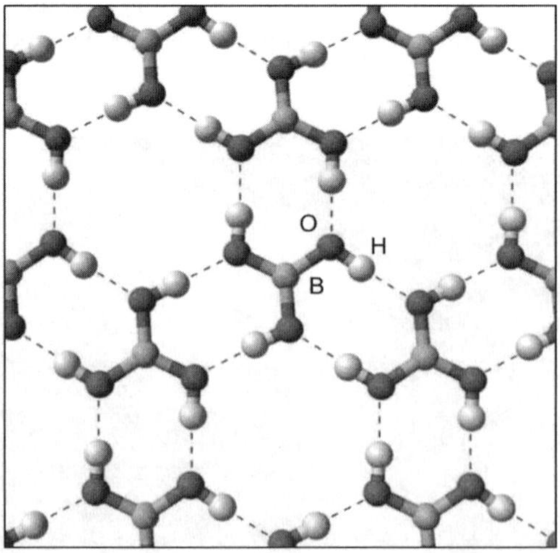

Figure 6.8 Planar configuration of boron, oxygen, and hydrogen atoms in layer of boric acid crystal (adopted from B. Mills, Odell-UK)

6.2.2.1 Lubricity of Boric Acid and Its Derivatives

As mentioned above, due to their layered-crystal structures, H_3BO_3 and α-HBO_2 are self-lubricating; hence, they afford very low friction and wear to sliding tribological surfaces. To demonstrate the lubricity of H_3BO_3, Erdemir performed extensive friction and wear tests with powders and solid compacts or pellets of H_3BO_3 on a pin-on-disk machine [24]. Sprinkling of the boric acid powders to the sliding interfaces of steel pins and disk samples dramatically reduced the friction coefficients. Figure 6.9 shows how feeding boric acid powders to a sliding contact interface of a steel test pair decreased friction from about 0.8 to less than 0.05. In the case of solid pellets or compacts, disk-shaped samples with a nominal diameter of 2.5 cm were compacted from 99.8 wt % H_3BO_3 powders by cold-pressing at about 35 MPa. These samples were then rubbed against a steel ball under 2 to 5 N loads. The friction coefficient of the pin/disk pair described above was measured as a function of sliding distance. The initial friction coefficient of this tribosystem was approximately 0.2; it then decreased steadily with distance and eventually reached a steady-state value of less than 0.1 after sliding about 20 m [24].

H_3BO_3 can spontaneously form on the outer surfaces of bulk B_2O_3 or its thin film forms. Erdemir *et al.* investigated the formation and tribological characteristics of such boric acid films formed on the surfaces of vacuum-evaporated B_2O_3 layers [26]. They found that H_3BO_3, which formed spontaneously on the surfaces of B_2O_3 coatings, is remarkably lubricious. For

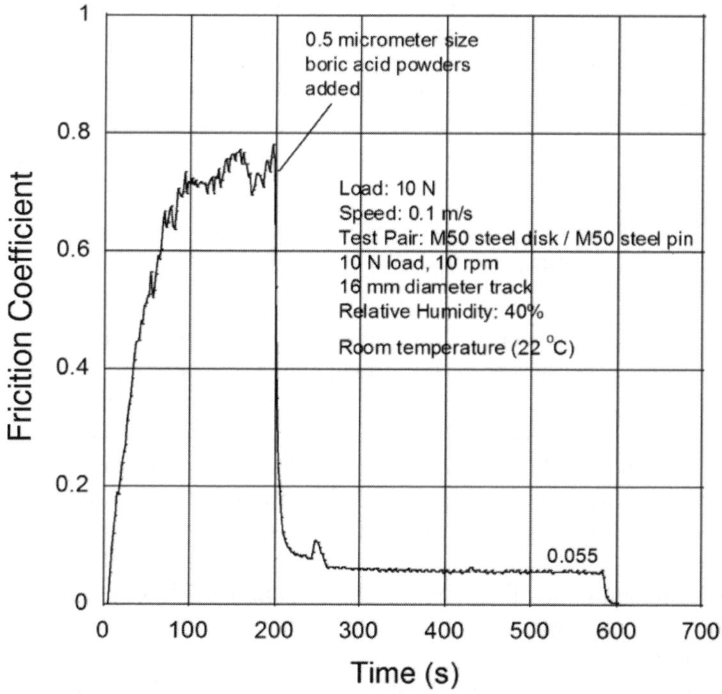

Figure 6.9 Effectiveness of boric acid powders in reducing friction between a steel pin and steel disk during test in a pin-on-disk machine. Up to 200 s, sliding was between pin and disk only; thereafter, 0.5 μm size boric acid powders were introduced to the sliding interface

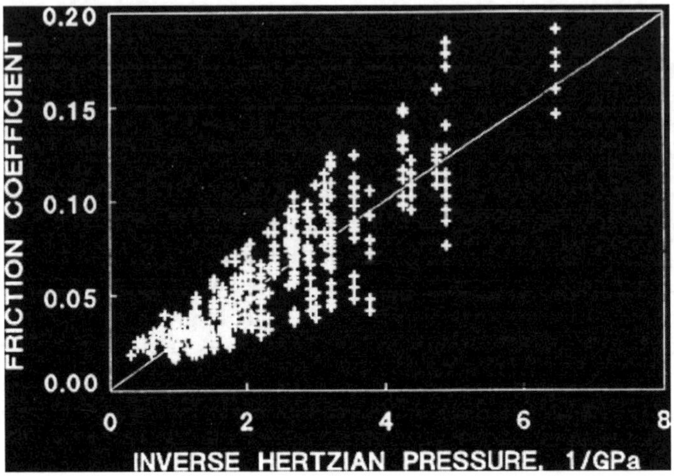

Figure 6.10 The relation of the friction coefficient of boric acid films to Hertzian contact pressure

a sliding pair consisting of a sapphire ball and B_2O_3-coated Al_2O_3 disk, they reported friction coefficients ranging from 0.02 to 0.05 in open air with 50 % relative humidity, depending on applied forces [30,31]. Figure 6.10 summarizes the relationship between friction coefficient of thin boric acid films and Hertzian contact pressure in a ball-on-disk type test setup. In general, the higher the contact pressure, the lower the friction coefficient.

Based on the results of surface and structure analytical studies, Erdemir et al. concluded that low friction is a direct consequence of the layered crystal structure of H_3BO_3 films forming on the exposed surfaces of B_2O_3 coating by the following spontaneous chemical reaction [31]:

$$1/2\ B_2O_3\ \text{(coating)} + 3/2\ H_2O\ \text{(environmental moisture)} \rightarrow H_3BO_3 \quad \Delta H_{298} = -45.1\ \text{kJ}\cdot\text{mol}^{-1}$$

The above reaction occurs spontaneously due to the negative heat of reaction, and a thin layer of H_3BO_3 forms everywhere on the surface of B_2O_3 exposed to humid air. Formation of such self-lubricating and self-replenishing films was later demonstrated on VB_2, B_4C, and borided steel surfaces, affording very low friction coefficients to these surfaces as well [32–34]. When such nano-scale boric acid films are tested under lubricated sliding conditions, extremely low friction coefficients (as low as 0.007) are attained, as shown in Figure 6.11. This particular film was produced on the surface of a boron-carbide-coated steel disk by a method described in Ref. [33]. Specifically, it was chemically extracted from the boron carbide coating by a high-temperature chemical conversion method. This finding suggests that if boric acid is formed on sliding surfaces as a strongly bonded boundary film, then extremely low friction coefficients should be feasible.

The lubrication mechanism of the bulk and thin film forms of boric acid is similar to that of the other known lamellar solids. Specifically, under shear stresses, plate-like crystallites of H_3BO_3 can align themselves parallel to the direction of relative motion and then slide over one another with relative ease to provide low friction, as shown in Figures 6.9 and 6.10 [24].

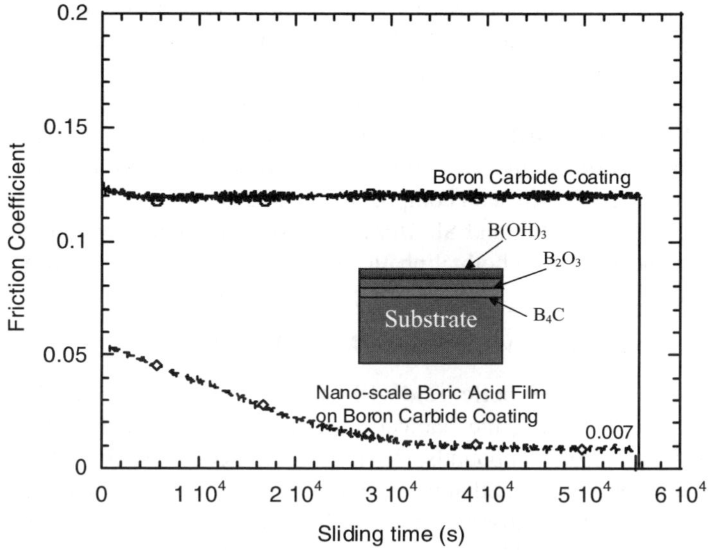

Figure 6.11 Lubricated sliding friction coefficient of a steel pin sliding against a B_4C coating and a nano-scale boric acid film formed on the B_4C coating, as illustrated in the inset

Such a lubrication mechanism was further substantiated by the electron microscopy image in Figure 6.3, where the crystalline layers of boric acid are indeed aligned parallel to the sliding surface. The image also shows strong evidence for interlayer shear. Boric acid films bond strongly to the sliding surfaces of magnesium and aluminum and their alloys and provide excellent lubricity when used as a spray coating or a colloidal lubricant [35]. Used as a filler in polymers, boric acid and boron oxide can substantially lower friction and increase the wear resistance of base polymers [7].

For applications involving elevated temperatures, the use of H_3BO_3 as solid lubricant is not recommended. Above about 170 °C, H_3BO_3 tends to decompose and gradually turns into other forms of metaboric acids. Upon further heating, it eventually turns into B_2O_3 (see Figure 6.7), thus losing its layered crystal structure and hence its lubricity. At temperatures greater than about 450 °C, B_2O_3 becomes a viscous liquid, and hence, low friction can be reinstated by a viscous-flow lubrication mechanism.

6.3 Lubrication by Colloidal Boric Acid Nano-particles and Other Boron Compounds

Boric acid can be prepared in powder forms ranging in size from tens of nanometers to several micrometers. Mixtures of both micro- and nano-scale powders with lubricating oils and/or greases have been prepared over a range of concentrations [7,36]. Because of their high surface-to-volume ratios and light weight, these nano-powders of boric acid disperse in oils quite well even without the use of additional surfactants. The low specific gravity of boric acid (i.e., $\sim$1.3 g/cm^3) further facilitates easy dispersion of such nano-powders in oils. When mixed with base mineral and synthetic oils, boric acid is known to be non-corrosive and to possess excellent anti-oxidant and anti-corrosion characteristics. In the past, other boron compounds such as borate esters, alkali-borates, and h-BN have been used as lubricant additives [6,37,38]. Borate esters are prepared by reacting boric acid or boron oxide with appropriate aliphatic or alkoxylated compounds. Other boron-based additives proposed for use in oils include borated derivatives of phosphorous, such as borated dihydrocarbyl hydrocarbylphosphonates. Certain organometallic boron-containing compounds have also been used as additives in oils in past years. Desired metals for use in such compounds include Na, K, Mg, Ca, Sr, Ba, Ti, Zr, V, Cr, Ni, Mn, Fe, Co, Cu, Zn, B, Pb, and Sb. Borated versions of such organometallic complexes are derived or synthesized from both aliphatic and heterocyclic organic compounds.

6.3.1 Preparation of Oils with Nano-Boric Acid Powders

Sub-micron size powders of boric acid (10–500 nm) can be produced by many well-established methods, including mechanical attrition, chemical precipitation, sonochemical synthesis from organoboron compounds, low-pressure gas condensation, and low-temperature evaporation of ethyl borates or methanol or ethanol solutions of boric acid into or through the lubricating oils. Mechanical ball-milling or crushing techniques may also be used to reduce the size of micro-size boric acid particles to the nano scale. Figure 6.12 shows the size distributions of typical nano-scale boric acid powders and their colloidal dispersion in a poly-alpha-olefin (PAO) oil. The nano-powders are milky white, and when mixed with oils, they make them look milky white as well. Because of a very high surface atom-to-bulk atom ratio, these nanometer-sized boric acid particles can be incorporated into various oil products, such as base and formulated vegetable, mineral, and synthetic oils; greases; and combinations thereof. Most of the atoms in these nanometer-sized particles reside on the surface of the particles, and they are chemically very active and are both physically and chemically attracted to

(a) (b)

Figure 6.12 (a) Electron microscopic image of nano-boric acid particles and (b) a colloidal mixture of these particles with base PAO oil

the hydrocarbon molecules in oils. Dispersion of these particles in oils with higher polarity (canola oil, polyolesters, etc.) appears to be much easier. Nanotubes, nanorods, and a range of other tubular nano/micro structures of boric acid have also been produced by plasma-based processes but have not yet been blended with oils or greases [39,40].

6.3.2 Lubrication Performance of Various Oils Containing Nano-boric Acid Particles

Figures 6.13–6.15 show the lubrication performance of nanometer-sized boric acid particles dispersed in base mineral and PAO oils. In these figures, lubrication performance is determined through friction coefficient measurements that were taken with a pin-on-disk machine, which consists of a stationary top-mounted steel pin that rubs against a unidirectional rotating steel disk or flat. The pins can be either flat or hemispherically tipped (typically, the pins have a 127-mm radius ground on one of the faces). The lubricants are applied to the disk surface, and the pins are rubbed against the disk. A load is applied to the pin by using dead weights. For the specific tests performed, 20 N loads were used, and the sliding velocity of the rotating disk was adjusted to give linear velocities of 0.01 and 0.1 m/s. Tests were run at room temperature and in open air whose relative humidity was varied between 30 and 60 %.

As shown in Figure 6.13, with the use of base PAO on sliding steel surfaces, the average friction coefficient is ~0.15. However, when PAO is blended with 5 wt % nano-scale boric acid particles, the friction coefficient is reduced to 0.04. At a concentration of 1 wt %, the friction coefficient was essentially the same.

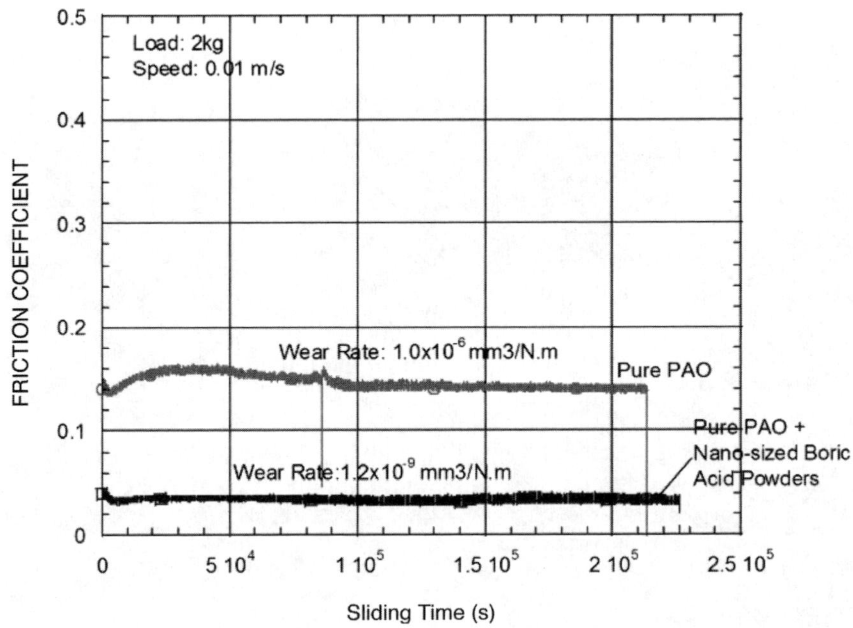

Figure 6.13 Friction coefficients of a steel pin-and-disk pair in PAO with and without 5 wt % nano-boric acid

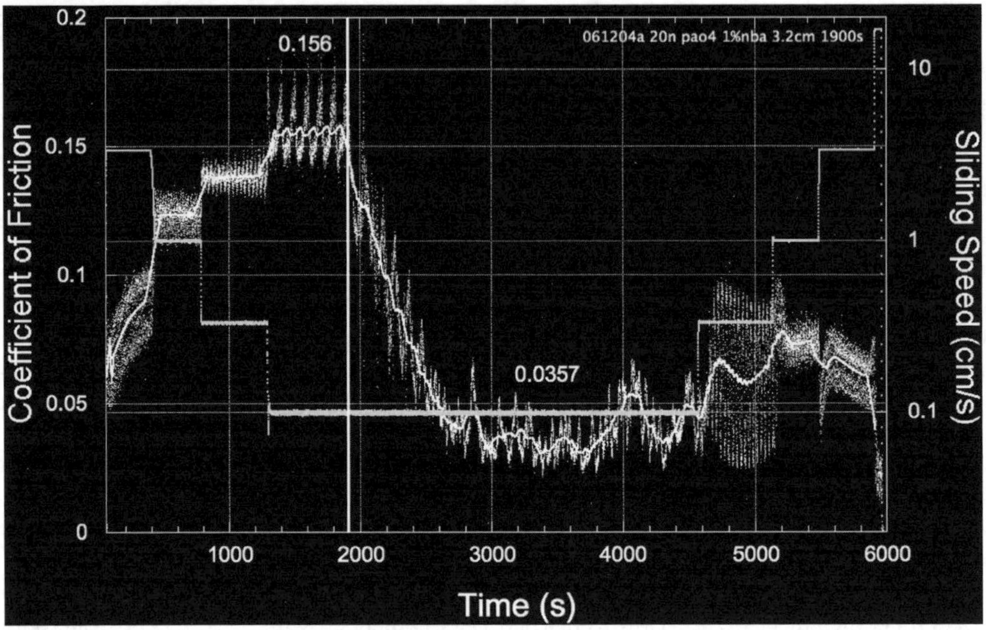

Figure 6.14 Friction coefficient as a function of speed and time for a steel pin-and-disk pair in pure PAO first and then in PAO containing nano-boric acid

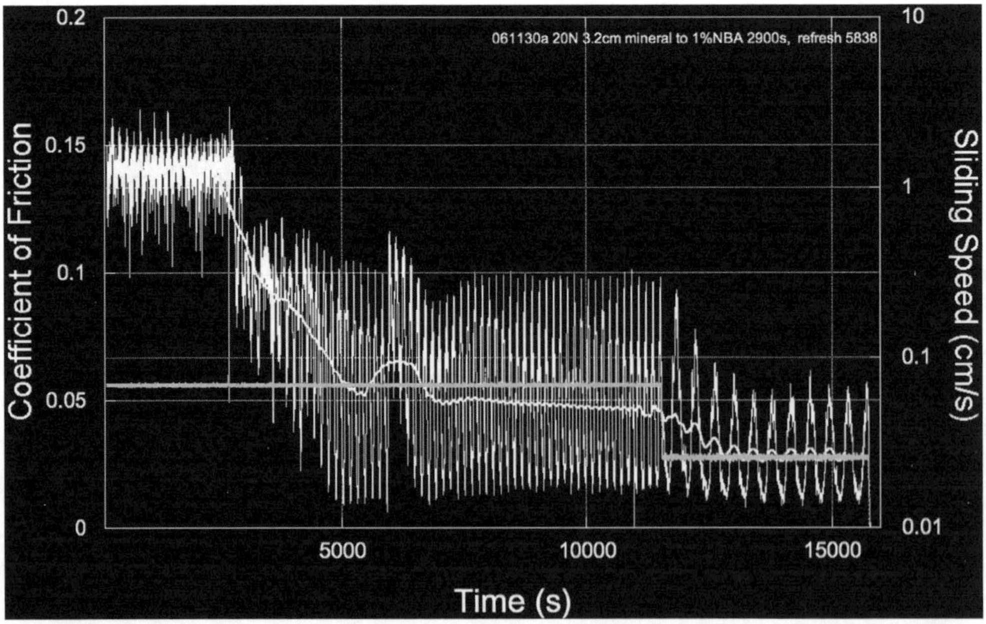

Figure 6.15 Variation of friction coefficient of a steel pin-and-disk pair in pure mineral oil first and then in mineral oil containing nano-boric acid for sliding velocities of 0.05 and 0.025 cm/s

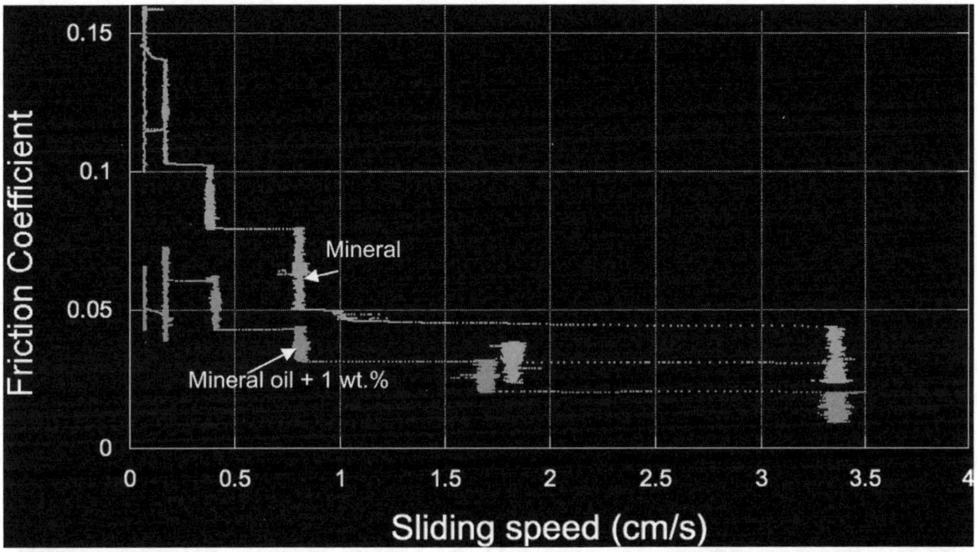

Figure 6.16 Variation of friction coefficient of a steel pin-and-disk pair with sliding speed when lubricated by mineral oil with and without 1 wt % nano-boric acid

Figure 6.14 shows the effect of nano-boric acid powder containing PAO on friction at different sliding velocities. Specifically, to illustrate the beneficial effect of nano-boric acid powders on friction, the sliding test was started with a pure PAO oil at a sliding speed of 5 cm/s, and then the speed was lowered stepwise to 0.1 cm/s. As expected, the friction coefficient increased to 0.15 under such low sliding velocity. However, when nano-boric acid powder containing PAO is added to the same sliding interface without changing the test conditions, the friction coefficient decreased rapidly and reached a value of about 0.04. Increasing speed stepwise back to 5 cm/s resulted in a gradual increase in friction to about 0.065.

Figure 6.15 shows the frictional behavior of nano-boric acid powder containing mineral oil. In this test, the sliding experiment was again started with a base mineral oil at a sliding velocity of 0.05 cm/s and let go until reaching 2000 s. Then, the mineral oil with nano-boric acid powder was introduced to the sliding contact interface. As is clear, the friction coefficient decreased steadily and eventually reached a value of ~0.05 after sliding for 7000 s. Upon further reduction of sliding velocity to 0.025 cm/s, the friction coefficient became even much lower, i.e., 0.02.

Figure 6.16 shows friction vs. sliding velocity for a mineral oil with and without nano-boric acid powders. As is clear, the friction coefficient is much lower for oil containing nano-boric acid. Note that the difference is particularly significant at low sliding velocities, where much severe metal-to-metal contact occurs. Overall, the addition of nanometer-sized boric acid powders significantly reduces the friction coefficients of both PAO and mineral base oils.

6.4 Lubrication Mechanism of Nano-Boric Acid Colloids in Oils

During sliding, solid nano-particles of boric acid are believed to be carried to the vicinity of the contact zone by liquid media and then dragged into the contact interface. Once they enter the contact zone, they are smeared on the real contact spots to form a very thin (a few nanometers thick) boundary film. Because of their high chemical affinity, these smeared particles may react with contact spots and hence establish a strongly bonded layer. Once bonded, they can protect these spots against further wear and reduce friction because of their easy shear. The molecules that make up these nano-particles may possibly rearrange themselves in a plate-like boric acid structure to enhance the anti-friction and -wear properties of the sliding surfaces.

Time-of-flight secondary-ion mass spectrometry (TOF-SIMS) and X-ray photoelectron spectroscopy (XPS) can provide complementary information on the existence and chemical nature of tribofilms that form on sliding surfaces. Accordingly, these techniques were used to determine the chemical nature of a tribofilm that formed on sliding steel surfaces during lubricated tribological tests with PAO oil containing nano-boric acid. These nanoparticles were produced by first mixing trimethyl borate and/or trimethoxyboroxin with PAO and then subjecting the mixture to rigorous ultrasonic agitation at 50 °C. During this sonochemical process, trimethyl borate and/or trimethoxyboroxin gradually react with environmental moisture and form nano-boric acid particles dispersed in oil. As shown in the TOF-SIMS images of the sliding-steel disk surfaces, high concentrations of various boron compounds were present within the wear tracks. However, outside the wear track, the concentration of boron was much lower. Note that these surfaces were subjected to brief sputtering in order to remove the environmental and/or contaminant layers prior to taking data. In support of the TOF-SIMS, XPS analysis of the same wear track (Figure 6.18) also indicated that boron is very rich within the wear track but poor outside of it.

Based on these surface analytical results, the anti-friction and -wear mechanisms of PAO oil containing nano-boric acid particles can be explained as follows. When nano-boric acid particles are carried into the sliding interface, they are deformed and smeared on these surfaces under the influence of contact load and sliding velocity. As is clear from Figures 6.17 and 6.18, they form a boron-rich boundary film on these surfaces. This film then shears easily to

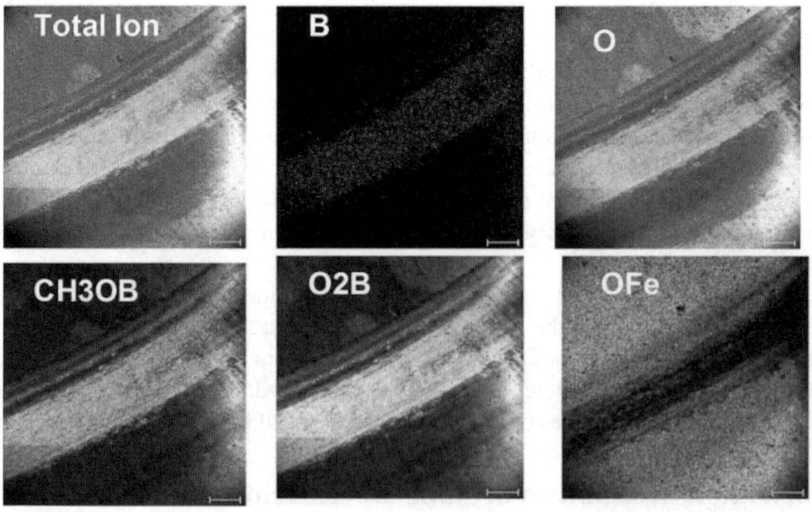

Figure 6.17 TOF-SIMS images of a wear track formed during testing of PAO with nano-boric acid. They reveal high concentrations for boron and other boron compounds within the wear track

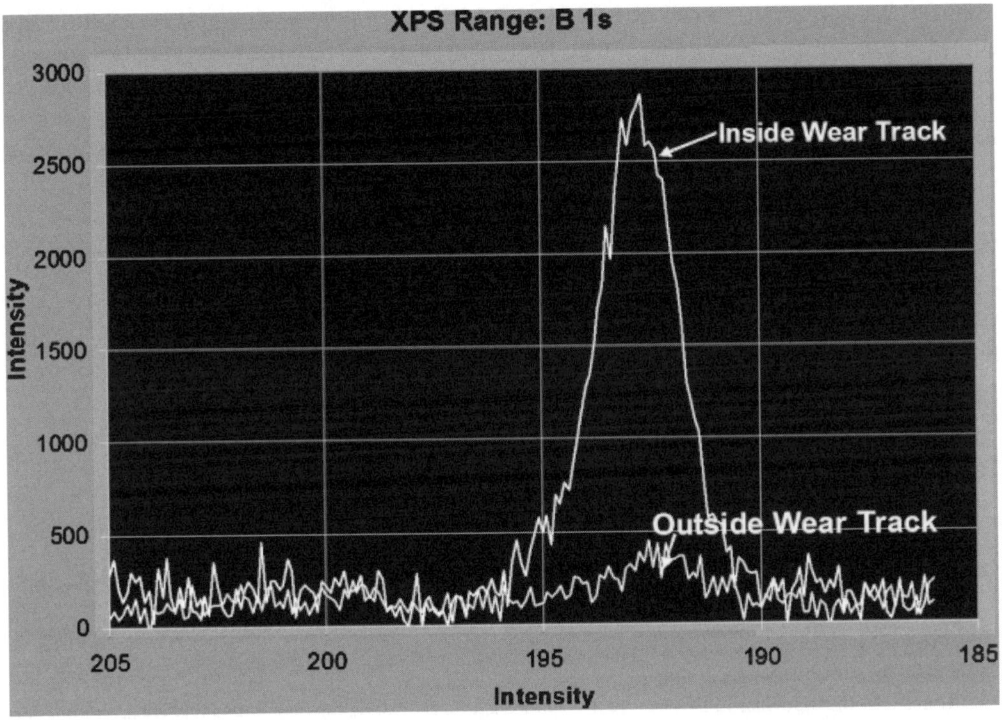

Figure 6.18 Comparison of the XPS B 1s peak heights of inside and outside regions of a wear track formed during test with PAO containing nano-boric acid. The greater peak height for B within the wear track suggests much higher boron concentration

provide low friction, as shown in Figures. 6.15 and 6.16. Because of its protective nature, the boron-rich boundary film can also reduce wear.

6.5 Summary

This chapter briefly reviewed recent developments in solid lubricants, in general, and their nano-lubrication potentials, in particular. The major emphasis of the chapter was on boric acid and its lubrication mechanism. As is clear from this and other chapters in this book, combining solid and liquid lubricants at nano-scales greatly improves lubrication performance; hence, such a hybrid lubrication strategy has much to offer for demanding tribological applications. With further refinement and realization of such lubrication strategies, higher energy efficiency and durability should be possible in future tribosystems, whose expected operating conditions are becoming more and more demanding.

Just like other lamellar solids, boric acid is also self-lubricating. Its lubrication mechanism is similar to that of the other solid lubricants. It can be manufactured as very fine nano-powders. Because of their light weight, these nano-powders can easily be dispersed in oils and used as colloidal anti-friction and -wear additives. Tribological studies have confirmed that they have very high capacity to reduce friction and wear. TOF-SIMS and XPS studies confirm that boric acid forms a strongly bonded boundary film on sliding surfaces. It is believed that this boron-rich film is very protective and slippery since it affords very low friction and wear to the sliding surfaces. In short, boric acid and other solid lubricants discussed in this book offer unique opportunities in meeting the increasingly more stringent operating conditions of future tribological systems. As they become available in very large quantities at reasonable costs, they are expected to gain significant market share.

Acknowledgement

This work was supported by the U.S. Department of Energy, Office of Energy Efficiency and Renewable Energy, FreedomCar and Vehicle Technology Program, under Contract No. DE-AC02-06CH11357. The author thanks Dr. O. L. Eryilmaz for help in surface analytical studies and Dr. Robert A. Erck for help in pin-on-disk experiments.

References

1. Erdemir, A., Solid lubricants and self-lubricating films, in 'Handbook of Modern Tribology,' B. Bhushan, ed., CRC Press, Boco Raton, FL, 2001, pp. 787–818.
2. Chinas-Castillo, F. and Spikes, H. A., Mechanism of action of colloidal solid dispersions, J. Tribology-Trans. ASME, 125 (2003) 552–557.
3. Pacholke, P. J. and Marshek, K. M., Improved worm gear performance with colloidal molybdenum-disulfide containing lubricants, Lubr. Eng., 43 (1987) 623–628.
4. Barnett, R. S., Molybdenum disulfide as an additive for lubricating greases, Lubr. Eng., 33 (1977) 308–313.
5. Broman, V. E., DeJovine, J., DeVries, D. L., and Keller, G. H. Testing of friction modified crankcase oils for improved fuel economy, SAE Reprint No. 780597 (1978).
6. Kimura, Y., Wakabayashi, T., Okada, K., Wada, T., and Nishikawa, H., Boron nitride as a lubricant additive, Wear, 232 (1999) 199–206.
7. Erdemir, A., Lubrication from mixture of boric acid with oils and greases, United States Patent #5,431,830 (1995).
8. Erdemir, A. and Donnet, C., Tribology of diamond and diamondlike carbon films: An overview, in 'Wear: Materials, Mechanisms and Practices,' G. W. Stachowiak, ed., John Wiley & Sons, 2006, pp. 191–222.
9. Wahl, K. J., Dunn, D. N., and Singer, I. L., Wear behavior of Pb-Mo-S solid lubricating coatings, Wear, 230 (1999) 175–183.
10. Martin, J. M., Pascal, H., Donnet, C., LeMogne, T., Loubet, J. L., and Epicier, T., Superlubricity of MoS_2: Crystal orientation mechanisms, Surf. Coat. Technol., 68–69 (1994) 427–432.
11. Winer, W. O., Molybdenum disulfide as a lubricant: A review of the fundamental knowledge, Wear, 10 (1967) 422–452.
12. Farr, L. P. G., Molybdenum disulfide in lubrication: A review, Wear, 35 (1975) 1–22.
13. Savage, R. H., Graphite lubrication, J. Appl. Phys., 19 (1948) 1–10.
14. Martin, J. M., LeMogne, T., Chassagnette, C., and Gardos, M. N., Friction of hexagonal boron nitride in various environments, Tribol. Trans., 35 (1992) 463–472.
15. Jamison, W. E., Electronic effects on the lubricating properties of molybdenum disulfide and related materials, ASLE Proc. 2nd International Conference on Solid Lubrication, ASLE SP-6, STLE, Park Ridge, IL, pp. 1–8 (1978).
16. Jamison, W. E., Structure and bonding effects on the lubrication properties of crystalline solids, ASLE Trans., 15 (1972) 296–305.
17. Donnet, C., and Erdemir, A., Historical developments and new trends in tribological and solid lubricant coatings, Surf. Coat. Technol., 180–181 (2004) 76–84.
18. Voevodin, A. A., Fitz, T. A., Hu, J. J., and Zabinski, J. S., Nanocomposite tribological coatings with 'chameleon' surface adaptation, J. Vac. Sci. Technol., A20 (2002) 1434–1444.
19. Tenne, R., Margulis, L., Genut, M., and Hodes, G., Polyhedral and cylindrical structures of tungsten disulphide, Nature, 360 (1992) 444–445.

20. Rapoport L., Fleischer N., and Tenne, R., Fullerene-like WS_2 nanoparticles: Superior lubricants for harsh conditions, Advanced Mat., 7–8 (2003) 651–655.
21. Joly-Pottuz, L., Dassenoy, F., Belin, M., Vacher, B., Martin, J. M., and Fleischer, N., Ultralow-friction and wear properties of IF-WS_2 under boundary lubrication, Tribo. Lett., 18 (2005) 477–485.
22. Watanabe, S., Miyake, S., and Murakawa, M., Tribological performance of cubic BN films, Diam. Films Technol., 7 (1997) 139–156.
23. Erdemir, A., Eryilmaz, O. L., and Fenske, G. R., Self-replenishing solid lubricant films on boron carbide, Surf. Eng., 15 (1999) 291–295.
24. Erdemir, A., Tribological properties of boric acid and boric-acid-forming surfaces. Part I: Crystal chemistry and mechanism of self-lubrication of boric acid, Lubr. Eng., 47 (1991) 168–172.
25. Fukunaga, O., Science and technology in the recent development of boron nitride materials, J. Phys.-Cond. Mat., 14 (2002) 10979–10982.
26. Erdemir, A., Fenske, G. R., Erck, R. A., Nichols F. A., and Busch, D. E., Tribological properties of boric acid and boric-acid-forming surfaces. Part II. Mechanisms of formation and self-lubrication of boric acid films on boron- and boric oxide-containing surfaces, Lubr. Eng., 47 (1991) 179–185.
27. Denton, R. M. and Fang, Z., Rock bit grease composition, U.S. Patent #5589443 (1995).
28. Zhi, C. Y., Bando, Y., Tan, C. C., and Golberg, D., Effective precursor for high yield synthesis of pure BN nanotubes, Sol. State Comm., 135 (2005) 67–70.
29. Zapol, P., Curtiss, L. A., and Erdemir, A., Periodic ab initio calculations of orthoboric acid, J. Chem. Phys., 113 (2000) 3338–3343.
30. Erdemir, A., Erck, R. A., and Robles, J., The relation of Hertzian contact pressure to friction behavior of self-lubricating boric acid films, Surf. Coat. Technol., 49 (1991) 435–438.
31. Erdemir, A., Fenske, G. R., and Erck, R. A., A study of the formation and self-lubricating mechanisms of boric acid films on boric oxide coatings, Surf. Coat. Technol., 43–44 (1990) 588–596.
32. Erdemir, A., and Bindal, C., Formation and self-lubricating mechanisms of boric acid on borided steel surfaces, Surf. Coat. Technol., 76–77 (1995) 443–449.
33. Erdemir, A., Bindal, C., and Fenske, G. R., Formation of ultralow friction surface films on boron carbide, Appl. Phys. Lett., 68 (1996) 1637–1639.
34. Erdemir, A., Halter, M., and Fenske, G. R., Preparation of ultralow-friction surface films on vanadium diboride, Wear, 205 (1996) 236–239.
35. Wei, J. J., Erdemir, A., and Fenske, G. R., Dry lubricant films for aluminum forming, Tribology Trans., 43 (2000) 535–541.
36. Erdemir A., Method to improve lubricity of low sulfur diesel and gasoline fuels, U.S. Patent # 6,783,651 (2004).
37. Komvopoulos, K., Chiaro, V., Pakter, B., Yamaguchi, E. S., and Ryason, P. R., Antiwear tribofilm formation on steel surfaces lubricated with gear oil containing borate, phosphorus, and sulfur additives, Tribol. Trans., 45 (2002) 568–575.
38. Adams, J. H., and Godfrey, D., Borate gear lubricant - EP film analysis and performance, Lubr. Eng., 37 (1981) 16–21.
39. Li, Y., Ruoff, R. S., and Chang, R. P. H., Boric acid nanotubes, nanotips, nanorods, microtubes, and microtips, Chem. Mater., 15 (2003) 3276–3285.
40. Enyashin, A. N., and Ivanovskii, A. L., Atomic and electronic structure of the orthoboric (H_3BO_3) and metaboric ($H_3B_3O_6$) acids nanotubes, Chem. Phys. Lett., 411 (2005) 186–191.

Appendix

Tribometers Used for the Studies of Chapters 2 and 3

Tribological tests were carried out on various pin-on-flat tribometers with a reciprocating motion. With this kind of tribometer, severe conditions of contact are obtained (contact pressure of approximately 1 GPa). Indeed, to characterize the effect of the addition of nanoparticles in oil, tribological tests must be performed in a boundary lubrication regime. In such a regime, lubricant film is too thin to be tribologically active and additives are necessary. Therefore, the effect of additives in oil can be studied using this lubrication regime, as well as the tribological properties of nanoparticles.

The contact obtained with a pin-on-flat tribometer is described in Figure A.1. Wear during the friction test is easily determined by measuring the wear scar diameter and comparing it with the theoretical Hertzian diameter. The Hertzian diameter depends on the experimental conditions used (pin diameter, pressure, constituents of the pin and flat).

A.1 Environmental Pin-on-Flat Tribometer

With this tribometer (Figure A.2) tribological tests can be carried out in ambient air or in controlled atmosphere (partial pressure of N_2, O_2, etc.). In this study, tribological tests were performed only in ambient air. A heating system can be used to perform tests up to 70 °C. Antagonistic surfaces used are made of AISI 52100 bearing steel. The dimensions of the flat are 10 mm × 8 mm × 2 mm. The diameter of the ball varies from 4.5 to 12 mm to obtain a contact pressure of 0.33 to 1.72 GPa (Table A.1). Flats and balls are mirror-polished with diamond solutions (of grain size 6, 3 and 1 μm). The surface roughness is approximately 25 nm.

The stroke length used was 2.5 mm and the sliding speed was 2.5 mm/s. Thus, friction tests were performed at a frequency of 0.5 Hz.

This tribometer allows measurements of the friction coefficient to be made, as well as the acquisition of triboscopic images of the friction coefficient or electrical contact resistance [1]. Such images show those data for each position of the pin along the wear scar and for a given number of cycles. Images of electrical contact resistance are particularly interesting as they

Nanolubricants Edited by J. M. Martin and N. Ohmae

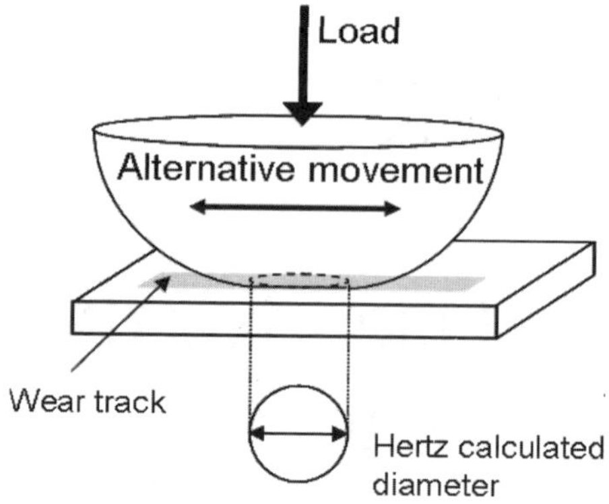

Figure A.1 Schematic representation of a pin-on-flat contact

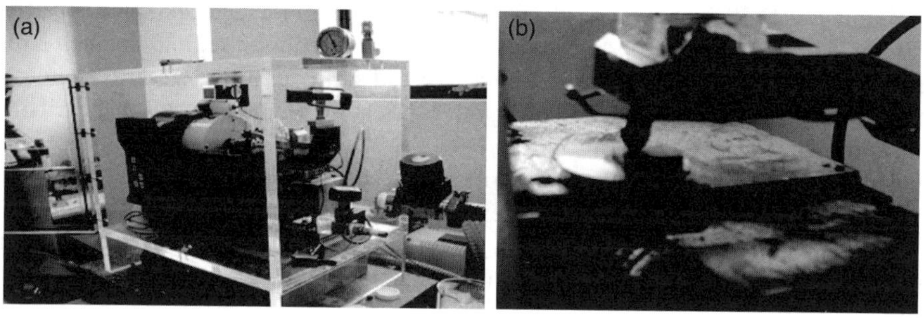

Figure A.2 (a) General sight of the environmental tribometer and (b) sight of the contact area

Table A.1 Experimental conditions used in this study

Ball diameter (mm)	Load (N)	Pressure (GPa)	Hertzian diameter (μm)
12	1	0.33	54
12	2	0.42	68
12	5	0.52	86
6	1	0.66	54
6	2	0.83	68
6	5	1.12	92
6	10	1.42	116
4.5	10	1.72	106

Table A.2 Capability of operating conditions for the environmental tribometer

Load	00.2 → 15N
Environment	Pressure partial of gas (N_2, O_2, H_2O)
Temperature	Ambiant → 70 °C
Sliding speed	0.05 → 13.6 mm/s
Stroke length	0.5 → 3.4 mm
Sizes of the flat ball	10 mm × 8 mm × 2 mm
	⊘ 4.5; 6; 9 or 12 mm

follow the formation of a tribofilm in the contact area on the pin or on the flat. The possible operating conditions of this tribometer are summarized in Table A.2.

A.2 Mobile Pin-on-Flat Tribometer

Another pin-on-flat tribometer was also used. It had two particular characteristics:

- its size, which enables it to be easily adapted to an optical or spectral measuring equipment, hence its name: mobile pin-on-flat tribometer;
- its transparent flat, which is made of sapphire ($e = 1.5$ mm), so *in situ* analyses of the contact area can be performed during friction tests (Figure A.3), where the image capture can be performed to follow the global flow of nanoparticles inside the contact area.

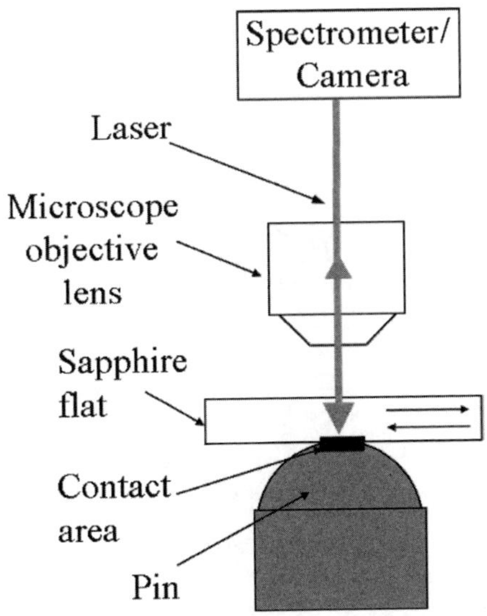

Figure A.3 Schematic representation of the alternative tribometer using a sapphire flat

Table A.3 Experimental conditions used in this study for the mobile pin-on-flat tribometer

Ball diameter (mm)	Load (N)	Pressure (GPa)	Hertz diameter (μm)
6	1	0.78	50
6	2	0.9	62
6	5	1.34	84

This tribometer can also be adapted to a Raman spectrometer. Raman spectroscopy is very useful when characterizing nanoparticles made of carbon or metal dichalcogenide. In this case, analyses are carried out *in situ*, but under no circumstances must the contact be opened during the friction tests whatever the time to acquire spectra, which can be long. Such a tribometer using a steel ball of 6 mm in diameter along with sapphire flats has led to several changes in the experimental conditions (Table A.3).

Table A.4 Possible operating conditions for the ultrahigh vacuum tribometer

Load	0.5 → 5N
Environment	➢Ultrahigh vacuum ➢Pressure partial of gas (N_2, O_2, H_2O)
Temperature	−150 °C → 550 °C
Sliding speed	0.01 → 3 mm/s
Stroke length	0.5 → 4 mm
Sizes of the flat & ball	10 mm × 8 mm × 2 mm Φ 8 or 16 mm

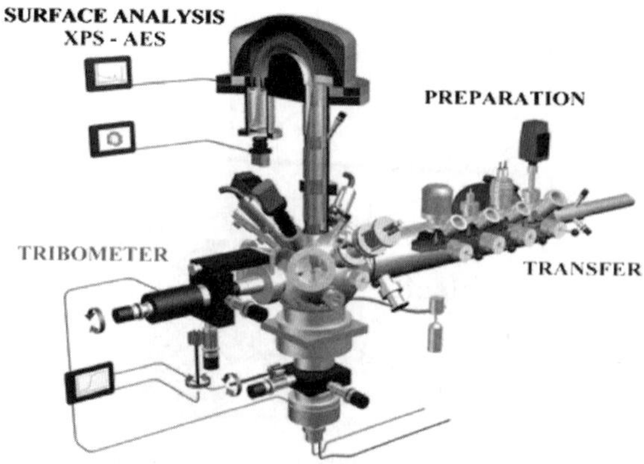

Figure A.4 Schematic representation of the ultrahigh vacuum tribometer

A.3 Ultrahigh Vacuum Tribometer

Experiments were carried out in ultrahigh vacuum (10^{-9} hPa) and in ambient temperature. Flats used were the same as those of the environmental pin-on-flat tribometer. The pin had a diameter of 16 mm. The various possible operating conditions for this tribometer are summarized in Table A.4.

Reference

1. Belin, M., Triboscopy: a new quantitative tool for microtribology, *Wear*, **168**, 1993, 7–12.

Index

Nanolubricants Edited by J. M. Martin and N. Ohmae